UNDERSTANDING THE NEW FIDIC RED BOOK

A CLAUSE-BY-CLAUSE COMMENTARY

AUSTRALIA
Law Book Co.
Sydney

CANADA AND USA
Carswell
Toronto

HONG KONG
Sweet & Maxwell Asia

NEW ZEALAND
Brookers
Wellington

SINGAPORE and MALAYSIA
Sweet & Maxwell Asia
Singapore and Kuala Lumpur

UNDERSTANDING THE NEW FIDIC RED BOOK

A CLAUSE-BY-CLAUSE COMMENTARY

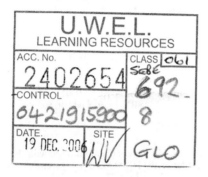

JEREMY GLOVER
PARTNER, FENWICK ELLIOTT SOLICITORS
WITH
SIMON HUGHES, KEATING CHAMBERS
AND AN INTRODUCTION BY
CHRISTOPHER THOMAS Q.C., KEATING CHAMBERS

LONDON
SWEET & MAXWELL
2006

Published in 2006 by
Sweet & Maxwell Limited of 100 Avenue Road,
http://www.sweetandmaxwell.co.uk
Typeset by J&L Composition, Filey, North Yorkshire
Printed and bound in Great Britain by
MPG Books Ltd, Bodmin, Cornwall

No natural forests were destroyed to make this product;
only farmed timber was used and re-planted.

British Library Cataloguing in Publication Data

A CIP catalogue record for this book
is available from the British Library

ISBN-10: 0421 915 900
ISBN-13: 978 042191590 9

CONTENTS

v

CLAUSE 4 – THE CONTRACTOR

CLAUSE 5 – NOMINATED SUBCONTRACTORS

CLAUSE 6 – STAFF AND LABOUR

CLAUSE 11 – DEFECTS LIABILITY

CLAUSE 12 – MEASUREMENT AND EVALUATION

CLAUSE 13 – VARIATIONS AND ADJUSTMENTS

CLAUSE 14 – CONTRACT PRICE AND PAYMENT

CLAUSE 20 – CLAIMS, DISPUTES AND ARBITRATION

GENERAL CONDITIONS OF DISPUTE ADJUDICATION AGREEMENT

FOREWORD
BY THE HON. SIR VIVIAN RAMSEY

The FIDIC form has been in common use for international construction projects since its inception in 1957. Whilst its format followed the Conditions of Contract published by the Institution of Civil Engineers, the two forms have now diverged. This has been most noticeable with the publication of the ICE 7th edn and in the development of the FIDIC suite of standard forms and, even more recently, the publication of the FIDIC Multilateral Development Bank Harmonised edn in 2005 with amendments in 2006.

International construction contracts necessarily adopt a proper law which reflects the choice of those international parties. Whilst one choice is always English law, the parties usually decide on a law which reflects the country of one of the participants in the project or the place of the project. In practice it has been found that the FIDIC form, despite being grounded in expressions derived from English common law, provides a comprehensive code which is easily applied in legal systems based on a civil code or, for instance, obligations under Islamic law. The familiarity of the FIDIC form within the international construction community has brought advantages both in tendering and in project management.

The common law parentage of the FIDIC form makes it appropriate for this book to have been produced by a team of specialist English construction lawyers drawing on the strength of Fenwick Elliott and Keating Chambers. Jeremy Glover, a partner at Fenwick Elliott with a wealth of practical knowledge has authored the majority of this book assisted by his colleague Yann Guermonprez. Simon Hughes at Keating Chambers has also contributed to the main commentary and Christopher Thomas, a Queen's Counsel with great experience in international construction law has provided the Introduction. That authorship would, in itself, be a sufficient recommendation. Its contents confirms that it provides authoritative and practical guidance. Through the use of an overview of key features it gives a helpful summary of the main points of each clause before embarking on a detailed commentary on a clause-by-clause basis. This is no dry academic work. It includes, for instance, the practical suggestion that a contractor should seek a copy of the contract between the employer and the engineer, whilst pointing out that there is no obligation on the employer to pass this on.

By citation of court decisions in different jurisdictions as well as articles by distinguished legal commentators it provides a well founded commentary which covers the ground for those who are working under the FIDIC form. It also provides guidance on legal arguments for those who are seeking to give advice when, as is inevitable, the complex requirements of construction projects give rise to questions on the scope and extent of obligations. The adoption of a Dispute Adjudication Board in the FIDIC form should avoid disputes and reduce the need for international arbitration. The commentary which this book provides will clearly be of great assistance in this area.

On visits to construction lawyers, arbitrators and judges around the world there are certain legal textbooks which always find a place on their bookshelves. I expect this book to be on those shelves in the near future.

Vivian Ramsey
Royal Courts of Justice
London

INTRODUCTION

1.   The inception of this book was the intent to provide a practical text aimed at lawyers and those advising contractors and employers by addressing and reviewing the 1999 edn of the "Red Book", published by the Fédération Internationale des Ingénieurs-Conseils (FIDIC). Surprisingly little commentary and review has taken place in respect of this edition, especially in the format of clause by clause consideration of the provisions. Subsequent to that inception FIDIC produced in May 2005 the Multilateral Development Bank Harmonised edition of the Conditions of Contract. FIDIC itself sought reviews of the MDB May 2005 version and produced an amended second version of the MDB Conditions of Contract in March 2006 set out and contrast the effect of the amendments to the clauses to be found in the MDB March 2006 version.

2.   In their original format, the familiarly termed FIDIC Conditions were the *"Conditions of Contract (International) for Works of Civil Engineering Construction"*. The first edn was published in August 1957 having been prepared on behalf of FIDIC and the Fédération Internationale du Bâtiment et des Travaux Publics (FIBTP). A subsequent edition was published in July 1969 recording a slight change of name of the FIBTP and the addition of the International Federation of Asian and West Pacific Contractors Associations as a sponsoring body, together with a supplementary section referable to clauses for dredging and reclamation work. This edition was reprinted in 1973 following the approval and ratification by two new sponsoring bodies, the Associated General Contractors of America, and the Inter-American Federation of the construction industry (Federación Interamericana de la Industria de la Construcción). The form of these editions of the FIDIC Conditions followed that of the English ICE Conditions of Contract and the 1973 version followed closely the fourth edn of the ICE Conditions, thus closing the provisions of the International Contract with specific terminologies and concepts referable to English Standard Form provisions and derivations. It is remarkable how little modification of an English domestic civil engineering contract was introduced into the FIDIC form which was promulgated for international use. Of the FIDIC 2nd edn it was written: *"as a general comment, it is difficult to escape the conclusion that at least one primary object in preparing the present international*

contract was to depart as little as humanly possible from the English conditions[1].

3.   The FIDIC form, as reflected in its third edn, consisted of 71 General Conditions including a heading for the fluctuations clause and uncompleted heads for separate agreement by the parties in latter parts. These were referred to as *"Conditions of Particular Application"* but were suggestions for subjects upon which the parties were required to make their own agreements. As with the English ICE Conditions there was also a Form of Tender and Appendix, and a Form of Agreement. The FIDIC Conditions themselves contemplated the existence of drawings and specification and bills of quantities as contract documents. The role of the engineer was assumed, with functions including certification and other determinations, and inherently supposed impartiality of the engineer as between contractor and employer in relation to decision making. The form also followed the English basis of re-measurement with quantities treated as approximate together with the system of nomination of sucontractors. This influence of English construction practices and legal concepts was extraordinary in respect of an organisation founded in 1913 by France, Belgium and Switzerland and remaining essentially based in Continental Europe, with the United Kingdom not becoming a member until 1949 and the United States membership in 1958.

4.   By the time of publication of the FIDIC 3rd edn in March 1977, the English ICE Conditions had reached their 5th edn (1973), but the revision committee of FIDIC introduced changes that were regarded as minimal[2]. From the origins of the first three editions of the FIDIC "Red Book", which provided a standard form for use in many projects undertaken in the Commonwealth countries, the premise had derived of the detailed design being provided to the contractor by the employer or his engineer. The civil engineering basis for the FIDIC Conditions had derived from the anticipated infrastructure projects of the nature of roads, bridges, dams, tunnels and water and sewage facilities. Thus the standard terms in the Red Book were less than satisfactory for contracts where major items of plant and alike were manufactured away from site. Focussing on this aspect led to the first edn of the "Yellow Book" produced by FIDIC for mechanical and electrical works in 1963, with its emphasis on testing and commissioning and more suitable for the manufacture and installation of plant. The second edition was published in 1980.

5.   Both the Red and Yellow Books were revised by FIDIC and new editions published in 1987. The 4th edn of the Red Book was the

---

[1] I.N. Duncan Wallace Q.C., The *International Civil Engineering Contract*, 1974.
[2] The FIDIC 3rd edn was commented upon in a supplement to the *International Civil Engineering Contract* by I.N. Duncan Wallace Q.C. in 1980.

product of a committee which had been established in 1983 and whose terms of reference had been premised on change only where change was necessary; maintain the basic role of the engineer; pay close attention to specific topics such as bonds, risk, insurance, claims procedures and dispute procedure and endeavour to update the language to be more understandable by those administering the contract on site. The 4th edn changed the title by the deletion of "International", reflecting a desire on the part of FIDIC that its form should be utilised domestically, albeit in domestic contracting where the effect of the law of the jurisdiction could give rise to different results from that which might otherwise have been intended by the continued English terminology and concepts[3].

6. Noticeably the 1987 4th edn introduced an express term which required the engineer to act impartially when giving a decision or taking any action which might affect the rights and obligations of the parties, whereas the previous editions had assumed this implicitly. In broad terms there was greater concentration on the allocation of risk, and with the significant addition of a ground for extension of time in the event of *"delay, impediment or prevention by the employer"*[4]. A Supplement was published in November 1996 which provided the user with the ability to incorporate alternative arrangements comprising an option for a Dispute Adjudication Board to go with modelled terms of appointment and procedural rules, and an option for payment on a lump sum basis rather than by reference to bills of quantities.

7. By this time FIDIC had responded to the increasing popularity of projects being procured on a design and build or turnkey premise. This resulted in the "Orange Book", namely the *Conditions of Contract for Design-Build and Turnkey* published in 1995[5].

8. The Orange Book reflected a significant move away from the FIDIC forms which had adopted the traditional role of the engineer. Whilst the 4th edn of the Red Book had introduced the obligation upon the engineer to consult with the employer and contractor, with the intended consequence of making the employer more visible, the Orange Book dispensed with the engineer entirely and provided for the "Employer's Representative". The express requirement to be impartial was also relinquished, albeit that when determining value, costs or extensions of

---

[3] Shalakani, "The Application of the FIDIC Civil Engineering Conditions in a Civil Code System" (1989) 6 ICLR 226; Frilet, "How certain provisions of the FIDIC Contract operate under French Laws" (1992) ICLR 121.

[4] E.C. Corbett, FIDIC 4th edn, *A Practical Legal Guide* (1991) N.G. Bunni, *The FIDIC Form of Contract* (1991).

[5] Jaynes, "The New Colour in FIDIC's Rainbow: The Trial Edition of the 'Orange Book'" (1995) ICLR 367.

time the Employer's Representative had to "*determine the matter fairly, reasonably and in accordance with the Contract*"[6].

9. Further the somewhat archaic provision for the submission of matters to the engineer for his "Decision" prior to an ability to pursue a dispute, was eliminated. In its place an Independent Dispute Adjudication Board was introduced consisting of either one or three members appointed jointly by the employer and the contractor at the commencement of the Contract, with the cost being shared by the parties[7]. In fact this replacement of the engineer's decision with a Dispute Board was a reflection of an amendment to the FIDIC Red Book which the World Bank had instituted in its standard bidding documents, and it was this together with the move towards Dispute Resolution Boards in the International Arena which led to the FIDIC 1996 Supplement to the Red Book.

10. It was however in 1994 that FIDIC established its task group to update both the Red and the Yellow Books in the light of developments in the international construction industry, and also to have regard to the work done in bringing about the preparation and publication of the Orange Book. The role of the engineer continued to be a topic upon which a nearly even balance of support or condemnation existed as to the requirement to act impartially in the circumstances of being employed and paid by the employer. Other aspects which the task group considered were the desirability that common definitions would be used in both Red and Yellow Books, and that the definitions which had been incorporated in the Orange Book would form a basis for the new standard forms and assist in harmonising the wording and thus eliminate the complexity that arises from differently formulated definitions. This aspect was reflected also in an attempt to utilise identical wording where considering the same topic in both documents. The group had also to balance the interests of familiarity with pre-existing terminologies and rendering the new documents more up to date. Equally the origins of the FIDIC forms in the contracts of the 19th and early 20th century English Building and Engineering Industry required to be simplified for the very reason that the FIDIC conditions were promulgated in English but in very many instances were being utilised by those whose language background was other than in English. This was especially important having regard to the intent that the new books would be suitable for use in both common law and civil law jurisdictions.[8]

---

[6] For a view at an earlier stage, "Barber, Rules of Conduct for the Engineer" (1988) 5 ICLR 290.
[7] Moulineaux, "Real Time Dispute Resolution: Updating FIDIC" (1995) ICLR 258, Seppala, "The New FIDIC Provisions for a Dispute Adjudication Board" (1997) ICLR 444.
[8] From the presentation of John Bowcock, Chairman, FIDIC Contracts Committee in Tallinn, Estonia, March 1997.

11.   One point that did become apparent to the task group was that the emphasis needed to move from the traditional aim of FIDIC forms towards civil engineering works of the nature of roads, bridges, dams and electrical and mechanical works of power generation and transmission equipment or hydro powered plants. Continuing development had led to more complexity and with projects requiring integration of different specialisms for which the existing forms were regarded as inadequate, albeit that the Orange Book had moved matters along considerably. This gave rise to the emergence of the view that the nature of the forms should depend more upon the parties' assumption of responsibility rather than upon the type of work, as between civil work and electrical and mechanical work. The distinction thus became one between works designed by the employer or his representative and works designed by the contractor.

12.   Test editions were promulgated in 1988 in order to benefit from comment and enhance the user-friendly aim.[9] Following these test editions publication was achieved in 1999 and this distinction was reflected in the titles:

> Conditions of Contract for Construction for Building and Engineering Works Designed by the Employer: The Construction Contract (the new Red Book).
>
> Conditions of Contract for Plant and Design-Build for Electrical and Mechanical Plant and for Building and Engineering Works, Designed by the Contractor – The Plant and Design/Build Contract (the new Yellow Book).

In addition with a change of colour FIDIC simultaneously brought out Conditions of Contract for EPC/Turnkey Projects: the *EPC Turnkey Contract* (Silver Book) and a *short form of contract* (the Green Book)[10].

13.   Each of the new books for major works includes General Conditions together with guidance for the preparation of the Particular Conditions, and a Letter of Tender, Contract Agreement and Dispute Adjudication Agreements. Whilst the Red Book refers to works designed by the employer, this reflects the main responsibility for design

[9] Booen, "FIDIC's Conditions of Contract for the Next Century: 1998 Test Editions" (1999) ICLR, 5. Corbett, "FIDIC's New Rainbow – An Overview of the Red, Yellow, Silver and Green Test Editions" (1999) ICLR, 39.
[10] The Silver Book is not considered further here, but it has been the subject of commentaries: Gaede, "The Silver Book: An Unfortunate Shift from FIDIC's Tradition of being Even-handed and of Focusing on the Best Interests of the Project" (2000) ICLR, 477; Kennedy, "EIC Contractor's Guide to the FIDIC Conditions of Contract for EPC Turnkey Projects (The Silver Book)" (2000) ICLR 504; Henchie: "FIDIC Conditions of Contract for EPC Turnkey Projects – The Silver Book Problems in Store?" (2001) ICLR 41.

and it is appropriate where the works include some contractor designed works whether civil, mechanical, electrical or construction work.

14. It is worthwhile to dwell upon the new 1999 books, and in particular upon the Red Book for construction[11]. The reason for this consideration is that the Multilateral Development Bank Harmonised edition derives from and utilises the work and considerations that went into the terms of the 1999 books.

15. The new Red Book has twenty clauses which are in reality chapters covering major topics by numerous sub-clauses. Clause 2 addresses the role of the employer, and interestingly, as well as reasonably, sub-cl. 2.4 renders it mandatory upon the employer following request from the contractor to submit *"reasonable evidence that financial arrangements have been made and are being maintained which will enable the employer to pay the contract price punctually . . ."* and *"Before the employer makes any material change to his financial arrangements, the employer shall give notice to the contractor with detailed particulars"*. Failure to submit such evidence provides the contractor with the entitlement to suspend work, *"or reduce the rate of work"*, unless and until the contractor has received the reasonable evidence. Further the employer is by sub-cl.2.5 required to give notice and particulars to a contractor *"if the employer considers himself to be entitled to any payment under any clause of these conditions or otherwise in connection with the Contract"*.

16. Clause 3 addresses the position of the engineer, making it clear at the outset that unless otherwise stated *"Whenever carrying out duties or exercising authority, specified in or implied by the Contract, the engineer shall be deemed to act for the employer"*. However, the express provision in the 1987 edition as to impartiality is foregone in favour of a provision that when the conditions provide that the engineer shall proceed in accordance with sub-cl.3.5 to agree or determine any matter:

> *". . . the engineer shall consult with each Party in an endeavour to reach agreement. If agreement is not achieved, the engineer shall make a fair determination in accordance with the Contract, taking due regard of all relevant circumstances."*

It is questionable as to whether this provision in fact moves the engineer to any different position of that reflected in the earlier editions when matters were subject to the engineers "decision". Nevertheless the rights and obligations of the parties are not made dependent upon opinions of the engineer.[12]

---

[11] Wade, "FIDIC's Standard Forms of Contract – Principles and Scope of the Four New Books" (2000) ICLR 5; Booen, "The Three Major New FIDIC Books" (2000) ICLR 24.

[12] For a consideration of some of the difficulties presented by the position of the engineer, Niklisch, "The Role of the Engineer as Contract Administrator and Quasi-Arbitrator in International Construction and Civil Engineering Projects" (1990) 7 ICLR 322.

17.   Clause 4 covers the contractor's general obligations including the requirement that in respect of contractor designed works:

> ". . . it shall, when the works are completed, be fit for such purposes for which the part is intended as are specified in the Contract."

18.   Despite Clause 4 addressing "*The Contractor*", sub-cl.4.10 requires the Employer to have made available to the contractor for his information prior to the date for submission of the tender "*all relevant data in the employer's possession on sub-surface and hydrological conditions at the Site, including environmental aspects*", together with an obligation to make available all such data which comes into the employer's possession subsequently. The Contractor is however rendered "*responsible for interpreting all such data*". Interestingly under sub-cl.4.12, which addresses "*unforeseeable physical conditions*", before any additional cost is finally agreed or determined the engineer "*may also review whether other physical conditions in similar parts of the Works (if any) were more favourable than could reasonably have been foreseen when the contractor submitted the tender*". Thus there is to be a balance between "adverse" and "more favourable" physical conditions.

19.   Clause 5 addresses Nominated Subcontractors, and the grounds upon which the contractor is given the opportunity to raise reasonable objection coincide neatly with the bases upon which nomination of sub-contractors as a mechanism may be regarded as unsatisfactory.

20.   Clauses 6 and 7 address the requirements for personnel, and for plant, materials and workmanship, incorporating at sub-cl.7.7 provision for the identification of the date when each item of plant and materials becomes the property of the employer. It is in cl.8 that provision is made for Commencement, Delay and Suspension and it contains the worthwhile requirement that the provision of programmes showing how the contractor proposes to execute the works is required to be supported by a report describing the methods which the contractor is to adopt. The terms as to entitlement to an extension of time are clear and unrelated to the formulation of any opinion for by sub-cl.8.4:

> ". . . the Contractor shall be entitled . . . to an extension of the Time for Completion if and to the extent that completion . . . is or will be delayed by any of the following causes . . ."

21.   The imposition of damages for delay is by sub-cl.8.7 brought within the purview of the engineer for the purposes of incorporation into payment certificates, and it is no longer an exception to the deductions which the engineer is required to implement. Inevitably linked with cl.8 is cl.10 providing for procedures for certification of completion and for the employer's taking over. Sub-clause 10.2 sensibly addresses

completion of parts of the works without the former reference to completion of substantial parts of the works.

22. Clause 12 deals with Measurement and Evaluation, and prescribes that a new rate or price *"shall be appropriate for an item of work if..."* the measured quantity of the item is changed by more than 10 per cent from the quantity in the bills of quantities, or the change in quantity multiplied by the rate exceeds 0.01 per cent of the accepted contract amount or the change in quantity directly changes the cost per unit quantity of the item by more than 1 per cent and the item is not specified as a *"fixed rate item"*.

23. Clause 13 addresses variations and incorporates adjustments for changes in legislation and in costs. However, provided the contractor notifies an inability to obtain the required goods a variation is not binding. Equally it is not binding in the case of contractor design if the proposed variation would have an adverse impact on safety, suitability or the achievement of performance criteria as specified.

24. The financial aspects are addressed in cl.14 together with procedures for payment. Clauses 15 and 16 provide for termination by the employer and suspension and termination by the contractor, with cl.17 dealing with Risk and Responsibility.

25. Clauses 17 and 18 concern Risk and Responsibility, and Insurance. As with the previous editions the contractor takes full responsibility for the care of the works, materials and plant from the commencement date until the issue of the Taking-Over Certificate, and thus if any loss or damage occurs other than due to an "employer's risk" the contractor is required to rectify that at his own cost. The definition of "employer's risks" thus becomes critical, and represents circumstances in which control by either party is not feasible and events or circumstances caused by the employer whether directly or indirectly. The extent of the contractor's responsibility before taking over now covers the additional element of "goods", meaning the contractor's equipment, materials, plant and temporary works as well as "Contractor's Documents" meaning the calculations, computer programmes and other software, drawings, manuals, models and other documents of a technical nature supplied by the contractor. In respect of delay or cost incurred by the contractor in rectifying loss or damage due to an employer's risk the extent of recovery on the part of the contractor is balanced as between previous positions adopted in FIDIC forms. The result is an entitlement only to recover cost save in the circumstances of use or occupation by the employer or designed by the employer where the contractor is provided with an entitlement to *"reasonable profit in addition"*. This reflects the nature of these two areas corresponding to default on the part of the employer.

26. Sub-clause 17.6 is a new provision as against the previous version of the Red Book. It is a limitation of liability on the part of the contractor.

The clause excludes the liability of both contractor and employer "*for loss of use of any works, loss of profit, loss of any contract or for any indirect or consequential loss or damage which may be suffered by the other party in connection with the contract*". This exclusion is expressly other than specifically provided for delay damages, cost of remedying defects, payment on and after termination, indemnities, the consequences of employer's risks and intellectual and industry property rights. Further, the total liability of the contractor to the employer is limited in money terms to the sum resulting from the application of a multiplier to the Accepted Contract amount, for which the Appendix to Tender must provide.

27.   Clause 19 is devoted to Force Majeure including its definition within sub-cl.19.1, namely meaning:

> "*An exceptional event or circumstance:*
>
> (a)   *which is a beyond a party's control,*
> (b)   *which such party could not reasonably have provided against before entering into the Contract, and*
> (c)   *which, having arisen, such a party could not reasonably have avoided or overcome, and*
> (d)   *which his not substantially attributable to the other party.*"

28.   Kinds of such exceptional events or circumstances are listed. This is a new provision referable to the previous editions of the Red Book, albeit that under the 4th edn some of the relief was available to the contractor for events now brought within the umbrella of this new provision, which includes aspects relating to the effect of events and the continuation of performance. The event of *Force Majeure* must prevent a party from performing any of its obligations, and as such is by definition not as severe as an event that prevents the performance of all of them. Notice is required following awareness of the event or circumstance and prevention from performance of an obligation on the part of the contractor by an event of *Force Majeure* may give rise to an extension of time and additional cost. It is only in the event of an execution of the substantially all the works in progress being prevented for a continuous period or multiple periods that the termination is provided for[13].

29.   The provisions conclude with cl.20, Claims, Disputes and Arbitration. This provides procedure and regulation for, and regulation of, claims not solely for additional payment but also for extension of time. It imposes the requirement on the contractor to give notice as soon as

---

[13] For a comprehensive analysis of the topic with a comparative consideration of legal systems, Treitel, *Frustration and Force Majeure*, 2nd edn, 2004.

practicable and not later than 28 days after the contractor became aware *"or should have become aware"* of the event or circumstance the subject of the claim. Failure to give notice renders time not being extended and the contractor not entitled to additional payment. The requirement to keep contemporary records is maintained but the novel feature of this provision is the mandatory requirement upon the engineer or employer to respond to the contractor and to any supporting particulars as to the basis of the claim *"with approval, or with disapproval and detailed comments"*. A request for further particulars may be generated yet the employer or engineer *"shall nevertheless give his response on the principles of the claim within . . ."* the 42 days.

30. It is apparent that the thinking of the Task Group was that there should be notice of a claim within 28 days for it to be regarded as valid, reflecting upon the bona fides of a claim and the ordinary ability on the part of a contractor to know whether a claim situation has arisen and on what basis[14]. Under this condition in Clause 20 disputes are to be adjudicated by a Dispute Adjudication Board, and it is in this area that there is a distinction in the 1999 Red Book from the system of Dispute Review Boards which was advocated by the World Bank.

31. Sub-clause 4.2 specifies that the contractor shall provide a Performance Security where the amount has been specified in the Appendix to Tender, and the sub-clause continues with provisions for extending the security and, importantly, an indemnity by the employer in favour of the contractor against damage, loss and expense resulting from a claim under the performance security *"to the extent to which the Employer was not entitled to make the claim"*. Accompanying the 1999 Red Book are sample forms of security which need to be carefully considered on the parts of both an employer and the contractor at the earliest possible stage.

32. The performance security is required to be issued by an entity approved by the employer *"and shall be in the form annexed to the Particular Conditions or in another form approved . . ."*. Thus the employer needs to have reached a decision on the document to comprise the performance security and its wording at the stage of the preparation of tender documentation. With the 1987 4th edn FIDIC provided two sample types of security comprising a Performance Guarantee and a Surety Bond. They were in conditional terms, co-extensive with the construction contract, and payable upon default. The desire for independence of the guarantee obligations and the assimilation of such guarantee obligations towards letters of credit allowed inextricably to the use of first

---

[14] Seppala, "FIDIC's new Standard Forms of Contract – Force Majeure, Claims, Disputes and other clauses" (2000) ICLR 235.

or on-demand bonds[15]. The Performance Guarantee, the Advanced Payment Guarantee and the Retention Money Guarantee represent a similar form and a payment mechanism. Their differences reflect different functions. Their nature is that of an on-demand guarantee but one which is payable upon the submission of identified documentation by the beneficiary. This is the written demand, but with a required statement as to the breach of obligations under the contract and the respect in which the contractor is in breach. These securities derive from the guidance of the International Chamber of Commerce and the Uniform Rules published by that body[16]. Certainly the introduction of the ICC Uniform Rules is a major step in providing clarity and certainty in this difficult area[17].

33. In general terms the first edition of the Rainbow series of 1999 was a welcome advance in clarity and in addressing areas of difficulty. The summary of the view of the European International Contractors Group was that the 1999 Construction Contract was an improvement on the 4th edn, albeit that the balance of the amendments would increase the risk to contractors, with the recognition that the new powers of the engineer could pose problems[18]. However, an experienced commentator's view is that the new forms would gain their places as leading international standard forms with an indication that they had been well received at the World Bank, and *"thus there is a likelihood that they will be included in standard bidding documents or in lists of forms approved for use in projects supported by multilateral funding agencies"*[19].

34. This was an extremely percipient comment because the banking community had for many years adopted the FIDIC Conditions as part of their standard bidding documents to which the banks required their borrowers and aid recipients to adhere. Whilst the banks utilised the FIDIC Conditions, it was through the Conditions of Particular Application that the banks introduced the terms and provisions which they particularly required and by which means amendments were effected to the FIDIC General Conditions. Thus not only did the procurement documents for a particular project require to repeat the amendments by this mechanism but there were inevitable differences between the amendments and conditions as between the banks. Those in charge of procurement at a number of Multilateral Development

[15] As described in *Edward Owen Engineering Ltd v Barclays Bank International Ltd and Umma Bank* (1978) QB 159, (CA) per Lord Denning M.R.
[16] Uniform Rules for Demand Guarantees (URDG, No. 458); 524 Uniform Rules for Contract Bonds (URCB No. 524).
[17] Bertrams, "The New Forms of Security in FIDIC's 1999 Conditions of Contract" (2000) ICLR 367.
[18] "EIC Contractors Guide to the FIDIC Conditions of Contract for Construction (the new Red Book)" (2003) ICLR 53.
[19] Corbett, "FIDIC's new Rainbow 1st Edition – An Advance?" (2000) ICLR 253.

Banks (MDB) were well aware of the problems created by these and the benefits that might accrue from uniformity. It was as a result of this that these banks wished to harmonise their bid documents by the creation of a modified form of the FIDIC Conditions of Contract for Construction 1st edn 1999 in which the General Conditions would provide in agreed terminology the effects of what previously had been incorporated by amendment.

35.   The resulting "harmonised edition" was the product of preparation by the FIDIC Contracts Committee and by a group of participating banks[20], and following comments from those to whom the draft was circulated, it was in May 2005 that the first harmonised edition of the 1999 Conditions was published.

36.   The harmonised edition carries a belief that it will simplify the use of the FIDIC Conditions of Contract not only for the MDBs and their borrowers but also for others involved with project procurement including engineers, contractors and contract specialists. It is intended for use on MDB financed projects only. It recognises that despite the use of the harmonised conditions which will reduce the number of additions and amendments in the Particular Conditions, nevertheless some special requirements will be generated by particular projects. It is for this reason that the harmonised edition contains provision for particular conditions, and it also contains sample forms for Contract Data, Securities, Bonds, Guarantees and Dispute Board Agreements. The harmonised edition is advanced on the basis that it follows FIDIC risk sharing principles, and whilst most of the amendments to reflect the harmonisation derived from the requirements of the banking community there are some minor changes introduced. There is however a significant change from the 1999 Red Book in connection with the dispute provisions contained in cll 20.2 to 20.8, which FIDIC considers to be an improvement.

37.   Accompanying the harmonised editions of May 2005 and amended in March 2006 is a Supplement prepared under the guidance of the FIDIC Contracts Committee[21]. The Supplement provides information and guidance for the preparation of contracts using the MDB harmonised construction contract, based on the Construction Contract 1st edn 1999 and elements of the MDB harmonised Master

---

[20] The banks involved were the African Development Bank, Asian Development Bank, Black Sea Trade and Development Bank, Caribbean Development Bank, European Bank for Reconstruction and Development, Inter-American Development Bank, International Bank for Reconstruction and Development (the World Bank), Islamic Bank for Development, Nordic Development Fund. However the Nordic Development Fund, because its activities are being wound up over the next few years, decided not to proceed.

[21] Comprising Christopher Wade, Nael Bunni, Axel-Volkmar Jaeger, Philip Jenkinson and Michael Mortimer-Hawkins, together with John Bowcock as special adviser and Christopher R. Sepala as legal adviser.

Procurement Document for Procurement of Works and User's Guide that were published in the *World Bank Standard Bidding Document for Works and User's Guide* May 2005.

38.   The Supplement is a most useful document containing a section on Changes to the Construction Contract General Conditions. This summarises the changes that were made to the FIDIC Conditions. One of the most noticeable changes is the addition in sub-cl.3.1 in connection with the engineer's duties and authority whereby the following additional provision is included:

> *"The Engineer shall obtain the specific approval of the Employer before taking action under the following sub-clauses of these conditions:*
>
> *(a)   Sub-clause 4.12: Agreeing or determining an extension of time and/or additional costs.*
> *(b)   Sub-clause 13.1: Instructing a Variation; except: (i) in an emergency . . .; or (ii) if such a variation would increase the Accepted Contract Amount by less than the percentage specified in the Contract Data.*
> *(c)   Sub-clause 13.3: Approving a proposal for variation submitted by the contractor in accordance with Sub-clauses 13.1 or 13.2.*
> *(d)   Sub-clause 13.4: Specifying the amount payable in each of the applicable currencies.*
>
> *Notwithstanding . . ."*

39.   Most critical, however, in relation to the engineer is the matter of his authority. Under the 1999 edn there is an express prohibition in sub-cl.3.1 in the following terms:

> *"The Employer undertakes not to impose further constraints on the Engineer's Authority, except as agreed with the Contractor".*

This is replaced in the harmonised version by the following:

> *"The Employer shall promptly inform the Contractor of any change to the authority attributed to the Engineer".*

Certainly the view of contractors is that this is a retrograde step permitting unilateral alteration of the engineer's authority, and thus potentially impacting upon the balance of risk[22]. This is a view apparently endorsed by FIDIC who accept that some of the changes have tilted the

---

[22] Appuhn and Eggink, "The Contractor's View on the MDP Harmonised Version of the new Red Book" (2006) ICLR 4.

balance of risk in favour of the Employer.[23] There are other notable features including the deletion of the four circumstances under which an employer is entitled to make a call under the performance security. There is no replacement, doubtless based on the view that the reference to the ICC Uniform Rules provide an adequate safeguard for both the employer, the financing institutions and the contractor.

40. In addition, the thresholds for variation in quantities and values of items in the bill of quantities for the purposes of determining whether a build rate may be changed have been increased in the harmonised version. Another notable feature lies in the terms of cl.20 and the replacement of Dispute Adjudication Boards with Dispute Boards.

41. A further useful section comprises Words and Phrases replaced in the General Conditions[24]. Perhaps the most useful is the Clause-by-Clause comparison between the provisions of the General Conditions in the Construction Contract and those in the harmonised edition which are set out side by side and with the amendments introduced by the latter specifically underlined.

42. This commentary had been prepared in the following way. First, the particular sub-clause has been set out in full. This is followed by an overview of the key features of that sub-clause. After that, we provide a deatield commentary explaining the effect and operation of the sub-clause. Where relevant we discuss case law and some of the problems encountered in practice. Finally, we set out any changes to be found in the MDB 2006 Harmonised Edition and comment on the effect of those changes.

43. The juridical basis of this commentary rests in English law, and it follows that the comments endeavour to interpret and reflect the contract according to the English rules of construction and interpretation. On occasion, we have provided the viewpoint of an alternative jurisdiction. It must however be recognised that the English rules are relatively strict as against interpretation and application under other legal systems, whether in Continental Europe or under Arab Civil Codes. Not only is this a caveat but it is a positive requirement that it is the approach and application of the applicable law which will govern the ultimate meaning and effect of the terms used[25]. Nevertheless it is hoped that a commentary on this basis and in the format adopted will be useful to English and international readers alike, with the latter then able to give or obtain advice as to whether the relevant proper law of the contract would modify the English view.

---

[23] Christopher Wade: the FIDIC Contract Forms and the new MDB Contract, paper given at the ICC-FIDIC conference, 17–18 October 2005.

[24] An example is the change from "Appendix to Tender" to "Contract Data".

[25] André-Dumont, "The FIDIC Conditions and Civil Law" (1988) 5 ICLR 43, Abrahamson, "checklist for Foreign Laws, FIDIC, Forensic Context" (1988) 5 ICLR 266.

FURTHER READING

There are two websites that provide a useful starting point:

(i)   The International Federation of Consulting Engineer's own website – *www.fidic.org* – provides a wealth of additional information about not only the conditions of contract or construction, but all aspects of international consulting engineering. This provides details of forthcoming courses and seminars as well as a number of useful articles:

(ii) There is a second website run by Fidic-NET – *www.fidic-net.org* – which is a worldwide inter-university collaboration set up to share experience and knowledge of Fidic contracts. In addition, Fidic-NET run annual workshops.

As will have been clear from the main text, there are only a limited number of judicial cases involving the FIDIC contracts, an inevitable consequence of the fact that any disputes are ultimately settled through (confidential) arbitration proceedings.

Any book on the latest FIDIC form of contracts must also owe a debt, as indeed the contracts do themselves, to the commentaries on the Old Red Book, FIDIC 4th edn prepared by Edward Corbett and Professor Nael G Bunni. They have also written a number of articles on the new FIDIC forms. Indeed, it would be fair to say that there are a number of articles that have been written on the new contracts, some on specific clauses, some of more general nature. For those interested in further reading, we would recommend starting with some of those listed below:

From the FIDIC Website:

Papers from the New Contracts Launch Seminar – 1998

1   The DRB/DAB: An attractive procedure if one takes certain precautions: Pierre M Genton, September 1999
2   FIDIC's New Contracts: Christopher Seppala, July 2000
3   New Edition of the Red Book: Impartiality of the Engineer: David Bateson, August 2000
4   The New FIDIC Contracts: Their Principles, Scope and Details: PL Booen, March 2001.
5   Termination, Risk and Force Majeure: Gordon L Jaynes, January 2001
6   Claims, Disputes and Arbitration: Gordon L Jaynes, January 2001
7   Claims and Adjustments of the Contract: Peter L Booen, 2001

8   Uniform Rules for Demand Guarantees: Christopher Seppala, June 2001
9   FIDIC Conditions of Contract: Christopher Wade, 2003
10   FIDIC Conditions of Contract and Dispute Adjudication Procedure: Christopher Seppala (2003)
11   Introduction to the FIDIC DAB Provisions: Gwen Owen, 2004
12   Clause 20, Dispute Resolution: Michael Mortimer-Hawkins, 2004
13   Dispute Boards & DAB: Peter H.J. Chapman, 2004
14   The Arbitration Clause In the FIDIC Contracts for Major Works: Christopher Seppala – Paper given at IBC Conference in London, December 2004
15   Overview of FIDIC Contracts: Christopher Wade, 2005
16   The FIDIC Contracts Guide: Christopher Wade (Oct. 2005)
17   The Gap in Clause 20.7: Noel G Bunni, 2005
18   Claims of the Employer: Christopher Wade, 2005
19   Force Majeure, Claims Disputed and Other Causes: Christopher Seppala
20   Dispute Boards and the MDB Contract: Gordon L Jaynes, April 2006.
21   Force Majeure: P. Goodwin and Dominic Roughton, 2006
22   Engineer or Dispute Adjudication Board: How to choose: Christopher Denny

From Other Sources:
1   "Contractor's Claims under the FIDIC contract for Major Works" Christopher Seppala, 2005 – 21 Const L.J. 278
2   "International Construction Contract Disputes – Commentary on ICC awards dealing with the FIDIC International Conditions of Contract" Christopher Seppala [1999] ICLR 339
3   "FIDIC's New Rainbow – An Overview of the Red, Yellow, Silver and Green Test Editions" Edward Corbett [1999] ICLR 39
4   "FIDIC's New Rainbow 1st Edition – an Advance?" Edward Corbett [2000] ICLR 253
5   "FIDIC's Standard Form of Contract – Principles and Scope of the Four New Books" Christopher Wade [2000] ICLR 5
6   "The Three Major New FIDIC Books" Peter L Booen [2000] ICLR 24
7   "EIC Contractor's Guide to the FIDIC Conditions of Contract for EPC Turnkey Projects (The Silver Book)" Frank M Kennedy [2000] ICLR 504
8   "FIDIC's New Standard Forms of Contract – Force Majeure, Claims, Disputes and other Clauses" Christopher Seppala [2000] ICLR 235
9   "Using FIDIC Contracts in Eastern Europe" Ilya Nikiforov [2000] ICLR 540
10   "FIDIC's New Suite of Contracts – Clauses 17 to 19" N.G. Bunni [2001] ICLR 523

11   Delivering Infrastructure – International Best Practice – FIDIC Contracts: A Developer's View. Michael Wahlgreen, August 2002, *www.scl.org.uk*

12   Delivering Infrastructure – International Best Practice – FIDIC Contracts: A Contractor's View. Corinna Osinski, August 2002, *www.scl.org.uk*

13   Delivering Infrastructure – International Best Practice – FIDIC's 1999 Rainbow – Best Practice? Edward Corbett, August 2002, *www.scl.org.uk*

14   "EIC Contractor's Guide to FIDIC Conditions of Contract for Construction" [2003] ICLR 53

15   "Dispute Boards: It's time to move on" Christopher Dering [2004] ICLR 438

16   "The Engineer in International Construction: Agent? Mediator? Adjudicator?" O/a Nisja [2004] ICLR 230

17   "Will the Silver Book become the World Bank's new gold standard? The interrelationship between the World Bank's infrastructure procurement policies and Fidic's Construction Contracts?" Matthew Bell [2004] ICLR 164

18   Overview of Fidic Contracts. Christopher Wade, 2005

19   FIDIC Contract Forms and the New MDB Contract. Christopher Wade, 2005

20   The FIDIC Contract's Guide. Christopher Wade, October 2005

21   Design Risk In FIDIC Contracts. Michael Black Q.C., March 2005 *www.scl.org.uk*

22   "The Contractor's view on the MDB Harmonised Version of the New Red Book" Richard Appuhn and Eric Eggink 2006 [ICLR] 4

23   "Dispute Boards Good News and Bad News: The 2005 "Harmonised" Conditions of Contract Prepare by Multilateral Development Banks and FIDIC" Gordon L Jaynes [2006] ICLR 102

24   Establishing Dispute Boards. Nicholas Gould May 2006 paper given to the 6th annual DRBF International Conference in Budapest *www.fenwickelliott.co.uk*

25   Engineering Ethics: Do engineers owe duties to the public? John Uff Q.C., Royal Academy of Engineering *www.raeng.org.uk*

ACKNOWLEDGEMENTS

The 1999 Edition of the Conditions of Contract for Construction for Building and Engineering Works Designed by the Employer (the new Red Book) and extracts of the MDB Harmonised General Conditions of Contracts and Sample Forms are the copyright of FIDIC and are reproduced with their kind permission. The 1999 edition of the Red Book can be obtained from FIDIC *www.fidic.org*. The original version of the MDB Harmonised General Conditions can be downloaded at *www.fidic.org/mdb*

TABLE OF CASES

TABLE OF STATUTES

CLAUSE 1 – GENERAL PROVISIONS

General Conditions

1. GENERAL PROVISIONS

1.1 Definitions   *In the Conditions of Contract ("these Conditions"), which*   **1–001**
include Particular Conditions, and these General Conditions,
the following words and expressions shall have the meanings
stated. Words indicating persons or parties include corpora-
tions and other legal entities, except where the context requires
otherwise.

1.1.1 The Contract   **1–002**

> *1.1.1.1 "Contract" means the Contract Agreement, the*
> *Letter of Acceptance, the Letter of Tender, these Con-*
> *ditions, the Specification, the Drawings, the Schedules,*
> *and the further documents (if any) which are listed in the*
> *Contract Agreement or in the Letter of Acceptance.*

> *1.1.1.2 "Contract Agreement" means the contract agree-*
> *ment (if any) referred to in Sub-Clause 1.6 [Contract*
> *Agreement].*

> *1.1.1.3 "Letter of Acceptance" means the letter of formal*
> *acceptance, signed by the Employer, of the Letter of*
> *Tender, including any annexed memoranda comprising*
> *agreements between and signed by both Parties. If there is*
> *no such letter of acceptance, the expression "Letter of*
> *Acceptance" means the Contract Agreement and the date*
> *of issuing or receiving the Letter of Acceptance means the*
> *date of signing the Contract Agreement.*

> *1.1.1.4 "Letter of Tender" means the document entitled*   **1–003**
> *letter of tender which was completed by the Contractor*
> *and includes the signed offer to the Employer for the*
> *Works.*

1

1.1.1.5 "Specification" means the document entitled specification, as included in the Contract, and any additions and modifications to the specification in accordance with the Contract. Such document specifies the Works.

1.1.1.6 "Drawings" means the drawings of the Works, as included in the Contract, and any additional and modified drawings issued by (or on behalf of) the Employer in accordance with the Contract.

1–004

1.1.1.7 "Schedules" means the document(s) entitled schedules, completed by the Contractor and submitted with the Letter of Tender, as included in the Contract. Such document may include the Bill of Quantities, data, lists, and schedules of rates and/or prices.

1.1.1.8 "Tender" means the Letter of Tender and all other documents which the Contractor submitted with the Letter of Tender, as included in the Contract.

1.1.1.9 "Appendix to Tender" means the completed pages entitled appendix to tender which are appended to and form part of the Letter of Tender.

1.1.1.10 "Bill of Quantities" and "Daywork Schedule" mean the documents so named (if any) which are comprised in the Schedules.

1–005 *1.1.2 Parties and Persons*

1.1.2.1 "Party" means the Employer or the Contractor, as the context requires.

1.1.2.2 "Employer" means the person named as employer in the Appendix to Tender and the legal successors in title to this person.

1.1.2.3 "Contractor" means the person(s) named as contractor in the Letter of Tender accepted by the Employer and the legal successors in title to this person(s).

1.1.2.4 "Engineer" means the person appointed by the **1–006**
Employer to act as the Engineer for the purposes of the
Contract and named in the Contract Data, or other
person appointed from time to time by the Employer and
notified to the Contractor under Sub-Clause 3.4
[Replacement of the Engineer].

1.1.2.5 "Contractor's Representative" means the person
named by the Contractor in the Contract or appointed
from time to time by the Contractor under Sub-Clause 4.3
[Contractor's Representative], who acts on behalf of the
Contractor.

1.1.2.6 "Employer's Personnel" means the Engineer, the
assistants referred to in Sub-Clause 3.2 [Delegation by
the Engineer] and all other staff, labour and other
employees of the Engineer and of the Employer; and any
other personnel notified to the Contractor, by the
Employer or the Engineer, as Employer's Personnel.

1.1.2.7 "Contractor's Personnel" means the Contractor's **1–007**
Representative and all personnel whom the Contractor
utilises on Site, who may include the staff, labour and
other employees of the Contractor and of each
Subcontractor; and any other personnel assisting the
Contractor in the execution of the Works.

1.1.2.8 "Subcontractor" means any person named in the
Contract as a subcontractor, or any person appointed as a
subcontractor, for a part of the Works; and the legal
successors in title to each of these persons.

1.1.2.9 "DAB" means the person or three persons so **1–008**
named in the Contract or other person(s) appointed
under Sub-Clause 20.2 [Appointment of the Dispute
Adjudication Board] or Sub-Clause 20.3 [Failure to
Agree on the Composition of the Dispute Adjudication
Board].

1.1.2.10 "FIDIC" means the Federation Internationale
des Ingenieurs-Conseils, the international federation of
consulting engineers.

1–009 *1.1.3 Dates, Tests, Periods*
and Completion

1.1.3.1 "Base Date" means the date 28 days prior to the latest date for submission and Completion and Completion of the Tender.

1.1.3.2 "Commencement Date" means the date notified under Sub-Clause 8.1 [Commencement of Works].

1.1.3.3 "Time for Completion" means the time for completing the Works or a Section (as the case may be) under Sub-Clause 8.2 [Time for Completion], as stated in the Appendix to Tender (with any extension under Sub-Clause 8.4 [Extension of Time for Completion]), calculated from the Commencement.

1–010

1.1.3.4 "Tests on Completion" means the tests which are specified in the Contract or agreed by both Parties or instructed as a Variation, and which are carried out under Clause 9 [Tests on Completion] before the Works or a Section (as the case may be) are taken over by the Employer.

1.1.3.5 "Taking-Over Certificate" means a certificate issued under Clause 10 [Employer's Taking Over].

1.1.3.6 "Tests after Completion" means the tests (if any) which are specified in the Contract and which are carried out in accordance with the provision of the Particular Conditions after the Works or a Section (as the case may be) are taken over by the Employer.

1–011

1.1.3.7 "Defects Notification Period" means the period for notifying defects in the Works or a Section (as the case may be) under Sub-Clause 11.1 [Completion of Outstanding Work and Remedying Defects], as stated in the Appendix to Tender (with any extension under Sub-Clause 11.3 [Extension of Defects Notification Period]), calculated from the date on which the Works or Section is completed as certified under Sub-Clause 10.1 [Taking Over of the Works and Sections].

1.1.3.8 "Performance Certificate" means the certificate issued under Sub-Clause 11.9 [Performance Certificate].

1.1.3.9 "day" means a calendar day and "year" means 365 days.

1.1.4 Money and Payments

1.1.4.1 "Accepted Contract Amount" means the amount accepted in the Letter of Acceptance for the execution and completion of the Works and the remedying of any defects.

1.1.4.2 "Contract Price" means the price defined in Sub-Clause 14.1 [The Contract Price], and includes adjustments in accordance with the Contract.

1.1.4.3 "Cost" means all expenditure reasonably incurred (or to be incurred) by the Contractor, whether on or off the Site, including overhead and similar charges, but does not include profit.

1.1.4.4 "Final Payment Certificate" means the payment certificate issued under Sub-Clause 14.13 [issue of Final Payment Certificate].

1.1.4.5 "Final Statement" means the statement defined in Sub-Clause 14.11 [Application for Final Payment Certificate].

1.1.4.6 "Foreign Currency" means a currency in which part (or all) of the Contract Price is payable, but not the Local Currency.

1.1.4.7 "Interim Payment Certificate" means a payment certificate issued under Clause 14 [Contract Price and Payment], other than the Final Payment Certificate.

1.1.4.8 "Local Currency" means the currency of the Country.

1.1.4.9 "Payment Certificate" means a payment certificate issued under Clause 14 [Contract Price and Payment].

1–015

1.1.4.10 *"Provisional Sum" means a sum (if any) which is specified in the Contract as a provisional sum, for the execution of any part of the Works or for the supply of Plant, Materials or services under Sub-Clause 13.5 [Provisional Sums].*

1.1.4.11 *"Retention Money" means the accumulated retention moneys which the Employer retains under Sub-Clause 14.3 [Application for Interim Payment Certificates] and pays under Sub-Clause 14.9 [Payment of Retention Money].*

1.1.4.12 *"Statement" means a statement submitted by the Contractor as part of an application, under Clause 14 [Contract Price and Payment], for a payment certificate.*

1–016 *1.1.5 Works and Goods*

1.1.5.1 *"Contractor's Equipment" means all apparatus, machinery, vehicles and other things required for the execution and completion of the Works and the remedying of any defects. However, Contractor's Equipment excludes Temporary Works, Employer's Equipment (if any), Plant, Materials and any other things intended to form or forming part of the Permanent Works.*

1.1.5.2 *"Goods" means Contractor's Equipment, Materials, Plant and Temporary Works, or any of them as appropriate.*

1.1.5.3 *"Materials" means things of all kinds (other than Plant) intended to form or forming part of the Permanent Works, including the supply-only materials (if any) to be supplied by the Contractor under the Contract.*

1–017

1.1.5.4 *"Permanent Works" means the permanent works to be executed by the Contractor under the Contract.*

1.1.5.5 *"Plant" means the apparatus, machinery and vehicles intended to form or forming part of the Permanent Works.*

1.1.5.6 "Section" means a part of the Works specified in the Appendix to Tender as a Section (if any).

1.1.5.7 "Temporary Works" means all temporary works of every kind (other than Contractor's Equipment) required on Site for the execution and completion of the Permanent Works and the remedying of any defects. **1–018**

1.1.5.8 "Works" mean the Permanent Works and the Temporary Works, or either of them as appropriate.

1.1.6 Other Definitions **1–019**

1.1.6.1 "Contractor's Documents" means the calculations, computer programs and other software, drawings, manuals, models and other documents of a technical nature (if any) supplied by the Contractor under the Contract.

1.1.6.2 "Country" means the country in which the Site (or most of it) is located, where the Permanent Works are to be executed.

1.1.6.3 "Employer's Equipment" means the apparatus, machinery and vehicles (if any) made available by the Employer for the use of the Contractor in the execution of the Works, as stated in the Specification; but does not include Plant which has not been taken over by the Employer.

1.1.6.4 "Force Majeure" is defined in Clause 19 [Force Majeure]. **1–020**

1.1.6.5 "Laws" means all national (or state) legislation, statutes, ordinances and other laws, and regulations and by-laws of any legally constituted public authority.

1.1.6.6 "Performance Security" means the security (or securities, if any) under Sub-Clause 4.2 [Performance Security].

1–021

1.1.6.7 "Site" means the places where the Permanent Works are to be executed and to which Plant and Materials are to be delivered, and any other places as may be specified in the Contract as forming part of the Site.

1.1.6.8 "Unforeseeable" means not reasonably foreseeable and against which adequate preventive precautions could not reasonably be taken by an experienced contractor by the date for submission of the Tender.

1.1.6.9 "Variation" means any change to the Works, which is instructed or approved as a variation under Clause 13 [Variations and Adjustments].

OVERVIEW OF KEY FEATURES

1–022   The definitions provided in cl.1.1 are extensive compared with many Standard Forms. The definitions for the defined terms, which are identical across all the FIDIC forms, are grouped not alphabetically but in six different categories, the contract, parties to the contract, dates, tests, periods and completion, money and payments, works and goods and other definitions.

As with any contract, it is important that these are reviewed by those in charge of running the project.[1]

COMMENTARY

1–023   A number of the defined terms warrant further consideration.

Sub-Clause 1.1

1–024   The reference to Particular Conditions is a reference to a number of optional additional contract conditions which have been prepared by FIDIC to supplement the General Conditions to be found at cll 1–20.

The Particular Conditions can be found at the end of the General Conditions and run to some 20 pages. Where appropriate, reference is made to the Particular Conditions in the text of this work. Indeed, as noted below, a number of the Particular Conditions have been adopted in the MDB Harmonised version.

---

[1] Indeed, this comment applies to the contract as a whole. It is surprising even today how often contracts are left in drawers and not reviewed by those who should know and understand them.

As with any standard contract, care should be used in adopting or using too many new or different clauses. The General Conditions are in a standard form which means that the parties are likely to be familiar with them and their meaning. The same may not be true of any amendments. Equally, care must be taken to avoid inconsistency and ambiguity between the existing and any new clauses.

Any Particular Condition which is adopted will be of some importance since in accordance with sub-cl.1.5, they will rate in order of priority above the General Conditions when it comes to interpreting the Contract.

Sub-Clause 1.1.1.3: *"Letter of Acceptance"*

In order for a document to become a "Letter of Acceptance" certain formalities must be observed (annexure of memoranda and signature by both parties). However, if these formalities are not observed, so that there is "no such letter of acceptance", or there is no letter of acceptance at all, then the parties are thrown back on the Contract Agreement itself.　　**1–025**

Sub-Clause 1.1.2: *"Parties to the Contract"*

Following the definition in sub-cl.1.1, parties can be firms, corporations, individuals or other legal entities.　　**1–026**

Sub-Clause 1.1.2.1: *"Party"*

The definition of "Party" only includes the Employer and the Contractor, not the Engineer, who is by implication, clearly not a party to the Contract.　　**1–027**

Sub-Clause 1.1.2.6: *"Employer's Personnel"*

This definition is wide-ranging and includes not just, as you would expect, those who work for the Employer but also the Engineer and its staff.　　**1–028**

Sub-Clause 1.1.2.7: *"Contractor's Personnel"*

The definition of "Contractor's Personnel"[2] is also very wide, as it includes "any other personnel assisting the Contractor in the execution of the Works". This definition may generate disputes as to what level of implication is necessary to constitute "assisting the contractor in the execution of the works". For example, would a technical consultant advising on a discrete issue fall inside the definition and so become one of the Contractor's Personnel? The　　**1–029**

---

[2] For further detail on the role of the Contractor's Representative, see sub-cl.4.3.

answer is probably yes, unless a legal dispute had arisen, when the consultant will not actually be assisting with works themselves.

Sub-Clause 1.1.2.9: *"DAB" [Dispute Adjudication Board]*

1–030 In terms of the dispute resolution procedures in the Contract, the most important change in this Form is the advent of the DAB or Dispute Resolution Boards whose decisions have temporary binding effect and replace the traditional Engineer's Decision. Those responsible for the drafting of standard forms have become increasingly attracted to Dispute Resolution Boards and this reflects changes within the construction and engineering industries both in the UK and internationally.

Typically, dispute resolution boards will often render quick (and usually temporarily binding) decisions on issues which arise during the currency of the contracts works. One reason for their adoption is concern for the cost of litigation and arbitration in the construction and engineering contexts.

In the UK engineering industry, the power to make temporarily binding decisions has traditionally been given to the Engineer, but the conventional view that the Engineer acts independently of his paymaster, the Employer, holding the requisite balance between the parties,[3] is the subject of widespread scepticism, even in the House of Lords.[4]

1–031 Finally, on May 1, 1998 the Housing Grants Construction and Regeneration Act 1996 came into force in the UK, which inter alia introduced into UK law the concept of a right to refer a dispute to adjudication under a construction contract (being a contract for the carrying out of defined "construction operations") with the aim of providing the contracting parties with a decision which is temporarily binding (binding unless and until it is altered by subsequent arbitration proceedings or litigation).

The UK construction and engineering industry have thus become accustomed to the concept of temporarily binding determinations, and this is a further reason for the rising popularity of DABs.

There is, however, a difference between DABs and adjudication in that if a party refuses to enforce a DAB decision, the other party will be unable to obtain immediate enforcement. This gap in cl.20 of the Contract has been identified by Professor Nael Bunni and is further discussed below in the commentary on cl.20.[5] In short, if a party refuses to enforce decision, he will be in breach of Contract but there will be no immediate consequences.

---

[3] *Sutcliffe v Thackrah* [1974] A.C. 727 at 737 per Lord Reid.
[4] *Beaufort Developments (N.I.) Ltd v Gilbert-Ash (N.I.) Ltd* [1999] 1 A.C. 266 per Lord Hoffman. Clearly the precise nature of the duty owed by any particular certifier will be a question of construction. However, for a recent review of the cases on the duties of the certifier, together with comments on this aspect of Lord Hoffman's speech, see the decision of Jackson J. in *Scheldebouw BV v St James Homes (Grosvenor Dock) Limited* [2006] EWHC 89, (2006) CILL 2317.
[5] See Nael Bunni "The Gap in Sub-Clause 20.7 of the 1999 FIDIC Contractors for Major Works" [2005] ICLR 272.

Clause 1.1.3.7: *"Defects Notification Period"*

This is an alteration from the Old Red Book, FIDIC 4th edn which used the phrase "Defects Liability Period". This change, and the replacement of the word "Liability", makes it clearer that the Contractor's liability does *not* end a year after taking-over.[6]   **1–032**

Clause 1.1.4.3: "Cost"

The definition should be noted since it is ". . . all expenditure reasonably incurred (or to be incurred) by the Contractor, whether on or off the Site, including overhead and similar charges, but does not include profit . . ."   **1–033**

Cost will need to be shown to have been incurred, and reasonably incurred, by the Contractor, but the entitlement to recover "Cost" does not automatically give rise to a right to an entitlement to an element of profit in addition.[7] That entitlement will depend on the actual wording of the individual sub-clause.

The FIDIC Contracts Guide ("the FIDIC Guide")[8] notes that overhead charges may include reasonable financing costs incurred by reason of payment being received after expenditure.

Sub-Clauses 1.1.5.4, 7–8

The reference to temporary and permanent works can create confusion. Some items (for example, tunnels, access ways) are potentially both. Alternatively, the permanent and temporary works may not make up the totality of the Works themselves. Therefore parties may want to consider agreeing specific definitions at the commencement of the project.   **1–034**

Sub-Clause 1.1.6.7: "Site"

Indeed this sub-clause requires that the site is clearly defined.   **1–035**

Sub-Clause 1.1.6.8: "Unforeseeable"

This is one of the more difficult and controversial sub-clauses.[9] As there have been two different versions of this definition in the two MDB Contracts, May 2005 and March 2006, it is important that the contracting parties check which version is being used. The definition of "unforeseeable" is of particular   **1–036**

---

[6] See Edward Corbett "FIDIC's New Rainbow 1st Edition – An Advance?" [2000] ICLR 253 at 254–255.
[7] See discussion on sub-clauses 1.2 and 8.4 below for further details.
[8] Published by FIDIC, 2000.
[9] Although perhaps not quite as controversial as it could have been. See discussion on the MDB version below.

relevance to sub-clauses 4.12 Unforeseeable Physical Conditions and 8.4 Extension of Time for Completion.

The concept of that which is unforeseeable – or that which is not reasonable to have foreseen by an experienced contractor – is familiar in the context of adverse physical conditions and artificial obstructions. The question of what the experienced contractor could and should have foreseen must take into account all the available sources of information, including, it is submitted, that which the Contractor *in fact* knew.

The reference in cl.1.1.6.8 to ". . . by the date of the submission of the Tender . . ." is an important reminder that the test of what was reasonably foreseeable is the time of submission of the tender. The definition is less clear than it might have been, however, on whether the definition is aimed at mere foresight, or whether there is an additional requirement that the Contractor was in a position to allow for the circumstance or event in his tender bid.

1–037   It is also important to note that events will only be regarded as "unforeseeable" when adequate precautions could not reasonably have been taken by an experienced contractor. This places an additional burden upon the Contractor since, when asserting conditions to have been "unforeseeable" he will, additionally, have to show that there were no adequate precautions which could reasonably have been taken by the experienced contractor.

The final part of the definition is that it must have been unforeseeable at the time of the date for the submission of the tender. This will presumably be the date for the return of priced tenders. However this is not a defined term.

MDB HARMONISED EDITION

1–038   There have been three words and/or phrases which have been replaced throughout in the MDB edition:

(i)   "Appendix to Tender" which has been replaced by "Contract Data".[9A] At one level, this is not a significant change as the information typically to be found in the Appendix to Tender now forms part of the Contract Data, defined in the new sub-cl.1.1.1.10. However the two are very different documents.

(ii)   "Dispute Adjudication Board" which has been replaced by "Dispute Board".
This is a change which reflects the World Bank terminology international usage in relation to dispute boards generally.

(iii)   "Reasonable profit" which has been replaced by "profit".

[9A] Just occasionally, see for example, sub-cl.14.5.

As discussed above, the definition of "cost" does not include profit.[10] However this does not mean actual profit. As set out below, there is a change to sub-cl.1.2, which states that where the expression, "*cost plus profit*", is used in the contract, the profit shall be one twentieth or 5 per cent of the Cost, unless otherwise indicated.

There are a number of other changes as follows:

Clause 1.1

The Particular Conditions in the MDB version have been divided into Part A and B. The detailed pages of Guidance from the 1999 Edition have been deleted. One reason for this is the incorporation of a number of the Particular Conditions into the MDB Version. The definition of profit set out above is one such example. Also, sub-cll 6.12–22 have been incorporated wholesale into the MDB Version from the original 1999 Edition Particular Conditions.   1–039

Part A relates to the "Contract Data" which is the information provided by the Employer. Essentially, this is the information which previously would have appeared in the "Appendix to Tender". For both the "Appendix to Tender" and "Contract Data", it is important that all the relevant information is filled in carefully.

Whilst there is no longer any need to fill in the Contractor's name and address, there are a number of new items which have been added to the Contract Data. These include details of the Bank's and Borrower's name (if any), and also a reference to limits on the Engineer's authority.[11]

If the relevant part of the "Contract Data" is filled in, variations which result in a percentage increase of the "Accepted Contract Amount" above a certain limit will require the approval of the Employer. There is also provision, if required, to set out the maximum total liability of the Contractor to the Employer.[12]   1–040

Part B relates to the Particular Conditions as set out in the 1999 Edition. However it is far more concise. The guidance notes that "*in a normal situation*", the General Conditions, taken with the information contained in Part A provide in the view of FIDIC, "*a perfectly satisfactory and legally sound basis for entering into the Contract*".

There are only two potential Particular Conditions mentioned. These relate to provisions taken from the Master Procurement Documents for Procurement of Works & Users Guide (March 2006):

(i) Sub-clause 6.23 – This recognises workers' rights to join workers' organisations;

---

[10] See sub-cl.1.1.4.3.
[11] See sub-cl.3.1 below for further information.
[12] See sub-cll 13 and 17.6 for further details.

(ii)   Sub-clause 14.1 – An additional sub-paragraph (e) is included relating to the importation of Contractor's Equipment.

Contractors and Employers should also be aware that it is likely that individual MDB's may develop their own specific conditions.

Sub-Clause 1.1.1

1–041 1.1.1.2: The words "(if any)" have been deleted.
1.1.1.4: The words "or letter of bid" have been added, which potentially serve to expand the meaning of "letter of tender".
1.1.1.9: The reference to "Appendix to Tender" has been deleted and replaced by the following:

"Bills of Quantities", "Daywork Schedule" and "Schedule of Payment Currencies" mean the documents so named (if any) which are comprised in the Schedules.

1.1.1.10: As the reference to "Bills of Quantities" and "Day Work Schedule" has been moved to 1.1.1.9, sub-cl.1.1.1.10 is new. It refers to "Contract Data" which means "the pages completed by the Employer entitled contract data which constitute Part A of the Particular Conditions".

Sub-Clause 1.1.2

1–042 1.1.2.2: The reference to "Appendix to Tender" has been replaced with "Contract Data".
1.1.2.9: The reference to "DAB" has been changed to "DB".
1.1.2.11: A new definition has been introduced: "Bank" means the financing institution (if any) named in the Contract Data. No definition of funder or lender appears in the 1999 Edition.
1.1.2.12: "Borrower" means the person (if any) named as the borrower in the Contract Data.

Sub-Clause 1.1.3

1–043 1.1.3.1: The words "and completion" have been added – presumably for certainty.
1.1.3.3: The words "Appendix to Tender" have been replaced by "Contract Data"
1.1.3.6: The words "provisions of the Particular Conditions" have been replaced by "specification".

1.1.3.7: The words "which extends over 12 months accept if otherwise" have been added, thereby clarifying the length of the Defects Notification Period.

The words "Appendix to Tender" have been replaced by "Contract Data".

Sub-Clause 1.1.4

There is no change. **1–044**

Sub-Clause 1.1.5

1.1.1.5   The following has been added to this clause: **1–045**

including vehicles purchased for the Employer and relating to the construction or operation of the Works.

The purchase of vehicles on behalf for the Employer, is something which the Contractor should keep a close eye on. If the Contractor is responsible for purchase (and as is often the case maintenance), costs can mount, sometimes unexpectedly.

1.1.5.6   The words "Appendix to Tender" have been replaced by "Contract Data".

Sub-Clause 1.1.6

1.1.6.7   The words "including storage and working areas" have been added **1–046** to the definition of site.

1.1.6.8   There is no change.

Note however that there was a controversial change in the first edition of the MDB Conditions in May 2005, which has now been withdrawn.

In this version, the words *"and against which adequate preventative precautions could not reasonably be taken"* were added to the definition of unforeseeable. This proposed change was met with widespread criticism, especially from contractors[13] who felt that the May 2005 definition increased their risk and would mean that the contractor would price their tender contingencies accordingly to take account of this risk. In particular, the contractors felt it was not reasonable to charge a contractor to have made adequate preventative precautions for events which were unforeseeable.

---

[13] See "The Contractor's View on the MDB Harmonised Version of the New Red Book", Richard Appuhn and Eric Eggink [2006] ICLR 4.

1.2 INTERPRETATION

1–047   *In the Contract, except where the context requires otherwise:*

  (a)   *words indicating one gender include all genders;*
  (b)   *words indicating the singular also include the plural and words indicating the plural also include the singular;*
  (c)   *provisions including the word "agree", "agreed" or "agreement" require the agreement to be record in writing;*
  (d)   *"written" or "in writing" means hand-written, type-written, printed or electronically made, and resulting in a permanent record;*

The marginal words and other headings shall not be taken into consideration in the interpretation of these Conditions.

OVERVIEW OF KEY FEATURES

1–048   • Any agreement must be recorded in writing.
       • Any agreement can be in writing if it is handwritten, typewritten, printed or electronic provided that a permanent record is made.

COMMENTARY

1–049   The first two parts of this sub-cl.(a) and (b) are entirely standard clauses.

   The statement that headings are not relevant to the interpretation of a contract is a useful reminder not to take them literally. Indeed a number of sub-clauses, including the MDB Version of sub-cl.1.2, relate to far more than the topic indicated by the heading. For example, sub-cl.4.8 is headed safety procedures but also deals with site security and the impact of the works on the surrounding areas. Whilst the heading of sub-cl.1.3 "Communications", rather surprisingly is the cover for the only place in the contract which sets any time requirement for the making of a determination for the issuing of certificates.

   The requirement at sub-para.(c) that any agreement made must be recorded in writing is important and can be easily forgotten.

1–050   This especially applies to any "electronically made" record which will be treated as a "written document" provided it results in a permanent record. Emails are taking on an increasing importance throughout the construction and engineering industry. Care must always be taken in drafting email and ensuring the email is sent to the correct addressee. It is recommended that emails are both printed out and saved to an electronic database.

The term permanent record applies to all four media. No definition is provided. With electronic documents, it should not be forgotten that, merely deleting an email will not prevent it from being recovered from a computer hard drive. Whether that would satisfy the need for a "permanent record" is unclear and has not yet been tested in the courts.

MDB HARMONISED EDITION

The following rather important paragraphs have been added: 1–051

(e)   *the word "tender" is synonymous with "bid", and "tenderer" with "bidder" and the words "tender documents" with "bidding documents".*

In these Conditions, provisions including the expression "Cost plus profit" require this profit to be one twentieth (5%) of this Cost unless otherwise indicated in the Contract Data.

The final addition is important and it seems strange that it has been inserted here and not somewhere more obvious, for example with the definition of "Cost" at subs.1.1.4. Nevertheless this represents an attempt to set a standard for what constitutes a reasonable profit. It is also an example of the adoption of one of the suggested Particular Conditions from the 1999 Edition.

1.3 COMMUNICATIONS

Wherever these Conditions provide for the giving or issuing of approvals,   1–052
certificates, consents, determinations, notices, requests and discharges, these communications shall be:

(a)   in writing and delivered by hand (against receipt), sent by mail or courier, or transmitted using any of the agreed systems of electronic transmission as stated in the Appendix to Tender; and

(b)   delivered, sent or transmitted to the address for the recipient's communications as stated in the Appendix to Tender. However:

   (i)   if the recipient gives notice of another address, communications shall thereafter be delivered accordingly; and

   (ii)   if the recipient has not stated otherwise when requesting an approval or consent, it may be sent to the address from which the request was issued.

Approvals, certificates, consents and determinations shall not be unreasonably withheld or delayed. When a certificate is issued to a Party, the certifier shall send a copy to the other Party. When a notice is issued to a Party, by the other Party or the sender, a copy shall be sent to the Engineer or the other Party, as the case may be.

OVERVIEW OF KEY FEATURES

1–053
- Approvals, Certificates, Consents, Determinations and the like shall be delivered as set out in the Appendix to Tender to the address as stated in the Appendix to Tender.
- Approvals, Certificates, Consents to Determination shall not be unreasonably withheld or delayed.

COMMENTARY

1–054 It is not just agreements which must be in writing. Indeed one of the few exceptions is instructions, which by sub-cl.3.3, are to be in writing, "wherever practical".

It is important that all parties are aware of the correct address to which communications should be sent.

Care must be taken by both parties to ensure that those working at the place to which communications are to be sent are also aware of this. For example, there is little point in giving a formal registered office address if that registered office is not used on a regular basis as this may mean that the notices and the like are not dealt with within either the contractually required or a reasonable time, if they are dealt with at all.

1–055 Care must also be taken to ensue that proper procedures are in place to monitor fax machines and computers. There have been two recent English authorities which demonstrate the potential problems which may arise. In one, a Judge held the delivery of contractual notice by fax constituted actual delivery for the purposes of the contract.[14] That notice had been served by fax on the morning of December 23, 2005. The office shut for Christmas that afternoon and owing to the Christmas break the fax was not seen until January 3, 2006.

In the second case[15] arbitration proceedings were served at an email address which appeared in the Lloyds Maritime Directory and on the

---

[14] *Construction Partnership UK Limited v Leek Developments Limited* 2006 CILL 2357.
[15] *Bernuth Lines v High Seas Shipping* 2006 CILL 2343.

company's website. The email was received, but then ignored by the clerical staff. The Judge held that the service was valid and the failings of the internal administration were the responsibility of the company concerned.

It should be noted that the European Union has adopted a community framework for electronic signatures under art.5 of Directive 1999/93 EC which provides that:

Member States shall ensure that advanced electronic signatures which are   **1–056**
*based on a qualified certificate and which are created by a secure-signature-
creation device:*

  (a)   *satisfy the legal requirements of a signature in relation to data in electronic form in the same manner as a hand-written signature satisfies
those requirements in relation to paper-based data; and*

  (b)   *are admissible as evidence in legal proceedings.*

The qualified certificate must be issued by a certification service provider meeting specific requirements laid down in the Directive.

A basic legal recognition of electronic signatures has now been largely   **1–057** implemented in the EU, which represents a success for the Directive. Nonetheless, the national interpretations of the Directive and the specific national regulations of details of e-signatures reveal significant differences.

Indeed it should also be noted that the spread of e-signatures is a world-wide phenomenon. For example, China recently promulgated its own "Electronic Signatures Law", which came into force on April 1, 2005.

The final part of this sub-clause is of particular significance. It is another example of a potentially important sub-clause being hidden under a fairly innocuous sounding heading. This is not something which is repeated else-where. The final paragraph notes that approvals, certificates, consents and determinations shall not be unreasonably, withheld or delayed. This is partic-ularly important because, for example, sub-cl.3.5 does not provide any time limit for the Engineer to make his determinations. Whilst this sub-clause provides some assistance, what is a reasonable period will depend on the facts of every particular case.

The case of *Neodox Limited v Swinton and Pendlebury Borough Council*[16]   **1–058** provides some guidance on this point. Here a civil engineering contractor entered into a contract with the Borough for the construction of sewerage works and sewers. The contractor alleged that a term could be implied that the engineer should supply:

*all the details and instructions necessary for the execution of the works in
sufficient time to enable the contractors to execute and complete the works in*

---

[16] 5 B.L.R. 38.

an economic and expeditious manner and/or in sufficient time to prevent the contractors being delayed in such execution and completion.

Diplock J. held that it was clear from the terms of the contract that instructions would be given from time to time during the course of the contract, and that what was a reasonable time did not depend solely on the convenience and financial interest of the contractor. The question of reasonableness had also to be considered from the point of view of the engineer and his staff and of the owners themselves. Other relevant matters affecting reasonableness would be the order in which the works were to be carried out as approved by the engineer, whether requests for particular details had been made by the contractor, whether the details related to variations or to the original works, and also the contract period.

1–059   What was a reasonable time was a question of fact having regard to all the circumstances of the case. Therefore, insofar as far as the late instructing of variations is concerned, any liability of the engineer depended on whether the reason for a later variation, and hence of any subsequent more detailed information, implied negligence or fault on the part of the engineer either in failing properly to pre-plan the work or due to a defective design on his part requiring subsequent correction. Instructions should be given at such times and in such manner as not to hinder or prevent the contractor from performing his duties under the contract. As it happened, here the engineer's decisions could not be regarded as requiring the performance of a variation or an addition to the works and the contractor was not entitled to recover a sum additional to the contract rates.

MDB HARMONISED EDITION

1–060   The words "Appendix to Tender" have been replaced by "Contract Data". The "Contract Data" does not provide for details of the Contractor's address and this will need to be clarified, particularly for the purposes of this sub-clause.

1.4 LAW AND LANGUAGE

1–061   *The Contract shall be governed by the law of the country (or other jurisdiction) stated in the Appendix to Tender.*

If there are versions of any part of the Contract which are written in more than one language, the version which is in the ruling language stated in the appendix to tender shall prevail.

The language for communications shall be that stated in the Appendix to Tender. If no language is stated there, the language for communications shall be the language in which the Contract (or most of it is written).

OVERVIEW OF KEY FEATURES

- The Appendix to Tender should set out the law of the Contract.
- The Appendix to Tender should set out the language of the Contract.
- This is important because the parties should choose which law and language governs the contract and the performance of the Works.

1–062

COMMENTARY

"Governing law" clauses, such as this one, should not be treated lightly in contract negotiations, especially if there is an international element to the contract in question. It is important to remember that the governing law clause dictates the law which is to be adopted when resolving disputes. It is therefore different from the jurisdiction clauses, to be found at clause 20 which dictates the forum in which any dispute is to be resolved.

1–063

This is particularly important here because, unlike with the language provisions, there are no fall back provisions, if the parties fail to choose a governing law. Therefore, the failure to choose a governing law may lead to a long and costly dispute. Particular problems arise in cases where there is a dispute under a contract with an international element. Typically, the parties in question may not be based in the same country, or the performance of the contract might take place in a different country and so different parties will have a preference for different laws.

In other words, the parties should be aiming to act differently to Apple Corps and Apple Computer Inc who in an attempt to resolve their differences in 1991 came to an agreement which contained no provisions as to:

(i)   the courts of which country should have jurisdiction over any dispute concerning the agreement;
(ii)   which country's laws should govern the agreement; or even
(iii) in which country the agreement was made.

When the courts, some 13 years later, almost inevitably, had to decide whether the law of England or California was the governing law, Mr Justice Mann noted:

1–064

The evidence before me showed that each of the parties was overtly adamant that it did not wish to accept the other's jurisdiction or governing law, and could reach no agreement on any other jurisdiction or governing law. As a result [the agreement] contains no governing law clause and no jurisdiction clause. In addition, neither party wanted to give the other an advantage in terms of where the agreement was finalised. If their intention in doing so was to create obscurity and difficulty for lawyers to debate in future years, they have succeeded handsomely.[17]

Plainly, wherever possible this should be avoided.

MDB HARMONISED EDITION

1–065   The words "Appendix to Tender" have been replaced by "Contract Data".
In addition, the second and third paragraphs have been simplified so that the ruling language is to be that stated in the Contract Data, or if no language is stated there, the language for communication shall be the ruling language of the contract. No definition is provided of "ruling language" of the contract. It is to be presumed that this is the language in which the majority of the contract is drafted.

1.5 PRIORITY OF DOCUMENTS

1–066   *The documents forming the Contract are to be taken as mutually explanatory of one another. For the purposes of interpretation, the priority of the documents shall be in accordance with the following sequence:*

(a)   *the Contract Agreement (if any),*
(b)   *the Letter of Acceptance,*
(c)   *the Letter of Tender,*
(d)   *the Particular Conditions,*
(e)   *these General Conditions,*
(f)   *the Specification,*
(g)   *the Drawings, and*
(h)   *the Schedules and any other documents forming part of the Contract.*

If an ambiguity or discrepancy is found in the documents, the Engineer shall issue any necessary clarification or instruction.

---

[17] *Apple Corps Limited v Apple Computer Inc* [EWHC] 768.

OVERVIEW OF KEY FEATURES

- The priority of documents is, unless stated otherwise as listed at items (a)–(h).[18]   **1–067**
- The Engineer shall be responsible for resolving any discrepancy.

COMMENTARY

The order of priority of the documents forming the Contract can be significant where complex questions of interpretation arise. Such clauses are typically included in case there is a conflict in meaning between the various contract documents.   **1–068**

It is to be noted that the General Conditions – the FIDIC Standard Form – is placed relatively low in order of priority, but this itself reflects the traditional English common law position that the written words chosen by the parties ought to have priority, on questions of interpretation, over the printed words (or standard wording) used by the parties.[19] Those advising parties in the negotiation of the Contract will need to think carefully about the matters of significance to be included in the Contract Agreement and the Letter of Acceptance, together with the Particular Conditions, which it should be noted also rank of higher importance than the General Conditions.

Although, the final paragraph of this sub-clause sensibly suggests that if necessary, the Engineer should issue clarification, there is no apparent sanction if no such instruction is issued.

The Particular Conditions suggest the following alternative if no order of precedence is to be prescribed: **1–069**

The documents forming the Contract are to be taken as mutually explanatory of one another. If no ambiguity or discrepancy is found, the priority shall be such as may be accorded by the governing law. The Engineer has authority to issue any instruction which he considers necessary to resolve an ambiguity or discrepancy.

However, it is thought better to use the original sub-clause.

---

[18] This order reflects the sequence of definitions set out in sub-clause 1.1.1.
[19] *Robertson v French* (1803) 4 East 130; and *Glynn v Margetson* [1893] A.C. 351, HL.

MDB HARMONISED EDITION

1–070   The list has been extended so that items (d) and (e) read Particular Conditions – Part A and Part B respectively.

1.6 CONTRACT AGREEMENT

1–071   *The Parties shall enter into a Contract Agreement within 28 days after the Contractor receives the Letter of Acceptance, unless they agree otherwise. The Contract Agreement shall be based upon the form annexed to the Particular Conditions. The costs of stamp duties and similar charges (if any) imposed by law in connection with entry into the Contract Agreement shall be borne by the Employer.*

OVERVIEW OF KEY FEATURES

1–072   • The Parties are to enter into the Contract Agreement within 28 days of the receipt of the Contractor's of the Letter of Acceptance.
        • The form of contract shall be that annexed to the Particular Conditions.
        • The Employer will be responsible for any stamp duties or other charges.

COMMENTARY

1–073   Lawyers frequently stress the importance, in terms of certainty if nothing else, of the parties actually signing up to and entering into a formal no contract. This sub-clause is designed to encourage this. However, there is and practically cannot be, any sanction for failing to do so. It is entirely possible that using the general rules of offer and acceptance that the Parties could find themselves under a binding Contract without having entered into any of the formalities set out here.

MDB HARMONISED EDITION

1–074   The words "unless they agree otherwise" have been changed to read "unless The Particular Conditions establish otherwise".

Note that under the Contract Data, one of the conditions governing the Commencement Data is the signing of the Contract Agreement by both parties. This thereby acts as some form of imperative to ensure the Contract is signed.

1.7 ASSIGNMENT

Neither Party shall assign the whole or any part of the Contract or any benefit   **1–075**
or interest in or under the Contract. However, either Party:

(a) *may assign the whole or any part with the prior agreement of the other Party, at the sole discretion of such other Party, and*
(b) *may, as security in favour of a bank or financial institution, assign its right to any moneys due, or to become due, under the Contract.*

OVERVIEW OF KEY FEATURES

Neither party can assign its interest in the Contract without the prior   **1–076**
agreement of the other.

COMMENTARY

This sub-clause includes a prohibition on assignment, save where (at his   **1–077**
absolute discretion) the other party gives its consent, or where the party's interest under the Contract is offered to secure finance (no consent required).

MDB HARMONISED EDITION

There is no change.   **1–078**

1.8 CARE AND SUPPLY OF DOCUMENTS

The Specification and Drawings shall be in the custody and care of the   **1–079**
Employer. Unless otherwise stated in the Contract, two copies of the Contract

and of each subsequent Drawing shall be supplied to the Contractor, who may make or request further copies at the cost of the Contractor.

Each of the Contractor's Documents shall be in the custody and care of the Contractor, unless and until taken over by the Employer. Unless otherwise stated in Contract, the Contractor shall supply to the Engineer six copies of each of the Contractor's Documents.

The Contractor shall keep, on the Site, a copy of the Contract, publications named in Specification, the Contractor's Documents (if any), the Drawings and Variations and other communications given under the Contract. The Employer's Personnel shall have the right of access to all these documents at all reasonable times.

If a Party becomes aware of an error or defect in a document of a technical nature which was prepared for use in executing the Works, the Party shall promptly give notice to the other Party of such error or defect.

OVERVIEW OF KEY FEATURES

1–080
- The Employer shall keep the Specification and Drawings.
- The Employer shall provide the Contractor with two copies of the Contract and every drawing.
- If the Contractor requires more than two copies, he shall have to pay for them.
- The Contractor shall look after the Contractor's Documents.
- The Contractor must supply the Engineer with six copies of each of the Contractor's Documents.
- At least one copy of the Contract, Contractor's Documents, Drawings and Variations must be kept by the Contractor on site.
- The Employer's Personnel shall have access to the documents kept on site at all reasonable times.
- If a party becomes aware of an error or defect of a technical nature in any of the documents prepared for the use in executing the works, they must promptly give notice of this to the other parties.

COMMENTARY

1–081 This is another example of a potentially significant clause being almost hidden away in the definitions part of the contract.

The first three paragraphs are self-explanatory, although the parties should take notice of the number of copies of documents that must be provided. For example, the Employer is entitled to charge the Contractor if it requires more than two copies of the contract or drawings; whereas the Contractor has to supply to the Employer six copies of its documents.

The most important part of this clause relates to the final paragraph. If a party becomes aware of an error or defect in a document of a technical nature[20] then it must promptly give notice to the other party of this error or defect. Whilst the Contract sets out who is responsible for the errors, the responsibility for those errors (or at least the degree of responsibility for the costs, in time and money, for rectifying those errors), might be substantially reduced if it can be proven that a party became aware of an error in a document, but did not bring it to the other's attention.

MDB HARMONISED EDITION

The words "of a *technical nature*" have been deleted, thereby extending the   **1–082** meaning of this clause. If a Party becomes aware of any defect at all then they must promptly give notice of this.

1.9 DELAYED DRAWINGS OR INSTRUCTIONS

The Contractor shall give notice to the Engineer whenever the Works are likely   **1–083** *to be delayed or disrupted if any necessary drawing or instruction is not issued to the Contractor within a particular time, which shall be reasonable. The notice shall include details of the necessary drawing or instruction, details of why and by when it should be issued, and details of the nature and amount of the delay or disruption likely to be suffered if it is late.*

If the Contractor suffers delay and/or incurs Cost as a result of a failure of the Engineer to issue the notified drawing or instruction within a time which is reasonable and is specified in the notice with supporting details, the Contractor shall give a further notice to the Engineer and shall be entitled subject to Sub-Clause 20.1 [Contractor's Claims] to:

(a) an extension of time for any such delay, if completion is or will be delayed, under Sub-Clause 8.4 [Extension of Time for Completion], and
(b) payment of any such Cost plus profit, which shall be included in the Contract Price.

After receiving this further notice, the Engineer shall proceed in accordance with Sub-Clause 3.5 [Determinations] to agree or determine these matters.

---

[20] The FIDIC Guide specifically notes that sub-clause 1.8 does not refer to errors or defects of a financial or other nature.

However, if and to the extent that the Engineer's failure was caused by any error or delay by the Contractor, including an error in, or delay in the submission of, any of the Contractor's Documents, the Contractor shall not be entitled to such extension of time, cost or profit.

OVERVIEW OF KEY FEATURES

1–084   Sub-clause 1.9 provides for the giving of notice by the Contractor upon the occurrence of delay and/or where information is not provided. There is the provision for a *further* notice, after which, if there is continued default, the Contractor is entitled to an extension of time and his Costs plus profit.

COMMENTARY

1–085   This is an important provision relating to claims that might easily be overlooked in this essentially introductory section of the Contract. It is suggested that the positioning of this important provision is not helpful. At all events, the provision in its terms places a significant practical burden upon the Contractor since, in a complex project, it will not always be clear that the lack of provision, or the late provision, of, say, steel erection drawings, is likely of itself to cause delay and/or disruption to the Works. However, the Contractor ought to have sufficient site resource and site experience to take a pragmatic, but cautious, approach to written notices, and provide the requisite notice where there is any realistic possibility of delay *or disruption* resulting from late release of information.

An important aspect of the notice requirements in sub-clause 1.9 is the requirement for the Contractor to specify a reasonable period by which the information is required (presumably the date being selected as one which will avoid any, or any further, delay or disruption to the Works). If this date is not met, then the Contractor is required to provide a further notice. The *effect* of such late provision is not to give rise to a claim for breach of implied terms as to hindrance and non-cooperation, but to give rise to claims for time and money under the Contract.

The two-stage process of notification by the Contractor – with the Contractor, at the first stage, identifying reasonable dates for the provision of information which is *ex hypothesi* already late – is a useful and welcome tool for recovering progress during the course of the works. From the Contractor's point of view, compliance with, and making practical use of, sub-cl.1.9, will require solid planning expertise during the course of the Works.

MDB HARMONISED EDITION

There are two changes; one small the other more significant.   **1–086**
    First, the words "details of which appear before the nature" in the first paragraph have been deleted.
    Second, the word "reasonable" has been deleted in sub-cl.(b). This is a deletion which occurs throughout the MBD version.[21]

1.10 EMPLOYER'S USE OF CONTRACTOR'S DOCUMENTS

As between the Parties, the Contractor shall retain the copyright and other   **1–087**
intellectual Contractor's Documents property rights in the Contractor's
Documents and other design documents made by (or on behalf of) the
Contractor.
    The Contractor shall be deemed (by signing the Contract) to give to the
Employer a non-terminable transferable non-exclusive royalty-free licence to
copy, use and communicate the Contractor's Documents, including making and
using modifications of them. This licence shall:

(a)   apply throughout the actual or intended working life (whichever is longer)
       of the relevant parts of the Works,
(b)   entitle any person in proper possession of the relevant part of the Works
       to copy, use and communicate the Contractor's Documents for the
       purposes of completing, operating, maintaining, altering, adjusting,
       repairing and demolishing the Works, and
(c)   in the case of Contractor's Documents which are in the form of computer
       programs and other software, permit their use on any computer on the
       Site and other places as envisaged by the Contract, including replace-
       ments of any computers supplied by the Contractor.

The Contractor's Documents and other design documents made by (or on   **1–088**
behalf of) the Contractor shall not, without the Contractor's consent, be used,
copied or communicated to a third party by (or on behalf of) the Employer for
purposes other than those permitted under this Sub-Clause.

---

[21] See also sub-cll 2.1, 4.7, 7.4, 10.2 and 10.3, 11.8, 12.3, 16.1 and 17.4.

OVERVIEW OF KEY FEATURES

1–089 • The Contractor retains copyright in its own documents.
 • The Employer has a right to use and copy the Contractor's documents during the life of the project provided that use relates to the project in question

COMMENTARY

1–090 Although copyright remains with the Contractor, the Employer for the purposes of the particular contract only, has the right to use the Contractor's documents.

MDB HARMONISED EDITION

1–091 There is no change.

1.11 CONTRACTOR'S USE OF EMPLOYER'S DOCUMENTS

1–092 *As between the Parties, the Employer shall retain the copyright and other intellectual Employer's Documents property rights in the Specification, the Drawings and other documents made by (or on behalf of) the Employer. The Contractor may, at his cost, copy, use, and obtain communication of these documents for the purposes of the Contract. They shall not, without the Employer's consent, be copied, used or communicated to a third party by the Contractor, except as necessary for the purposes of the Contract.*

OVERVIEW OF KEY FEATURES

1–093 • The Employer retains copyright in its own documents.
 • The Contractor may, for the purpose of the project, copy and use the employer's documents.

COMMENTARY

As with sub-cl.1.10, the Contractor may use, for the purposes of this contract only, documents belonging to the Employer.

1–094

MDB HARMONISED EDITION

There is no change.

1–095

1.12 CONFIDENTIAL DETAILS

The Contractor shall disclose all such confidential and other information as the Engineer may reasonably require in order to verify the Contractor's compliance with the Contract.

1–096

OVERVIEW OF KEY FEATURES

If the Engineer requires information, which is confidential, to check up on the Contractor's performance of the Contract, that information must be disclosed by the Contractor.

1–097

COMMENTARY

This is a one-sided provision and no guidance is provided on what is to happen to this information. In particular, the Contract does not provide any specific obligation on the Engineer to keep any such information disclosed confidential.

1–098

Equally there is no requirement on either party to keep details of the contract private and confidential. This has been addressed in part in the Particular Conditions which suggest the following addition:

The Contractor shall treat the details of the Contract as private and confidential, except to the extent necessary to carry out obligations under it or to comply with applicable Laws. The Contractor shall not publish, permit to be published or disclose any particulars of the Works in any trade or technical paper or elsewhere without the previous agreement of the Employer.

This proposed addition is one sided in that it only imposes an obligation of confidentiality upon the Contractor and not the Employer.

MDB HARMONISED EDITION

There have been substantial changes to this clause. The words "the Engineer may reasonably require" and "the Contractor's compliance with the Contract" have been deleted.

1–099   Following on from the Particular Conditions, a new paragraph has been added. Therefore cl.1.12 reads as follows:

> *The Contractor's and the Employer's Personnel shall disclose all such confidential and other information as may be reasonably required in order to verify compliance with the Contract and allows its proper implementation.*
>
> *Each of them shall treat the details of the Contract as private and confidential, except to the extent necessary to carry out their respective obligations under the Contract or to comply with applicable Laws. Each of them shall not publish or disclose any particulars of the Works prepared by the other Party without the previous agreement of the other Party. However, the Contractor shall be permitted to disclose any publicly available information, or information otherwise required to establish as qualifications to compete for other projects.*

1–100   Thus the MDB wording is not one-sided and means both parties must, if it is reasonable to do so, disclose confidential information.

The final sentence will be a particular relief to Contractors when it comes to looking and bidding for new projects to work on.

1.13 COMPLIANCE WITH LAWS

1–101   *The Contractor shall, in performing the Contract, comply with applicable Laws. Unless otherwise stated in the Particular Conditions:*

> *(a)   the Employer shall have obtained (or shall obtain) the planning, zoning or similar permission for the Permanent Works, and any other permissions described in the Specification as having been (or being) obtained by the Employer; and the Employer shall indemnify and hold the Contractor harmless against and from the consequences of any failure to do so; and*
>
> *(b)   the Contractor shall give all notices, pay all taxes, duties and fees, and obtain all permits, licences and approvals, as required by the Laws in rela-*

tion to the execution and completion of the Works and the remedying of any defects; and the Contractor shall indemnify and hold the Employer harmless against and from the consequences of any failure to do so.

OVERVIEW OF KEY FEATURES

- The Contractor must comply with the applicable laws.   **1–102**
- The Employer is responsible for obtaining planning and other permissions for the permanent works.
- If the Employer fails to obtain the correct or proper planning permission, it must indemnify the Contractor for any consequence of this failure.
- It is the responsibility of the Contractor to give all notices and pay all taxes and other duties in relation to the completion of the Works.
- If the Contractor fails to do this, it is responsible for indemnifying the Employer for the consequences of this failure.

COMMENTARY

This sub-clause is tied up with sub-cl.1.4 and the extent of the Contractor's   **1–103**
obligations will depend on the law chosen by the parties to govern the Contract. This sub-clause also needs to be read in conjunction with sub-cl.2.2 which deals with permits, licences and approvals. Under sub-cl.2.2, the Employer must provide reasonable assistance to the Contractor by obtaining copies of the laws of the country and also with the application for any permits and licences which are required.

It is suggested that this sub-clause and in particular the differences in responsibility between the Employer and Contractor could be clearer. For example what does "similar permission" mean? To what extent, if at all, is it different from the need to obtain permits and licences? As a consequence, it might be beneficial for the contract to include a detailed scheme of the permits required which identifies the people responsible for obtaining them.

MDB HARMONISED EDITION

Sub-paragraph (a) the words "building permit" have been added after   **1–104**
"zoning". In addition, the words "(or being)" have been changed to "(or to be)".

The following has been added to para.(b):

Unless the Contractor is impeded to accomplish these actions and shows evidence of its diligence.

Therefore it is important that the Contractor keeps proper records of steps it takes so that it can, if necessary, justify its performance.

1.14 JOINT AND SEVERAL LIABILITY

1–105 *If the Contractor constitutes (under applicable laws) a joint venture, consortium or other unincorporated grouping of two or more persons:*

(a) *these persons shall be deemed to be jointly and severally liable to the Employer for the performance of the Contract;*
(b) *these persons shall notify the Employer of their leader who shall have authority to bind the Contractor and each of these persons; and*
(c) *the Contractor shall not after its composition or legal status without the prior consent of the Employer.*

OVERVIEW OF KEY FEATURES

1–106
  • If the Contractor is a joint venture, or other consortium, then all members of the joint venture must be jointly and severely liable to the Employer.
  • The joint venture, or other consortium, shall nominate a leader or other direct point of contact.

COMMENTARY

1–107 Contractors typically form consortiums for major international projects. One reason for this is the grouping of expertise. Projects can be of a wide-ranging nature and often companies simply do not have the necessary expertise. It is of no surprise that the Employer would require all members of the joint venture to be jointly and severely liable to it. Typically, the Employer may also require parent company guarantees or other forms of security from each member of the consortium. Indeed, the prudent Employer might also want to review the terms of any joint venture agreement.

It is therefore important that every member of the joint venture or other consortium, should fully understand that it could be held liable for defaults of their fellow members and equally obviously, it is important that the relationship between the members of the joint venture, or other consortium, is itself the subject of a clearly defined contract.

It should be noted that there is no similar provision relating to the Employer. Again, typically, Employers are joint ventures. Often, umbrella companies are formed for specific projects. This is one reason why it is important that contractors are as clear as they can be as to the status and likely financial security and stability of the Employer.

MDB HARMONISED EDITION

There is no change. 1–108

1.15 INSPECTIONS AND AUDIT BY THE BANK

The Contractor shall permit the Bank and/or persons appointed by the Bank to   1–109
inspect the Site and/or the Contractor's accounts and records relating to the
performance of the Contract and to have such accounts and records audited by
auditors appointed by the Bank if required by the Bank.

This sub-clause only appears in the MBD version. There is no specific requirement of confidentiality and it is not thought that the Bank is necessarily caught by the provisions of sub-cl.1.13. Therefore the Contractor might want to clarify this before granting the access required by this sub-clause.

CLAUSE 2 – THE EMPLOYER

2.1 RIGHT OF ACCESS TO THE SITE

The Employer shall give the contractor right of access to, and possession of, all **2–001**
parts of the Site within the time (or times) stated in the Appendix to Tender.
The right and possession may not be exclusive to the Contractor. If, under the
Contract, the Employer is required to give (to the Contractor) possession of
any foundation, structure, plant or means of access, the Employer shall do so in
the time and manner stated in the Specification. However, the Employer may
withhold any such right or possession until the Performance Security has been
received.

If no such time is stated in the Appendix to Tender, the Employer shall give
the Contractor right of access to, and possession of, the Site within such times
as may be required to enable the Contractor to proceed in accordance with the
programme submitted under Sub-Clause 8.3 [Programme].

If the Contractor suffers delay and/or incurs Cost as a result of a failure by
the Employer to give any such right or possession within such time, the
Contractor shall give notice to the Engineer and shall be entitled subject to
Sub-Clause 20.1 [Contractor's Claims] to:

(a)   an extension of time for any such delay, if completion is or will be
delayed, under Sub-Clause 8.4 [Extension of Time for Completion], and
(b)   payment of any such Cost plus reasonable profit, which shall be included
in the Contract Price.

After receiving this notice, the Engineer shall proceed in accordance with **2–002**
Sub-Clause 3.5 [Determinations] to agree or determine these matters.

However, if and to the extent that the Employer's failure was caused by any
error or delay by the Contractor, including an error in, or delay in the submis-
sion of, any of the contractor's Documents, the Contractor shall not be entitled
to such extension of time, cost or profit.

OVERVIEW OF KEY FEATURES

- The Employer shall give the Contractor right of access to and possession   **2–003**
  of all parts of the site as stated in the Contract.
- Such right to access to and possession of the site may not be exclusive.

37

- Access or possession to the site may be withheld until the Contractor provides the Performance Security required by sub-cl.4.2.
- If the Contractor suffers delay or incurs costs due to (a) the failure by the Employer to give access as set out in the Contract or, (b) if no time is specified, within such time as may be required to enable the Contractor to proceed in accordance with his programme, provided he gives notice to the Engineer the Contractor may be entitled to an extension of time pursuant to sub-cl.8.4 and payment of any costs (including reasonable profit).
- No extension of time will be granted if the Employer's failure to give possession or access was caused by an error or delay by the Contractor, including delay in the submission of the Contractor's Documents.

COMMENTARY

2–004   This sub-clause provides that the Employer must give the Contractor access to enter the site within the time set out in the Contract. This is a key obligation on the part of the Employer. A failure to do so might lead to the Contractor being entitled to an extension of time and consequential costs. As recommended in the Particular Conditions, the prudent Employer if it recognises that it may not be able to grant access early or exclusively, should make this clear in the Specification.

Access

2–005   The definition of site is a wide one. Under sub-cl.1.1.6.7, the "Site" is defined as the place not only where the permanent works are to be executed, and to which plant and materials are to be delivered, but also "any other places as may be specified in the Contract as forming part of the site". Therefore, it is important that all parties to the Contract understand where these "other places" may be.

The right of access referred to here is the right to enter the site. This is not the same as the "Access Route" which is referred to at sub-cl.4.15. It is the Contractor's responsibility to satisfy himself as to the adequacy of any access routes to the site.

The time within which the Employer must provide access is usually set out in the Appendix to Tender, where there is provision to fix the number of days within which the Employer must give access to the site. This timescale is calculated by using the Commencement Date as a base date. By sub-cl.8.1, the Contractor will be given seven-days Notice of the Commencement Date. The Commencement Date will be no later than 42 days after the Contractor receives the Letter of Acceptance.

If nothing is set out in the Appendix to Tender, then the Employer must **2–006** provide access not within a reasonable time, but so as to enable the Contractor to proceed in line with the programme that has been submitted in accordance with sub-cl.8.3. By sub-cl.8.3, the Contractor's programme must be provided to the Engineer within 28 days after receiving the Notice, pursuant to sub-cl.8.1, of Commencement of Works. In addition, the reference in the second paragraph to times in the plural suggests that the provision of access in these circumstances need not be to the entire site in one go.

Sub-clause 2.1 is carefully qualified in that it expressly states that the Contractor will not necessarily have exclusive rights of access and possession of the site. It is important that the Contractor appreciates this. In contrast, under English (or common law) jurisdictions, where there is no express term, a term would be implied into most contracts to provide the Contractor with such reasonable access and possession as is required to enable him to carry out his contractual rights and obligations.

The Contractor should try to establish as part of the tender process to what extent his access will not be exclusive. If there are areas of site over which the Contractor has only a shared control or if the Contractor is unable to have free and unfettered possession and is required to work with or around other contractors, then this will have an obvious impact on both the likely tender figure and also the project programme. In addition, by virtue of sub-clause 4.6, the Contractor must co-operate with other contractors working for the Employer on or near the site.

This sub-clause also refers to the possibility of the Contractor being given **2–007** access to parts of the site. It is important that the Contractor is aware of exactly which parts of the site will be available and when. The Contractor should check that this is made clear in the Appendix to Tender.

As soon as the Contractor takes possession of the site, he becomes responsible for a number of obligations. Among the most important of these are provided by sub-cl.4.8 and 4.22 whereby the Contractor becomes fully responsible for safety and site security. These are important considerations in themselves but again the Contractor will need to understand the likely effect on these obligations if his access and/or possession are not exclusive.

This sub-clause does not deal with the question of the Employer's right of access to the site. Therefore, although it could be said that this question is dealt with by sub-cl.7.3, for the avoidance of any doubt[1], the Employer might want to consider inserting a clause confirming his rights of access:

"The Contractor will allow the Employer and/or the Employer's Personnel access to the site as and when required by the Employer for any purpose whatsoever."

---

[1] As sub-cl.7.3 primarily deals with rights of inspection.

Performance Security

2–008   Sub-clause 4.2 deals with the bond or Performance Security which the Contractor may be required to provide under the Contract. By sub-cl.4.2, the Contractor must provide the Performance Security within 28 days of receiving the Letter of Acceptance. The parties to the contract must therefore check the appropriate timescales set out in the Contract for the provision of access to the site to ensure that there is no conflict between these two requirements.

By sub-cl.14.6, an Employer may withhold payment if the Performance Security has not been provided within the contractual timescale. Here sub-cl.2.1 provides an additional tool for the Employer to encourage the provision of the necessary Performance Security. The Employer *may* withhold possession where no Performance Security has been provided. Therefore it is important that the parties keep each other informed of any delays in obtaining this security. An Employer will need to balance the importance of obtaining the financial security provided by the Performance Security against the need to commence work on site promptly.

It is not unusual for contracts to try and make the provision of a performance bond and/or guarantee a condition precedent to liability or obligation on the part of the Contractor.

2–009   In *Chiemgauer Membran und Zeltbau GMBH (formerly t/a Koch Hightex GMBH) v New Millennium Experience Co Ltd (Formerly Millennium Central Ltd) (1999)* [2] a claim for damages following an alleged wrongful termination was met by the defence that as the Claimant had failed to fulfil the terms of a condition precedent to provide the bond, the Defendant was under no liability. This defence failed because the English Court of Appeal concluded that the real issue was whether the parties had intended that the provision of the documents was of such fundamental importance that the result of the failure to provide them would be that the Claimant would lose all rights to payment, even under the letter of intent. That question was answered quite sharply on the basis that it clearly represented a "commercial nonsense".

Whilst that is not the situation here and the Employer has a choice about whether to withhold access or not, it is submitted that a Contractor would not be able to make a successful claim for an extension of time in circumstances where the reason that site access was withheld was a failure to provide a bond.

Claims

2–010   If a Contractor wants to claim an extension of time, costs and profit as a consequence of any failure of the Employer to provide access to the site in

---

[2] (2000) CILL 1595.

due time, the Contractor must give notice in accordance with the provisions of cl.20.

In the English case of *Rapid Building Group Ltd v Ealing Family Housing Association Ltd*[3], the Contractor, where the contract stated that possession of the site should be given on a specified date, found that part of the site was occupied by squatters. Consequently the employer was not able to give the Contractor possession of the whole of the site on the due date and the Contractor was entitled to an extension of time for the 19 days it took to evict the squatters. Whilst, under sub-cl.2.1 a similar result would happen here, the Contractor should remember that if, for example, protestors came on to the site once the Contractor had possession (and was therefore responsible for site security), then they would be the Contractor's and not the Employer's responsibility.

A Contractor will not be entitled to an extension of time, profit or cost if the Employer's failure to give possession or access was caused by an error or delay by the Contractor. The example given in the sub-clause is any delay in the submission of the Contractor's Documents. This is confirmed by sub-cl.4.6, which requires that where the Employer is required to give possession of any structure, plant or means of access in accordance with the Contractor's Documents, then the Contractor is responsible for submitting that information in a timely and accurate manner.

MDB HARMONISED EDITION

As noted above, the words "Appendix to Tender" have been replaced by **2–011** "Contract Data". In addition, the word "reasonable" has been deleted, so that the Contactor is entitled to payment of its Cost plus profit not reasonable profit.

Further, the words "without disruption" have been added to the second paragraph. These words serve to reinforce that the requirement that the Employer must give the Contractor access in such a way that there is no impediment on the Contractor working in accordance with its programme.

2.2 PERMITS, LICENCES OR APPROVALS

The Employer shall (where he is in a position to do so) provide reasonable **2–012** *assistance to the Contractor at the request of the Contractor:*

---

[3] 29 BLR 5.

(a)   by obtaining copies of the Laws of the Country which are relevant to the Contract but are not readily available, and

(b)   for the Contractor's applications for any permits, licences or approvals required by the Laws of the Country:

   (i)   which the Contractor is required to obtain under Sub-Clause 1.13 [Compliance with Laws],

   (ii)   for the delivery of Goods, including clearance through customs, and for the export of Contractor's Equipment when it is removed from the Site.

OVERVIEW OF KEY FEATURES

2–013   • To the extent that the Employer can, the Employer shall assist the Contractor:

   (i)   in obtaining copies of the relevant laws which are not readily available; and

   (ii)   in the making of any applications for permits, licences and approvals which the Contractor will need to make in accordance with those laws.

   • The Employer is only required to provide this assistance if so requested by the Contractor.

COMMENTARY

2–014   This sub-clause requires the Employer to assist the Contractor in obtaining any permits, licences and copies of the Laws of the Country relevant to the Contract which the Contractor may need. However this obligation arises only if the Contractor specifically requests this assistance.

The sub-clause does not refer to the obtaining of planning permission. As noted in sub-cl.1.13(a), this is the responsibility of the Employer.

Although the clause does not specifically say so, as the obligation on the Employer is one to assist,[4] it is presumed that the Contractor has the primary obligation of actually preparing whatever documentation may be required to obtain any necessary permits.

---

[4] No definition of "reasonable assistance" is provided, although the FIDIC Guide suggests that it would not be reasonable for the Contractor to expect the Employer to do anything he can do himself.

By sub-cl.1.1.6.5, "Laws" means all national (or state) legislation, ordinances and other laws and regulations and by-laws of any legally constituted public authority. It is possible that the Laws of the Country will not necessarily be the governing law under which the contract operates and this should be checked. **2–015**

However, this is a sub-clause which is of considerable potential benefit to the Contractor (depending on the extent to which a Contractor has any local representatives), as the Employer is more likely than the Contractor to be based in the Country where the contract is taking place and therefore should be better placed to make use of local knowledge and contacts.

The obligation to assist is qualified in that the Employer shall only provide reasonable assistance where he is in a position to do so. Nevertheless, a Contractor would be well advised to consider taking advantage of the opportunities this clause provides.

By sub-cl.1.13, the Contractor, in carrying out his obligations under the Contract, must comply with all applicable laws. Sub-clause 1.13 also specifically requires the Contractor to give all necessary notices, pay taxes and obtain all permits and licences in relation to the execution and completion of the works and remedying of any defects. By sub-cl.1.13(b), the Contractor shall indemnify the Employer for the consequences of any failure to do so. It is important that the Contractor acts promptly. Licences may well be required to import plant, equipment and materials. The Contractor needs to be aware of any likely fees for handling and/or freight charges as well as the time periods required for the giving of any notices or processing any applications. **2–016**

In the Canadian case of *Ellis-Don Limited v The Parking Authority of Toronto,*[5] the contractor was prevented from commencing works because an excavation permit was not issued until after it had commenced work on site. It was clear under the contract that the employer had to apply for the permit. The employer contended that the contractor should have been aware at the time of tender that no excavation permit had been issued and that such a permit would not be issued until detailed shoring drawings had been provided. Further, the employer continued that since the contractor was responsible for the shoring and bracing, these drawings should have been submitted by that contractor. The court rejected the employer's argument. The wording of that contract was clear. Under the FIDIC conditions, the delay would be entirely the responsibility of the Contractor. However, the Contractor, would have been able to seek the Employer's assistance in applying for the excavation permit.

There is one proviso to this which illustrates a potential difficulty with this sub-clause. If the acquiring of the "excavation permit" could be said to be part of the process of obtaining planning permission, then the Contractor

---

[5] (1985) BLR 98.

could argue that any delay was the responsibility of the Employer as laid down by sub-cl.1.13(a). However, it is more likely that the need for an excavation permit would fall under the auspices of sub-cl.1.13(b) being an approval required in relation to the execution of the Works. Both parties to the contract would therefore be advised to consider carefully where the responsibility for obtaining the necessary approvals lie.

2–017   Unlike other sub-clauses, no specific sanction is provided in respect of a failure by the Employer to provide reasonable assistance or from a failure by the Employer to provide accurate information. For example, sub-cl.2.1 specifically refers to the potential entitlement of the Contractor to an extension of time and/or payment of any costs plus profit for failure by the Employer to provide access to the site.

However, sub-cl.8.5 provides that a Contractor may be entitled to an extension of time under sub-cl.8.4(b) as a result of any delays caused by authorities, provided the Contractor has diligently followed the procedures laid down by the relevant legally constituted public authorities in the country. Therefore, if the information which is supplied by the Employer is out of date or otherwise inaccurate, this may give the Contractor grounds to make such an application. An alternative and possibly easier route would be to claim under sub-cl.8.4(e) which refers to any delay, impediment or prevention caused by or attributable to the Employer.

This sub-clause does not mention payment for any assistance given by the Employer, but it is to be presumed that a reasonable payment will be required.

2–018   Finally, a word of caution. It is only by fully understanding both the provisions of the contract and the potential impact that the applicable law might have on the proposed works that a Contractor can fully understand and allocate the risks inherent in pricing and programming its Tender. We mentioned above the importance of local knowledge, sub-cl.2.2 is a sub-clause of the contract. The prudent Contractor will have assessed the risks and sought to obtain details of the necessary laws and procedures necessary to obtain licences prior to entering into that contract.

MDB HARMONISED EDITION

2–019   There are two changes, both of which could be said to favour the Contractor by clarifying and it could be argued adding to what the Employer must do to assist the Contract or in obtaining permits and other approvals.

First, the words "where he is in a position to do so", which might have provided a ground for the Employer to argue that it could not assist, have been deleted from the first line. The first paragraph now reads:

The Employer shall provide, at the request of the Contractor, such reasonable assistance as to allow the Contractor to obtain properly.

Second, the words "by obtaining" are now longer needed at the beginning of sub-paragraph (a) and so have been deleted, as have the words "for the Contractor's applications" from the beginning of sub-paragraph (b).

2.3 EMPLOYER'S PERSONNEL

The Employer shall be responsible for ensuring that the Employer's Personnel   **2–020**
and the Employer's other contractors on the Site:

(a)   *co-operate with the contractor's efforts under Sub-Clause 4.6 [Co-Operation], and*
(b)   *take actions similar to those which the Contractor is required to take under sub-paragraphs (a), (b) and (c) of Sub-Clause 4.8 [Safety Procedures] and under Sub-Clause 4.18 [Protection of the Environment].*

OVERVIEW OF KEY FEATURES

The Employer shall be responsible for ensuring that the Employer's   **2–021**
Personnel and his other contractors:

(i)   co-operate with the Contractor to the limited extent provided for in sub-clause 4.6;
(ii)   comply with the health & safety requirements set out in paragraphs (a)–(c) of sub-clause 4.8;
(iii)   take all reasonable steps to protect the environment both on and off the Site as required by sub-clause 4.18.

COMMENTARY

This sub-clause firstly requires that the Employer take responsibility for   **2–022**
ensuring its personnel and other contractors co-operate with the Contractor to the limited extent required by sub-cl.4.6. Second, it imposes safety and environmental obligations on the Employer.

    The definition of Employer's Personnel is a wide one. Under sub-cl.1.1.2.6, the Employer's Personnel include the Engineer and the Engineer's assistants, all staff and other employees of the Engineer and the Employer together with any other personnel notified to the Contractor by either the Engineer or the Employer as being members of the Employer's personnel.

Co-operation

2–023   Sub-clause 4.6 provides that the Contractor shall allow "appropriate opportunities" for carrying out work to the Employer's Personnel together with any other contractor employed by the Employer and personnel of any legally constituted public authority. The obligation on the Employer here is intended to complement that. Of course, it is in the interests of the Employer's Personnel and other contractors to assist, since the requirement imposed by sub-cl.4.6 is on the Contractor to allow these other parties the opportunity to carry out their own Works.

One potential difficulty for the Contractor is the extent to which other contractors are employed to carry out works on the Site at the same time as he is. Although the Contractor is under an obligation to co-operate with these other contractors and the Employer is under an obligation to ensure those contractor's co-operate, that seems to be the extent of the obligations. Neither sub-cl.2.3 nor sub-cl.4.6 indicate or confirm that there is any similar obligation imposed on these other potential contractors.

It is important that the Contractor understands and is fully aware of the extent to which he would have to work on a site with other contractors. It is the Employer and not the Contractor who enters into contracts with those other contractors and the Employer should of course take steps to ensure that these provisions are reflected in any other contracts. It is the Employer and not the Contractor who has control over those other contractors. The Contractor needs to know exactly how the work of any other contractor might impact on his own planned programme. The Employer should also appreciate the potential difficulties this may cause the Contractor, as it is in the Employer's interests to ensure that the project runs smoothly.

2–024   Sub-clause 2.3 provides an express term that the Employer must ensure that its personnel co-operate with the Contractor's efforts in allowing others to carry out work. There is no express obligation to co-operate in any other way.

The extent to which a Contractor might be able to imply other terms depends on the jurisdiction of the Contract. When preparing the tender, the prudent contractor would be advised to assume that no terms can be implied. The following examples, taken from the English Courts, are provided to give a flavour of the factors which both parties ought to take into account.

By sub-cl.2.1, the Employer must give the Contractor possession of the site. This obligation in fact could be said to be part of an overall implication that the Employer agrees to do all that is necessary on his part to bring about completion of the contract.[6] In *London Borough of Merton v Stanley Hugh*

---

[6] *MacKay v Dick* (1881) 6 App Cas 251.

Leach Ltd,[7] the Court when considering the JCT 63 contract found for, amongst others, the following implied terms:

(i)   that the employer would not hinder or prevent the contractor from carrying out its obligations in accordance with the terms of the contract and from executing the works in a regular and orderly manner; and

(ii) that the employer and the architect would do all things necessary to enable the contractor to carry out the work, and the employer would be liable for any breach of this duty on the part of the architect.

The English Courts have considered whether a term for co-operation on the part of an Employer where no standard form conditions were agreed can be implied. In *Allridge (Builders) v Grandactual*,[8] a number of different contractors were engaged to carry out conversion works in parallel. The contractor asked for a term to be implied that: the employer would co-operate with the contractor so as to enable the contractor to carry out the works in a regular and orderly manner and would not hinder or prevent the contractor from so doing.   **2–025**

Although the court found in favour of the contractor, it did not use the word co-operate in its final summary of conclusions saying that there were implied terms that the employer would:

(i)   not hinder or prevent the contractor from carrying out its obligations in accordance with the terms of the contract and from executing the works in a regular and orderly manner; and

(ii) take all steps reasonably necessary to enable the contractor to discharge its obligations and execute the works in a regular and orderly manner.

However the Court would not imply terms: **2–026**

(i)   that all preceding works to be carried out by other contractors engaged by the employer and which necessarily had to be completed for the contractor to carry out his works in a regular and ordinary manner, would be so completed by those other contractors prior to the contractor commencing on site;

(ii) that the employer would ensure that the site would be either entirely or reasonably free from rubbish and other debris as at the contractor's commencement on site so as to enable the works to proceed; nor

(iii) that the employer would ensure that the site would comply with basic health and safety requirements so as to allow the contractor to carry out the works in a non hazardous environment.

---

[7] (1985) 32 BLR 51.
[8] (1996) CILL 1225.

Of these three terms, it is the first which might cause the Contractor the most concern if there are other contractors on the site, whose work might directly impact upon the Contractor's own work. The FIDIC conditions do not contain any of these three terms either, and the Contractor should take this into account when preparing its tender.

Health & Safety and the Environment

2–027   Sub-clause 4.8 provides that the Contractor shall comply with all applicable safety regulations, keep the site clear of unnecessary obstruction, provide fencing and other guarding and provide any temporary works which may be necessary.

Sub-clause 4.18 provides that the Contractor shall take all reasonable steps to protect the environment.

By sub-cl.2.3, the Employer is therefore under an obligation to ensure that its own personnel and any other contractors on site take similar efforts. Accordingly, it is important that the Employer undertakes suitable supervision of its own personnel to ensure that this obligation is carried out and that any contract with any other contractors contain similar obligations.

MDB HARMONISED EDITION

2–028   There is no change.

2.4 EMPLOYER'S FINANCIAL ARRANGEMENTS

2–029   *The Employer shall submit, within 28 days after receiving any request from the Contractor, reasonable evidence that financial arrangements have been made and are being maintained which will enable the Employer to pay the Contract Price (as estimated at that time) in accordance with Clause 14 [Contract Price and Payment]. If the Employer intends to make any material change to his financial arrangements, the Employer shall give notice to the Contractor with detailed particulars.*

OVERVIEW OF KEY FEATURES

- If the Contractor so requests, the Employer shall provide within 28 days reasonable evidence that financial arrangements are in place to pay the Contract Price in accordance with cl.14.   **2–030**
- The Employer must give notice to the Contractor if he intends to make any material change to his financial arrangements.
- If that evidence is not provided, the Contractor may, subject to the provision of the necessary notices, suspend work or even terminate the contract.

COMMENTARY

This is an entirely new provision. It provides a mechanism whereby the Contractor can obtain confirmation that sufficient funding arrangements are in place to enable him to be paid. This is something which may be of particular importance if the Employer is a company which has been specifically set up to carry out the project in question and this is therefore, typically backed by loan finance. It is also a clause that typically the Employer will seek to delete.   **2–031**

Simple consideration of the clause in isolation suggests that there is little sanction available to the Contractor as a result of any failure by the Employer to provide the reasonable evidence requested. However, in accordance with sub-cl.16.1, should the Employer fail in the 28-day period to comply with sub-cl.2.4, the Contractor may, after giving not less than 21 days' notice to the Employer, suspend work.

This potentially gives a period of 49 (i.e. 28 plus 21) working days before a Contractor is entitled to suspend.

If the Contractor does not receive reasonable evidence within 42 days of giving notice under sub-cl.16.1 then, pursuant to sub-cl.16.2, the Contractor is entitled to terminate the Contract. This of course gives a total of 105 days (42 plus 49 plus the need to give 14 days notice as required by sub.cl.16.2(a)) from the date of the initial request for reasonable evidence to termination of the Contract. A Contractor might consider this to be a long time, particularly if costly works are being carried out during the 91-day period. Companies or even national economies can collapse in that time.   **2–032**

One potential difficulty with this clause is that no definition of "reasonable evidence" has been provided. All that is said is that the evidence must show that the Employer is able to pay the contract Price in accordance with cl.14. Given that a Contractor has the potential right to terminate if reasonable evidence is not provided, it is easy to see how a dispute could arise as to the sufficiency of any evidence supplied. Also the sub-clause does not either provide how often or with what frequency a Contractor can make such a request. The timing of a request might be considered to have an impact on the reasonableness of it.

This sub-clause deals with matters arising in the course of the contract. It might therefore be particularly important if a significant and thereby costly variation is instructed mid-way through the project.

2–033 However the sub-clause does not deal with funding questions that might arise before and during the tender period prior to the Contract being entered into. Depending on the nature of the Employer and the way in which the project will be funded, a prudent Employer might be well advised to provide evidence of its funding as part of the tender documentation. Equally, the prudent Contractor should try and establish prior to submitting the tender, if not evidence of how the Employer is funding the work, then at the very least a definition of what "reasonable evidence" might be provided.

If the Employer intends to make a "material change" to his financial arrangements he must give detailed notice to the Contractor. No definition again is provided of "material change". It seems largely up to the Employer to decide what a material change is. The most obvious example of a "material change" is a change in backer or part of the funding team. Indeed the Employer might consider that there is always the chance that, if a change was made, that the Contractor, in theory, may never find out. However that would clearly be unwise given that the Contractor has the power, through this sub-clause to ask for evidence of the financial arrangements that are in place.

MDB HARMONISED EDITION

2–034 The following changes have been introduced:

The first paragraph has been expanded so that the Employer must now submit reasonable evidence of its financial arrangements "*before the Commencement Date and thereafter*" within 28 days of any request by the Contractor.

This means that this information must be provided initially by the Employer even if it has not been requested by any of the potential Contractors.

2–035 Then, the final sentence of the sub-clause has been deleted and replaced by the following:

Before the Employer makes any material change to his financial arrangements, the Employer shall give notice to the Contractor with detailed particulars.

The change here is a minor but significant one, namely the replacement of the words "If the Employer intends to make" with "Before the Employer makes". Now the Employer only has to give notice if he actually makes any material change to his financial arrangements. This will, it is submitted, have the effect of pushing back the time when the Employer needs to inform the Contractor of any such change.

Finally, the following new paragraph has been added at the end of the sub-clause: **2–036**

In addition, if the Bank has notified to the Borrower that the Bank has suspended disbursements under its loan, which finances in whole or in part the execution of the Works, the Employer shall give notice of such suspension to the Contractor with detailed particulars, including the date of such notification, with a copy to the Engineer, within 7 days of the Borrower having received the suspension notification from the Bank. If alternative funds will be available in appropriate currencies to the Employer to continue making payments to the Contractor beyond a date 60 days after the date of Bank notification of the suspension, the Employer shall provide reasonable evidence in such notice of the extent to which such funds will be available.

This second addition is potentially of particular benefit to the Contractor as it provides that the Employer must notify the Contractor within seven days if a Bank has suspended payment of any loan which may be financing the project, thereby giving the Contractor early warning of any potential financial problem.

2.5 EMPLOYER'S CLAIMS

If the Employer considers himself to be entitled to any payment under any **2–037** *Clause of these Conditions or otherwise in connection with the Contract, and/or to any extension of the Defects Notification Period, the Employer or the Engineer shall give notice and particulars to the contractor. However, notice is not required for payments due under Sub-Clause 4.19 [Electricity, Water and Gas], under Sub-Clause 4.20 [Employer's Equipment and Free-Issue Material], or for other services requested by the Contractor.*

The notice shall be given as soon as practicable after the Employer became aware of the event or circumstances giving rise to the claim. A notice relating to any extension of the Defects Notification Period shall be given before the expiry of such period.

The particulars shall specify the Clause or other basis of the claim, and shall include substantiation of the amount and/or extension to which the Employer considers himself to be entitled in connection with the Contract. The Engineer shall then proceed in accordance with Sub-Clause 3.5 [Determinations] to agree or determine (i) the amount (if any) which the Employer is entitled to be paid by the Contractor, and/or (ii) the extension (if any) of the Defects Notification Period in accordance with Sub-Clause 11.3 [Extension of Defects Notification Period].

This amount may be included as a deduction in the Contract Price and **2–038** *Payment Certificates. The Employer shall only be entitled to set off against or* *make any deduction from an amount certified in a Payment Certificate, or to otherwise claim against the Contractor, in accordance with this Sub-Clause.*

OVERVIEW OF KEY FEATURES

2–039
- If the Employer considers himself entitled to either any payment or an extension of the Defects Notification period under the Contract, the Employer or Engineer shall give notice and particulars to the Contractor.
- The notice relating to payment should be given as soon as practicable after the Employer has become aware of the event or circumstance which gives rise to the claim.
- Any notice relating to the extension of the Defects Notification Period should be given before the expiry of that period.
- The Employer must also provide substantiation including the basis of the claim and details of the relief sought.

2–040
- Once notice has been given, the Engineer shall make a determination in accordance with sub-cl.3.5.
- Any amount payable under sub-cl.2.5 may be included as a deduction in the Contract Price and Payment Certificates.
- The Employer cannot make any deduction by way of set-off or any other claim unless it is in accordance with the Engineer's determination.
- Notice is not required for payments due to the Employer for services under sub-cl.14.19 or equipment under sub-cl.4.20.

COMMENTARY

2–041   Sub-clause 2.5 is another new clause. It is designed to prevent an Employer from summarily withholding payment or unilaterally extending the Defects Notification Period. The final paragraph of the sub-clause specifically confirms that the Employer no longer has a general right of set-off. The Employer can only set-off sums once the Engineer has agreed or certified any amount owing to the Contractor following a claim.

The Employer should remember that in accordance with sub-cl.14.7, he must pay any amount certified, even if he disagrees with the Engineer's decision. By sub-cl.14.8, were the Dispute Adjudication Board to decide that the Employer had not paid the amount due, the Contractor would be entitled to finance charges.

Accordingly, sub-cl.2.5 imposes a specific notice procedure on any Employer who considers that it has any claims against the Contractor. Unless the Employer follows the procedure laid down by this sub-clause, he cannot withhold or otherwise deduct any sums due for payment to the Contractor.

There are a number of different clauses throughout the Contract which **2-042** provide the Employer with a right to claim payment from the Contractor. These are as follows:

Sub-clause 4.19 – Electricity, Water and Gas
Sub-clause 4.20 – Employer's equipment and free-issue material
Sub-clause 7.5 – Rejection
Sub-clause 7.6 – Remedial Work
Sub-clause 8.1 – Commencement of Works
Sub-clause 8.6 – Rate of Progress
Sub-clause 8.7 – Delay Damages
Sub-clause 9.4 – Failure to pass tests on completion
Sub-clause 10.2 – Taking over of parts of the works
Sub-clause 11.3 – Extension of defects notification period
Sub-clause 11.4 – Failure to remedy defects
Sub-clause 13.7 – Adjustments for changes in legislation
Sub-clause 15.3 – Valuation at date of termination
Sub-clause 15.4 – Payment after termination
Sub-clause 17.1 – Indemnities
Sub-clause 18.1 – General requirements for insurances
Sub-clause 18.2 – Insurance for works and contractor's equipment

The Employer must give notice "as soon as practicable" of him **2-043** becoming aware of a situation which might entitle him to payment. Therefore unlike sub-cl.20.1, where a Contractor has 28 days to give notice, there is no strict time limit within which an Employer must make a claim, although any notice relating to the extension of the Defects Notification Period must of course be made before the current end of that period. In addition it is possible that the Applicable Law might just impose some kind of limit.

It might have been thought that one option would have been to suggest that the Employer should be bound by the same 28 day limit as the Contractor. Instead sub-cl.2.5 provides a simpler claims mechanism with no time bar. However the rationale for the difference in treatment is presumably that in the majority of, if not all, situations, the Contractor will be (or should be) in a better position to know what is happening on site and so will be much better placed to know if a claims situation is likely to arise than an Employer.

The notice must be in writing and delivered in accordance with the requirements of sub-cl.1.3. It is unclear as to whether the particulars are required to be provided at the same time as the notice is served. The sub-clause does not require that the particulars are provided at the same time as no time limit or frame is imposed on either.

The particulars that the Employer must provide are details of the clause **2-044** (or basis) under which the claim is made together with details of the money is time relief sought.

Details of any notices served by the Employer are also required by sub-cl.4.21(f) to form part of the regular progress reports.

The Employer does not need to give the Contractor notice where, he has supplied electricity, water gas or equipment to the Contractor, in accordance with sub-cll 4.19 and 4.20.

2–045   If either party is not satisfied with the determination made by the Engineer under sub-clause 3.5, then the resulting dispute could be referred to the Dispute Adjudication Board under cl.20. An Employer would therefore be advised not to deduct the amount to which he is believed to be entitled, before any such determination of the Dispute Adjudication Board, as to do so would leave the Employer liable to a claim from the Contractor.

MDB HARMONISED EDITION

2–046   The reference to Free-Issue Material has been changed to Free-Issue Materials.

The following change has been introduced to the sentence which details when the Employer must give notice. The second paragraph now reads as follows:

The notice shall be given as soon as practicable <u>and no longer than 28 days</u> after the Employer became aware, <u>or should have become aware</u>, of the event or circumstances giving rise to the claim. A notice relating to any extension of the Defects Notification Period shall be given before the expiry of such period.

The first impression given by the addition of the underlined words is that they serve to tighten up the period in which the Employer must notify any claim – an impression reinforced by the apparent 28-day time limit. However, the new words introduce an additional subjective reasonableness test. Whereas before all that mattered was when the Employer actually became aware of the circumstances giving rise to a claim, now some consideration needs to be given to when the Employer should have realised that a claims situation had arisen. However in reality, save for extreme cases, little has changed. There is still no time limit to serve as a condition precedent which might serve to deprive the Employer of the opportunity to make a claim.

CLAUSE 3 – THE ENGINEER

3.1 ENGINEER'S DUTIES AND AUTHORITY

The Employer shall appoint the Engineer who shall carry out the duties   **3–001**
*assigned to him in the Contract. The Engineer's staff shall include suitably
qualified engineers and other professionals who are competent to carry out these
duties.*

   The Engineer shall have no authority to amend the Contract.

   *The Engineer may exercise the authority attributable to the Engineer as
specified in or necessarily to be implied from the Contract. If the Engineer is
required to obtain the approval of the Employer before exercising a specified
authority, the requirements shall be as stated in the Particular Conditions. The
Employer undertakes not to impose further constraints on the Engineer's
authority, except as agreed with the Contractor.*

   However, whenever the Engineer exercises a specified authority for which the   **3–002**
*Employer's approval is required, then (for the purposes of the Contract) the
Employer shall be deemed to have given approval.*

   Except as otherwise stated in these Conditions:

*(a)   whenever carrying out duties or exercising authority, specified in or
implied by the Contract, the Engineer shall be deemed to act for the
Employer;*

*(b)   the Engineer has no authority to relieve either Party of any duties,
obligations or responsibilities under the Contract; and*

*(c)   any approval, check, certificate, consent, examination, inspection,
instruction, notice, proposal, request, test, or similar act by the Engineer
(including absence of disapproval) shall not relieve the Contractor from
any responsibility he has under the Contract, including responsibility for
errors, omissions, discrepancies and non-compliances.*

OVERVIEW OF KEY FEATURES

- The Engineer is appointed by the Employer.
- The Engineer is obliged to carry out the functions described in the   **3–003**
  Contract but he has no power to *amend* the Contract.
- The Engineer's authority to act derives in the first place from the Contract
  documents, whether by express words and by virtue of powers necessarily
  implied from the express words.

- For certain further acts, specified authority is required, and that permission required by the Engineer from the Employer should be set out in the Particular Conditions.
- The Engineer shall be deemed to act on behalf of the Employer.
- Neither any act nor omission by the Engineer shall relieve either the Employer or Contractor of any obligation or responsibility under the Contract.

COMMENTARY

3–004   The Engineer is not a party to the Contract. Indeed the Engineer is not defined as a Party at sub-cl.1.1.2. His contract will be with the Employer. Nevertheless he essentially acts as agent for the Employer and is expressly stated to be acting for the Employer whenever he carries out his duties under the Contract. The prudent Contractor might want to consider asking for a copy of the contract between the Employer and Engineer, although there is no express obligation on the Employer to pass this on.

There has been considerable debate in recent years about the role of the Engineer. The traditional understanding of the role of the Engineer is that he is required to be independent and fair, holding a balance between the Employer and the Contractor. In his judgment in *Amec Civil Engineering Limited v Secretary of State for Transport*[1], Rix L.J. (at para.81 of the Judgment) provided the following summary of the obligations of the Engineer by reference to the various decided cases:

(a)   the Engineer must "retain his independence in exercising [his skilled professional] Judgment";[2]

(b)   he must "act in a fair and unbiased manner" and "reach his decisions fairly, holding the balance";[3]

(c)   he must "act fairly";[4]

(d)   if he hears representations from one party, he must give a similar opportunity to the other party to answer what is alleged against him; [5] and

(e)   he must "act fairly and impartially" where fairness is "a broad and even elastic concept" and impartiality "is not meant to be a narrow concept".[6]

[1] [205] CILL 2288.
[2] Megarry J. in *Hounslow LBC v Twickenham Garden Developments* [1971] 1 Ch 233 at 259G.
[3] Lord Reid in *Sutcliffe v Thackrah* [1974] A.C. 727 at 737D.
[4] Lord Morris in *Sutcliffe v Thackrah* at 744G.
[5] *AC Hatrick (NZ) Ltd v Nelson Carlton Construction Co Ltd* [1964] NZLR 72.
[6] Cooke J. in *Canterbury Pipe Lines Ltd v Christchurch Draining Board* [1979] 16 BLR 76 NZCA.

However, in the case of *Amec Civil Engineering Limited v Secretary of State*   **3–005**
for Transport, the Court of Appeal decided that the Engineer (unlike adjudicators acting either under the Housing Grants Construction and
Regeneration Act 1996 or under a contractual scheme) did not have to
observe the rules of natural justice in reaching decisions such as those
required under cl.66 of the ICE Conditions of Contract.

In recent years, the traditional view of the Engineer – typified by the
speeches in *Sutcliffe v Thackrah* – has been subject to a degree of scepticism.
In context of clause 3, there are perhaps further complications for the
Engineer and for an understanding of his role.[7]

Under sub-cl.2.6 of the Old Red Book FIDIC 4th edn, wherever the
Engineer was required to exercise his discretion, he was required to act
"impartially within the terms of the Contract and having a regard to all
circumstances". That has now changed and the word impartial no longer
appears.

One reason for this change is likely to be the introduction of the Dispute   **3–006**
Adjudication Boards. Under the Old Red Book, cl.67 provided for the
Engineer to adjudicate on disputes between the Contractor and Engineer.
Now, although the Engineer still rules on, for example, extension of time
claims at first instance, a new layer has been added and the Dispute
Adjudication Board stands between the parties and arbitration.

Sub-clause 3.1 makes it clear that the Engineer is expressly the agent of the
Employer. Under the first stage envisaged by sub-cl.3.5, the Engineer is to
"... endeavour to *reach agreement* ..." between the parties. Under the
second stage envisaged by sub-cl.3.5, the Engineer is obliged to make "... *a
fair determination* ...".

There is scope for further complication where, as is often the case, the
Engineer has been responsible for the preparation of the *design* for the
Works. In a case where there are complaints about alleged defects, and
the Contractor relies upon defects in the underlying design, then of course
the Engineer must make any necessary determination under the second stage
envisaged by sub-cl.3.5, but the Engineer is unlikely to be able to broker an
agreement between the parties under the first limb of sub-cl.3.5.

The identity and competence of the Engineer is of vital importance to the   **3–007**
project and may also have an influence on a potential Contractor's approach
to tendering for the job. However, cl.3 does not say when the Employer must
appoint the Engineer. The assumption must be that at the very least the
Engineer will have been named in the tender documentation. If he has not,
then the prudent Contractor must ask legitimate questions about the nature
of the project. The reputation and technical ability of the Engineer will be
key to the success of any project.

[7] For further detail, see Ola O Nisja "The Engineer in International Construction: Agent?
Mediator? Adjudicator?" [2004] ICLR 230.

Under English and Common law, there is some disagreement as to whether the Engineer will owe a duty of care in tort, and if so to what extent, to the Contractor. This debate relates to two issues, first certification and second design, both of which have been touched on in the above discussion.

The reason for this is the case of *Pacific Associates v Baxter*.[8] Here, it was held that an engineer, while certifying an interim payment, does not owe the contractor any duty of care in negligence in relation to the content of tender documents it had prepared and supplied to the contractor. One of the reasons for the debate on this matter is that there was a disclaimer[9] in the contract between the contractor and the employer in relation to the work to be carried out by the engineer and how that work affected the contractor.

3–008   This it is thought, may have influenced the Court in arriving at its decision and from this, it may follow that a duty of care in tort owed by one party in a contractual framework to another party, in the same framework, may be qualified or negated by provisions in the latter's contract, although the former is not privy to it. However, the question as to whether or not a construction professional can be held to be negligent to a contractor in tort for certification has still to be finally decided and the general applicability of the decision is not entirely clear.[10]

For example, it has also been suggested that this decision was part of a 1980s trend in which the Courts were unwilling to allow certificates to be challenged,[11] a trend which has now been reversed.[12] In addition, over the recent years there has been a slow expansion of the doctrine of negligent misstatement, all of which cast doubt as to whether or not *Pacific Associates v Baxter* would be followed today.

In relation to certification, the English case of *John Mowlem & Co v Eagle Star Insurance & Ors*[13] indicated that the certifier (in this case the architect) might be held to be liable to the contractor for the tort of procuring a breach of contract if the certifier interfered with the contract. In that case it was suggested that the architect had wrongfully refrained from issuing extension of time awards and had wrongfully issued default notices.

---

[8] (1998) 44 BLR 33.

[9] This stated that "neither . . . the engineer nor any of his staff . . . shall be in any way personally liable for the actual obligation under the contract".

[10] See for example the debate between Ian Duncan Wallace and Nicholas Lane in the Construction Law Journal (2003) Vol 19, No.6.

[11] See for example paragraph 2–90 of Building Contract Disputes: Practise and Precedents Edited by Robert Fenwick Elliott and Jeremy Glover and the cases of *Northern Regional Health Authority v Crouch* [1984] 2 All E.R. 175 where the Court of Appeal refused to review a certificate as between employer and contractor, and *Lubenham v South Pembrokeshire District Council* (1986) 33 BLR 39 where the Employer was able to rely upon an architect's certificate that had been issued in good faith notwithstanding that it was actually wrong.

[12] See the House of Lords decision in *Beaufort Developments Ltd v Gilbert-Ash NI Ltd* (1998) CILL 1386.

[13] (1992) 62 BLR 126.

The situation is slightly different in Canada where on the one hand, the **3–09** Canadian Supreme Court[14] overturned a decision of the lower courts holding that an engineer owed a duty of care to a contractor who suffered loss as a result of errors in preparing the tender documents.

Against that, the Supreme Court has also ruled that an engineer might be liable for negligent misrepresentation in respect of economic loss caused by the reliance on drawing prepared for the tender stage.[15] The reason for the Court Decision was set out in the Judgment of McLauchlin who said:

> *The Engineers undertook to provide information (the tender package) for use by a definable group of persons with whom it did not have any contractual relationship. The purpose of supplying information was to allow the Tenderers to prepare a price to be submitted. The Engineers knew this. The Plaintiff Contractor was one of the Tenderers. It relied on information prepared by the Engineers in preparing its bid. Its reliance upon the Engineers' work was reasonable. It alleges it suffered loss as a consequence. These facts establish a prima facie case of action against the engineering firm.*

Another area where the Engineer might be liable to the Contractor is if approvals or determinations are unreasonably delayed. We have discussed the case of *Neodox Limited v Swinton and Pendlebury Borough Council*[16] above and although this case deals with instructions, similar principles will arise. Approvals must be issued without unreasonable delay.

Engineers should note that the final paragraph of this sub-clause, in a clear **3–010** attempt to limit liability on the part of the Engineer, states that any act carried out by the Engineer shall not relieve the Contractor from any responsibility he has under the Contract, including responsibility for errors, omissions, discrepancies and non-compliances.

The final matter that needs to be taken into consideration when reviewing the potential liability of the engineer to the contractor, at least as regards cases subject to English jurisdiction, relates to the impact of the Contracts (Rights of Third Parties) Act 1999. This Act provides that where two parties enter into a contract intended to benefit a third party, that third party can acquire direct rights under it. Thus it may serve to give a third party the right to enforce a contractual term where the parties to the contract itself intended such a term to be enforceable by the third party in question. Whilst in theory this may serve to alter the position in favour of the contractor, in practice it is easily excluded.

---

[14] *Auto Concrete Curb UB Limited v South Nation River Conservation Authority & Ors* (1994) 10 Const. L.J. 39 and (1995) 11 Const. L.J. 155.
[15] *Edgeworth Construction Ltd v N.D. Lea & Associates Ltd and Others* 66 BLR 56.
[16] See sub-cl.1.3.

Finally it should be noted that this sub-clause does potentially put one limit on the power of the Engineer and that is in the statement that the Particular Conditions must identify situations where the Engineer is required to obtain the approval of the Employer before exercising a specified authority. If this is not done at the outset then it will be necessary for the Employer to seek the Contractor's consent for any changes.

MDB HARMONISED EDITION

3–011   There have been a number of changes.

First, the final sentence of the third paragraph has been deleted and replaced with:

The Employer shall promptly inform the Contractor of any change to the authority attributed to the Engineer.

Second, a new sub-paragraph has been added to the list in paragraph 5:

(d)   Any act by the Engineer in response to a Contractor's request except otherwise expressly specified shall be notified in writing to the Contractor within 28 days of receipt.

3–012   Finally, a substantial section has been added at the end of the sub-clause:

The following provisions shall apply:

The Engineer shall obtain specific approval of the Employer before taking action under the following Sub-Clauses of these Conditions

(a)   Sub-Clause 4.12: Agreeing or determining an extension of time and/or additional cost.

(b)   Sub-Clause 13.1: Instructing a Variation, except;

(i)   in an emergency situation as determined by the Engineer, or

(ii)   if such a Variation would increase the Accepted Contract Amount by less than the percentage specified in the Contract Data.

(c)   Sub-Clause 13.3: Approving a proposal for Variation submitted by the Contractor in accordance with the Sub Clause 13.1 or 13.2.

(d)   Sub-Clause 13.4: Specifying the amount payable in each of the applicable currencies.

Notwithstanding the obligation, as set out above, to obtain approval, if, in the opinion of the Engineer, an emergency occurs affecting the safety of life or of the

Works or of adjoining property, he may, without relieving the Contractor of any of his duties and responsibility under the Contract, instruct the Contractor to execute all such work or to do all such things as may, in the opinion of the Engineer, be necessary to abate or reduce the risk. The Contractor shall forthwith comply, despite the absence of approval of the Employer, with any such instruction of the Engineer. The Engineer shall determine an addition to the Contract Price, in respect of such instruction, in accordance with Clause 13 and shall notify the Contractor accordingly, with a copy to the Employer.

Perhaps the first of these changes is the most controversial. Under the 1999   **3–013** Edition, the Employer actually undertook not to change the basis of the Engineer's authority without the agreement of the Contractor. This has been changed to give the Employer the right to make whatever changes it likes to the basis of the Engineer's authority. The only restriction is that it must inform the Contractor of these changes. There is no longer any requirement that the Contractor agrees to these changes.

The second change might be viewed as fettering the Engineer particularly in the fact that the new clause says that the Engineer cannot agree or determine any extension of time or cost consequence of said extension without the Employer's approval.

3.2 DELEGATION BY THE ENGINEER

The Engineer may from time to time assign duties and delegate authority to   **3–014** *assistants, and may also revoke such assignment or delegation. These assistants may include a resident engineer, and/or independent inspectors appointed to inspect and/or test items of Plant and/or materials. The assignment, delegation or revocation shall be in writing and shall not take effect until copies have been received by both Parties. However unless otherwise agreed by both Parties, the Engineer shall not delegate the authority to determine any matter in accordance with Sub-Clause 3.5 [Determinations].*

Assistants shall be suitably qualified persons, who are competent to carry out these duties and exercise this authority, and who are fluent in the language for communications defined in Sub-Clause 1.4 [Law and Language].

Each assistant, to whom duties have been assigned or authority has been delegated, shall only be authorised to issue instructions to the Contractor to the extent defined by the delegation. Any approval, check, certificate, consent, examination, inspection, instruction, notice, proposal, request, test, or similar act by an assistant, in accordance with the delegation, shall have the same effect as though the act had been an act of the Engineer. However:

3–015   *(a)*   *any failure to disapprove any work, Plant or Materials shall not consti-tute approval, and shall therefore not prejudice the right of the Engineer to reject the work, Plant or Materials;*

          (b)   *if the Contractor questions any determination or instruction of an assis-tant, the Contractor may refer the matter to the Engineer, who shall promptly confirm, reverse or vary the determination or instruction.*

OVERVIEW OF KEY FEATURES

3–016
- The Engineer can delegate functions to assistants or to a Resident Engineer.
- These delegations must be in writing and must define the extent of the functions so delegated.
- Unless agreed by both the Employer and Contractor, the Engineer may not delegate his Determination function under sub-cl.3.5.

COMMENTARY

3–017   This is an important clause for the Engineer, since it is unlikely that he will be able to do everything himself.

    The sub-clause clearly envisages through the reference of a site engineer, that the Engineer need not be full-time resident on the project. Any staff to whom authority is delegated will fall under the definition of Employer's Personnel to be found at sub-cl.1.1.2.6.

    There are limits on this delegation and the sub-clause makes it clear that the Engineer cannot delegate his authority to determine under sub-cl.3.5 without the agreement of both the Contractor and Employer.

MDB HARMONISED EDITION

3–018   There is no change.

3.3 INSTRUCTIONS OF THE ENGINEER

3–019   *The Engineer may issue to the Contractor (at any time) instructions and addi-tional or modified Drawings which may be necessary for the execution of the Works and the remedying of any defects, all in accordance with the Contract.*

The Contractor shall only take instructions from the Engineer, or from an assistant to whom the appropriate authority has been delegated under this Clause. If an instruction constitutes a Variation, Clause 13 [Variations and Adjustments] shall apply.

The Contractor shall comply with the instructions given by the Engineer or delegated assistant, on any matter related to the Contract. Whenever practicable, their instructions shall be given in writing. If the Engineer or a delegated assistant:

(a) gives an oral instruction,

(b) receives a written confirmation of the instruction, from (or on behalf of) the Contractor, within two working days after giving the instruction, and

(c) does not reply by issuing a written rejection and/or instruction within two working days after receiving the confirmation.

then the confirmation shall constitute the written instruction of the Engineer or delegated assistant (as the case may be).

OVERVIEW OF KEY FEATURES

- The Contractor is obliged to comply with instructions given by the Engineer (or his assistants to whom proper authority under sub-cl.3.2 has been delegated). **3–020**
- Where possible those instructions should be in writing.
- Where the instructions are oral, if they are confirmed in writing by the Contractor within two days, then they shall be deemed to be a written instruction unless the Engineer responds to the contrary within two days.
- If the Contractor considers that the instruction in fact constitutes a variation then it should respond in accordance with sub-cl.13.3.

COMMENTARY

This sub-clause enables the Engineer to issue instructions and additional **3–021** and/or modified drawings at any time. Whilst the Engineer should try and issue instructions in writing, this sub-clause recognises that this might not always be possible. The Contractor should note the two day time limit in responding to oral instructions.

The Contractor must comply with these. In addition, if the Contractor considers that any of these drawings or instructions in reality constitutes variations, then the Contractor should follow the variation procedures laid down by sub-cl.13.3.

MDB HARMONISED EDITION

3–022 There is no change.

3.4 REPLACEMENT OF THE ENGINEER

3–023 *Notwithstanding Sub-Clause 3.1, if the Employer intends to replace the Engineer Employer shall, not less than 42 days before the intended date of replacement, give notice to the Contractor of the name, address and relevant experience of the intended replacement Engineer. The Employer shall not replace the Engineer with a person against whom the Contractor raises reasonable objection by notice to the Employer, with supporting particulars.*

OVERVIEW OF KEY FEATURES

3–024
- The Employer must give 42 days notice of any intention to replace the Engineer.
- If the Contractor raises a reasonable objection against the proposed replacement, then the Employer must find someone else.

COMMENTARY

3–025 The Employer may replace the Engineer at any time.

Where the Employer proposes to replace the Engineer, the Contractor is entitled to receive notice 42 days in advance together with details of the intended replacement. The Employer is not obliged to provide reasons for the intended replacement. The Contractor may raise objections if he considers the intended replacement to be unsuitable. The objections must be detailed and presumably in writing, although this is not specifically stated.

The Employer must give reasonable consideration to any objections made by the Contractor. If those objections are reasonable, an alternative must be found.

MDB HARMONISED EDITION

The time limit for the notice of any change to be given to the Contractor has been reduced from 42 to 21 days.   **3–026**

In addition, the final sentence has been substantially changed and now reads as follows:

If the Contractor considers the intended replacement Engineer to be unsuitable, he has the right to raise reasonable objection against him by notice to the Employer, with supporting particulars, and the Employer shall give full and fair consideration.

This is a significant change.

The MDB version does not say that the Employer "shall not replace" the Engineer where the Contractor has reasonable objections. In other words, the ability of the Contractor to object to the replacement has been considerably weakened and thereby the Employer has far more freedom in its choice of a replacement Engineer.

3.5 DETERMINATIONS

Whenever these Conditions provide that the Engineer shall proceed in accordance with this Sub-Clause 3.5 to agree or determine any matter, the Engineer shall consult with each Party in an endeavour to reach agreement. If agreement is not achieved, the Engineer shall make a fair determination in accordance with the Contract, taking due regard of all relevant circumstances.   **3–027**

The Engineer shall give notice to both Parties of each agreement or determination, with supporting particulars. Each Party shall give effect to each agreement or determination unless and until revised under Clause 20 [Claims, Disputes and Arbitration].

OVERVIEW OF KEY FEATURES

* Where the Engineer is required by the Contractor to reach a determination, the Engineer is bound to consult with each party in order to reach agreement.   **3–028**
* If no agreement is reached, the Engineer's duty is to provide a fair determination in accordance with the Contract and having regard to all the circumstances.

- The Engineer must provide a reasoned notice of his determination.
- That determination is binding unless revised in accordance with clause 20.

COMMENTARY

3–029 The Engineer may be required to make determinations in respect of the following sub-clauses:

1.9   Delayed Drawings or Instructions
2.1   Right of Access to the Site
2.5   Employer's Claims
4.7   Setting Out
4.12 Unforeseeable Physical Conditions
4.19 Electricity, Water and Gas
4.20 Employer's Equipment and Free-issue Materials
4.24 Fossils
7.4   Testing
7.5   Rejection
7.6   Remedial Work
8.4   Extension of Time
8.5   Delays caused by Authorities
8.6   Rate of Progress
8.9   Consequences of Suspension
9.4   Failure to Pass Tests on Completion
10.2 Taking Over Parts of the Works
10.3 Interference with Tests on Completion
11.4 Failure to Remedy Defects
11.8 Contractor to Search
12.3 Evaluation
12.4 Omissions
13.2 Value Engineering
13.3 Variation Procedure
13.7 Adjustments for Changes in Legislation
14.4 Schedule of Payments
14.8 Delayed Payment
15.3 Valuation of Date of Termination
15.4 Payment after Termination
16.1 Contractor's Entitlement to Suspend Work
16.4 Payment on Termination
17.4 Consequences of Employer's Risks
18.1 General Requirements for Insurances
18.2 Insurance for Works and Contractors Equipment
19.4 Consequences of Force Majeure

19.6   Optional Payment
20.1   Contractor's Claims

The two key areas where the Engineer is required to make a determination   **3–030**
are in respect of the claims by the Contractor for extension of time and/or
reimbursement of cost. In proceeding to make his determination, the
Engineer must first try and consult with the Employer and Contractor to
see if an agreement can be reached. One reason for this consultation
process is the importance of trying to maintain a good working relationship over the course of a potentially lengthy project. The FIDIC Guide
notes that:

> *Complying with these procedures and maintaining a co-operative approach to*
> *the determination of all adjustments should enhance the likelihood of*
> *achieving a successful project.*

It will be up to the Engineer to determine how this is done. Depending on the
nature of the dispute, the Engineer might try informal discussions; alternatively the Engineer might want to consider adopting a more formal approach,
such a mediation or another form of alternative dispute resolution.

If no agreement is reached, the Engineer must reach a fair determination.   **3–031**
It is important that the Engineer proceeds not to make a determination which
is simply fair, but one which is fair according to the requirements of the
Contract.

The sub-clause does not impose any time limit within which the Engineer
must reach his decision. However sub-clause 1.3 provides that any *"Approvals,*
certificates, consents and determinations shall not be unreasonably withheld or
delayed."

MDB HARMONISED EDITION

The following time-limit has been added to the first sentence of the second   **3–032**
paragraph: *"within 28 days from the receipt of the corresponding claim or*
request except when otherwise specified."

This deals with the omission mentioned above. Whilst 28 days might be
considered to be tight, this is the time limit for decisions to be found in UK
adjudications under the Housing Grants Construction & Regeneration Act
1996. It should be noted that there is no provision for this time period to be
extended, although presumably the parties could chose to extend the time
period if they so chose.

We discuss above the fact that an Engineer's determination must be fair
according to the requirements of the Contract. This can lead to different
results under the two different versions. For example, where profit is

allowed,[17] under sub-cl.1.2 of the MDB version this means the Engineer must assess profit at 5 per cent of the cost; under the 1999 Edition version, the basis of that assessment will be the "reasonable profit".

3.6 MANAGEMENT MEETINGS

3–033 *The Engineer or the Contractor's Representative may require the other to attend a management meeting in order to review the arrangements for future work. The Engineer shall record the business of management meetings and supply copies of the record to those attending the meeting and to the Employer. In the record, responsibilities for any actions to be taken shall be in accordance with the Contract.*

OVERVIEW OF KEY FEATURES

3–034 This is a Particular Condition only. Therefore it will only appear in the Contract if the Parties chose to incorporate it.[18]

- Either the Engineer or the Contractor's Representative may call meetings to review future progress.
- The Engineer must take formal minutes of any meeting that takes place.

COMMENTARY

3–035 This is a potentially useful clause for both the Employer and the Contractor. Its usefulness can be judged by the thought that there will be times, depending on progress, where either party might be reluctant to attend such a meeting.

As noted below in the discussion of sub-cl.4.21, there is no contractual requirement to hold even regular site meetings. Sub-clause 3.6 envisages extraordinary site meetings. Therefore one middle-way solution might be simply to impose a requirement for there to be regular site meetings.

---

[17] And remember that under sub-cl.8.4 an entitlement to Cost does not always mean an entitlement to profit on top.
[18] Self-evidently this comment applies to each and every clause. However, it is suggested that the General Conditions are far more likely to form the basis of any FIDIC Contract. The Particular Conditions will be treated differently.

However, the question of site meetings is something to which both the Employer and Contractor should give some consideration to prior to work commencing.

MDB HARMONISED EDITION

The Particular Conditions have effectively been removed from the MDB Version so that this sub-clause will not feature in the standard version of the MDB Edition. Practically this may mean it is less likely to be incorporated. **3–036**

CLAUSE 4 – THE CONTRACTOR

4.1 CONTRACTOR'S GENERAL OBLIGATIONS

The Contractor shall design (to the extent specified in the Contract), execute **4–001**
and complete the works in accordance with the Contract and with the Engineer's instructions, and shall remedy any defects in the Works.

The Contractor shall provide the Plant and Contractor's Documents specified in the Contract and all Contractor's Personnel, Goods, consumables and other things and services, whether of a temporary or permanent nature, required in and for this design, execution, completion and remedying of defects.

The Contractor shall be responsible for the adequacy, stability and safety of all Site operations and of all methods of construction. Except to the extent specified in the Contract, the Contractor (i) shall be responsible for all Contractor's Documents, Temporary Works, and such design of each item of Plant and Materials as is required for the item to be in accordance with the Contract, and (ii) shall not otherwise be responsible for the design or specification of the Permanent Works.

The Contractor shall, whenever required by the Engineer, submit details of **4–002**
the arrangements and methods which the Contractor proposes to adopt for the execution of the Works. No significant alteration to these arrangements and methods shall be made without this having previously been notified to the Engineer.

If the Contract specifies that the Contractor shall design any part of the Permanent Works, then unless otherwise stated in the Particular Conditions:

(a) The Contractor shall submit to the Engineer the Contractor's Documents for this part in accordance with the procedures specified in the Contract;

(b) These Contractor's Documents shall be in accordance with the Specification and Drawings, shall be written in the language for communications defined in Sub-Clause 1.4 [Law and Languages], and shall include additional information required by the Engineer to add to the Drawings for co-ordination of each Party's designs;

(c) The Contractor shall be responsible for this part and it shall, when the works are completed, be fit for such purposes for which the part is intended as are specified in the Contract; and

(d) Prior to the commencement of the Tests on Completion, the Contractor shall submit to the engineer the "as-built" documents and operation and maintenance manuals in accordance with the Specification and in sufficient detail for the Employer to operate, maintain, dismantle, reassemble, adjust and repair this part of the Works. Such part shall not be considered

to be completed for the purposes of taking-over under Sub-Clause 10.1 [Taking Over of the Works and Sections] until these documents and manuals have been submitted to the Engineer.

OVERVIEW OF KEY FEATURES

Contractor's Primary Obligation

4–003   The Contractor's primary obligation is to "execute and complete the Works" and "remedy any defects" in accordance with the Contract and with the Engineer's instructions.

Provision of Plant

4–004   The Contractor is responsible for ensuring the supply of the necessary plant, personnel, and other goods and services necessary for the completion of the Works in good time in accordance with the requirements of the Contract.

Site Operations

4–005   The Contractor is responsible for all site operations and construction methods. This includes all design of plant and materials to the extent required by the Contract.

Method Statements

4–006   • A Contractor must submit whenever required by the Engineer details of the methods which the Contractor intends to adopt in carrying out the Works.
   • The Contractor must notify the Engineer of any alterations to that method statement.

Design

4–007   • The Contractor only has design responsibility to the extent specified in the Contract.
   • If the Contractor undertakes any element of design, that design must be fit for the purpose for which the designed part is intended as specified in the contract.
   • The four sub-clauses (a) to (d) provide clear guidelines for the Contractor to follow where he is required to undertake design.
   • The Contractor will be fully responsible for any element of design which he undertakes.

72

COMMENTARY

General

Clause 4 is the longest clause in the Contract. It is also one of the most impor-   **4-008**
tant. Sub-clause 4.1 serves as a general wrap up for the basic obligations of the
Contractor as set out in cl.4 as a whole.

In short the Contractor's primary obligation is to carry out the Works in
accordance with the Contract, namely within the time set out in the
Appendix to Tender (as may be extended in accordance with cl.8) and for the
agreed price (which again may be subject to change in accordance with cl.14
in particular).

The Works are defined at sub-cl.1.1.5.8 as being "the Permanent Works
and the Temporary Works". The obligation here is therefore an obligation to
"complete" any item of work which is necessary to complete the Works as a
whole. The Contractor should consider this obligation carefully. It might not
be enough simply to carry out the work shown on the contract documents.
Equally, there may be items of ether the permanent or temporary works for
which the Contractor is not fully responsible.

This sub-clause is based on clause 8 of the Old Red Book, FIDIC 4th edn.   **4-009**
However there are a number of changes to the general wording. These
changes serve to tighten the responsibilities of the Contractor. Under cl.8 of
the Old Red Book, FIDIC 4th edn, the general obligation to execute the
works was qualified with the words "with due care and diligence". These
words have been taken out. Therefore the sub-clause must be read, it is
submitted, as an absolute obligation.

The requirements of sub-cl.4.1, mirror cl.3 of the Contract Agreement
whereby the Contractor "covenants" with the Employer to "execute and
complete the Works and remedy any defects therein in conformity with the
provisions of the Contract".

Accordingly, the second paragraph of sub-cl.4.1 requires the Contractor to
provide everything necessary to carry out his Works. This includes personnel
and plant, as well as the Contractor's Documents which are widely-defined at
sub-cl.1.6.1.1 as including calculations, computer programs and other software, drawings, manuals models and other technical documents. The
Contractor's Documents must be written in the language for communications
as defined in sub-cl.1.4.

It is important that the Contractor considers carefully what personnel and   **4-010**
plant he will need and when, in order to keep to his programme.

Sub-clause 4.1 provides that the Contractor shall submit details of the
methods of work to the Engineer. The only time requirement put on this obligation is that the Engineer can request such information whenever he
chooses. Once the Contractor has submitted the information, he cannot
make any "significant" alteration to his methods without notifying the

Engineer of his intentions. There is no indication given either that the Engineer must approve the alteration or that he could reject the proposed alteration.

It is likely that an Employer would want the Engineer to consider the proposals. If the Engineer did have to approve them, in accordance with sub-cl.1.3, his approval could not be unreasonably withheld. However, the Contractor should remember that sub-clause 3.1(e) holds that any approval given by the Engineer will not relieve him of his responsibilities under the Contract.

Design[1]

4–011   The Contract is intended for use where it is the Employer who provides the design and where it is the Contractor who constructs according to that design. Therefore, whilst it is unlikely that a Contractor will be required to carry out any design, it is important that this is carefully checked. This means that it is important that the Contract as a whole (including the specification and any drawings or bills of quantities) is carefully checked to see if there are any design obligations contained therein.

Where the Contract specifies that the Contractor shall design any part of the permanent[2] works, then the Contractor must submit to the Engineer the Contractor's Documents in relation to the designed portion of the Works and the Contractor shall be responsible for this design. Any permanent works designed by the Contractor must be fit for the purpose for which the Works were intended as specified in the Contract. Where a Contractor undertakes design, the Contractor will be responsible for the fitness for purpose of that design. This absolute obligation obviously depends on the nature of the definition of the purposes of the Works contained in the Contract.

However, the Contract does not define the "intended purpose" and there is surprisingly no contractual obligation on the Employer to provide such a definition. The prudent Employer would include such details as part of the tender documentation. In many jurisdictions such an obligation to provide a definition would be implied, although if the purpose is not clear, the simplest solution would be for the Contractor to ask.

4–012   To the extent that the Contractor undertakes any design obligation, that obligation will mirror the obligation of the design and build contractor. The general rule under English jurisdictions, is that a contractor who agrees to be responsible for both design and build of a structure assumes a duty to ensure that it will be fit for its purpose as communicated to that contractor.

The introduction of this fitness for purpose obligation is new, and it is important to remember that under English or Common law, the fitness

---

[1] For further discussion see Design Risk in FIDIC Contracts by Michael Black Q.C. – a paper given to the Society of Construction Law in November 2004.
[2] The sub-clause does not relate to the design of any temporary works.

for purpose duty is stricter than the ordinary responsibility of an architect or other consultant carrying out design where the implied obligation is one of reasonable competence to "exercise due care, skill and diligence".

In *Greaves v Baynham Meikle*,[3] Lord Denning said this of the fitness for purpose obligation:

> "Now, as between the building owners and the Contractors, it is plain that the owners made known to the Contractors the purpose for which the building was required, so as to show that they relied on the Contractors skill and judgement. It was therefore, the duty of the Contractors to see that the finished work was reasonably fit for the purpose for which they knew it was required. It was not merely an obligation to use reasonable care, the Contractors were obliged to ensure that the finished work was reasonably fit for the purpose."

The duty is, therefore, absolute.   **4–013**

In *IBA v EMI and BICC*[4] Lord Scarman said:

> "In the absence of any term (express or to be implied) negating the obligation, one who contracts to design an article for a purpose made known to him undertakes that the design is reasonably fit for the purpose."

Further, in *Viking Grain Storage v T.H. White Installations Ltd*,[5] Judge John Davies said:

> "The virtue of an implied term of fitness for purpose is that it prescribes a relatively simple and certain standard of liability based on the "reasonable" fitness of the finished product, irrespective of considerations of fault and of whether its unfitness derived from the quality of work or materials or design."

Where an Employer can be seen to rely on a Contractor for the design, the Contractor's legal responsibility is to produce (in the absence of express provision in the contract) a final work which is reasonably suitable for its purpose. Given that there is express provision in the FIDIC conditions, the absence of negligence in the design, will not therefore be a defence for the Contractor.

In England since it can be difficult for contractors to obtain insurance for   **4–014** a fitness for purpose obligation, many standard forms or bespoke contracts limit a contractor's design responsibility to that of a consultant. There is no such limit here.

---

[3] [1975] 1 W.L.R. 1095.
[4] (1980) 14 BLR.
[5] (1986) 33 BLR.

The obligation to provide Works that are fit for their purpose will only be effective if elsewhere in the documentation the purpose of the plant has been clearly made known to the Contractor. For example, it is not enough for a Contractor to assume from the Tests on Completion which may need to be carried out under cl.9 that it knows what the Employer requires. A Contractor should ensure that he has been provided with a general description of any outputs that the Employer intends to achieve, or an indication of how the Employer expects the plant to perform in a given number of years.

Sub-clause 4.1 attempts to clearly define where design responsibility lies. However, a Contractor should always take care to consider the implications of the design even if he thinks he has no design responsibility whatsoever. In some jurisdictions a Contractor is under a duty to warn the Employer of any problems with the design.

4–015   It has been held by the Supreme Court of Canada in *Brunswick Construction v Nowlan*[6] that a contractor executing work in accordance with plans of the employer's architect is under a duty to warn the employer of obvious design defects. The situation in England is slightly different. Prior to 2000, it was not clear whether there was a duty to warn.[7] In *Plant v Adams*,[8] whilst the Court of Appeal expressly reserved its position as to such an obligation where there was a design defect which the sub-contractor knew or ought to have known about which was not dangerous, where there was potential danger, it held that it was clear that a sub-contractor owed a duty of care to point out design faults and was required to protest vigorously and even walk off site, unless a safe design was produced.

A slightly different approach was suggested by the subsequent case of *Aurum Investments Limited v Avonforce Limited (in liquidation)*,[9] where an underpinning subcontractor was held not to be liable under the duty to warn principle when part of the excavation work collapsed. The subcontractor could not know of the design and build contractor's method or work. Mr Justice Dyson said that:

"the law is moving with caution in this area . . . a court should not hold a contractor to be under a duty to warn his client unless it is reasonable to do so."

Therefore the prudent Contractor would be advised to alert the Engineer to any obvious design defects which he comes across.

[6] (1974) 21 BLR 27.
[7] There were conflicting authorities. See *Victoria University of Manchester v Hugh Wilson and Lewis Womersley and Pochin (Contractors) Limited* [1984] CILL 126 and *University of Glasgow v W Whitfield and John Laing Construction Limited* [1988] 42 BLR 66.
[8] (2000) BLR 205.
[9] [2001] CILL 1729.

MDB HARMONISED VERSION

A new third paragraph has been added as follows: **4–016**

> *All equipment, material, and services to be incorporated in or required for the Works shall have their origin in any eligible source country as defined by the Bank.*

This amounts to a straightforward prohibition on the use of any materials from those countries deemed by the World Bank not to be an "eligible source country". As set out in the IBRD Guidelines for Procurement Under IBRD Loans and IDA Credit, a list is maintained of countries from which bidders, goods and services are not eligible to participate in procurement which is financed by the World Bank.[10] The list is regularly updated and can be obtained from the Public Information centre of the World Bank.

The Contractor should also be aware, under both editions of the Contract,   **4–017** of the possibility that local law or regulation might impose sanctions or otherwise prohibit using equipment from certain Countries. It is also possible that there might be similar restrictions imposed by the United Nations.

Finally, the words "if applicable" have been added to sub-para.(d) to go before "operation and maintenance manual".

4.2 PERFORMANCE SECURITY

The Contractor shall obtain (at his cost), a Performance Security for proper   **4–018** *performance, in the amount and currencies stated in the Appendix to Tender. If an amount is not stated in the Appendix to Tender, this Sub-Clause shall not apply.*

The Contractor shall deliver the Performance Security to the Employer within 28 days after receiving the Letter of Acceptance, and shall send a copy to the Engineer. The Performance Security shall be issued by an entity and from within a country (or other jurisdiction) approved by the Employer, and shall be in the form annexed to the Particular Conditions or in another form approved by the Employer.

The Contractor shall ensure that the Performance Security is valid and enforceable until the Contractor has executed and completed the Works and remedied any defects. If the terms of the Performance Security specify its expiry date, and the Contractor has not become entitled to receive the

---

[10] As of January 1998, representatives from the following countries were not eligible to participate in procurement financed by the World Bank: Andorra, Cuba, North Korea, Liechtenstein, Monaco, Nauru, San Marino and Tuvalu.

Performance Certificate by the date 28 days prior to the expiry date, the Contractor shall extend the validity of the Performance Security until the works have been completed and any defects have been remedied.

4–019   *The Employer shall not make a claim under the Performance Security, except for amounts to which the Employer is entitled under the Contract in the event of:*

(a)   *failure by the Contractor to extend the validity of the Performance Security as described in the preceding paragraph, in which event the Employer may claim the full amount of the Performance Security,*

(b)   *failure by the Contractor to pay the Employer an amount due, as either agreed by the Contractor or determined under Sub-Clause 2.5 [Employer's Claims] or Clause 20 [Claims, Disputes and Arbitration], within 42 days after this agreement or determination,*

(c)   *failure by the contractor to remedy a default within 42 days after receiving the Employer's notice requiring the default to be remedied, or*

(d)   *circumstances which entitle the Employer to termination under Sub-Clause 15.2 [Termination by Employer], irrespective of whether notice of termination has been given.*

The Employer shall indemnify and hold the Contractor harmless against and from all damages, losses and expenses (including legal fees and expenses) resulting from a claim under the Performance Security to the extent to which the Employer was not entitled to make the claim.

The Employer shall return the Performance Security to the Contractor within 21 days after receiving a copy of the Performance Certificate.

OVERVIEW OF KEY FEATURES

4–020   • The Contractor must provide Performance Security if the amount of the security required is set out in the Appendix to Tender.
   • If Performance Security is required, then the Contractor must, at his own cost, provide that security to the Employer no later than 28 days after receiving the Letter of Acceptance.
   • At the same time as the Performance Security is supplied to the Employer, a copy must be sent to the Engineer.
   • The Performance Security must be in the form annexed to the Particular Conditions. If it is not, it must be in a form agreed by the Employer.
   • The Performance Security must be issued by an entity and within a jurisdiction approved by the Employer.
   • The Performance Security must remain valid until the Contractor has completed the Works and remedied any defects.

- By sub-clause 14.6, if the Performance Security has not been received and accepted by the Employer, the Engineer will not issue an Interim Payment Certificate.
- The Employer may only make a call on the Performance Security under the four listed circumstances:

    (a)   Failure by the Contractor to extend the Performance Security beyond the expiry date if the Works are not complete at that date.   **4–021**

    (b)   Failure by the Contractor to pay an amount to the Employer under sub-cl.2.5 or cl.20.

    (c)   Failure by the Contractor to remedy a default within 42 days of being notified of the defect.

    (d)   When an Employer can terminate the Contract under sub-clause 15.2.

- If the Employer makes a claim when he is not entitled to do so, then the Contractor is entitled to an indemnity.

COMMENTARY

This sub-clause, which replaces clause 10 of the Old Red Book, FIDIC   **4–022** 4th edn provides that the Contractor must, if required by the Employer, obtain at his own cost, Performance Security in the amount specified by the Employer. The Contractor must maintain that Performance Security until his Works have been completed and any defects remedied.

The Contractor only has 28 days from receiving the Letter of Acceptance to provide the Performance Security. This is a relatively short period of time in which to obtain the security and it is important that the Contractor moves swiftly to obtain it. By sub-cl.2.1, the Employer may withhold access to the Site until the Performance Security has been obtained. Any delay in obtaining the Performance Security might have an adverse impact upon the Contractor's programme. In addition, by sub-cl.14.6, the Employer may withhold payment if the Performance Security has not been provided within the contractual timescale.

The Performance Security must be in a form approved by the Employer. The Employer is, however, free to choose his own form of security, although care must be taken to ensure that it complies with the Contract (as amended). Given the relatively tight timescale, if the Employer's preferred form is not one of those set out in the Contract Annex, then this should be made clear in the tender documentation.

Care must also be taken to ensure that the Performance Security applies   **4–023** with any applicable law. The Particular Conditions recommend that consideration is given to inserting the following to go after the second paragraph:

If the Performance Security is in the form of a bank guarantee, it shall be issued either (a) by a bank located in the Country, or (b) directly by a foreign bank acceptable to the Employer. If the performance security is not in the form of a bank guarantee, it shall be furnished by financial entity registered, or licensed to do business in the country.

The Annexes to the New Red Book set out seven recommended forms of security of which six relate to different types of security which the Contractor might be required to provide. These are:

(i)   Annexe A   Parent Company Guarantee
(ii)   Annexe B   Tender Guarantee
(iii) Annexe C   Performance Guarantee
(iv)   Annexe D   Performance Surety Bond
(v)   Annexe E   Advance Payment Guarantee
(vi)   Annexe F   Retention Money Guarantee
(vii) Annexe G   Payment Guarantee in favour of the Contractor

4–024   Annexes B, C, E, F and G are securities which are callable on demand. Annexes A and D are surety bonds payable on default.

An on-demand security bond is an unconditional obligation to pay when a demand has been made. A surety bond or performance guarantee requires certain conditions to be met before payment is made.

The most important are Annexes C and D, which are designed primarily to provide protection against any failure to complete the Works as a consequence of default on the part of the Contractor for example through insolvency. The Contractor only has to maintain the Performance Security until he has completed his works and remedied any defects. They do not provide protection in relation to latent defects.

4–025   These standard securities incorporate the following Rules produced by the International Chamber of Commerce[11]:

(i)   ICC Publication No.458, Uniform Rules For Demand Guarantees (1992);
(ii)   ICC Publication No.524, Uniform Rules For Contract Bonds (1993).

One advantage of incorporating these rules is that it will mean that there can be no argument over what laws govern the security or which jurisdiction will be competent to hear disputes in connection with it. However the rules are not identical.

By art.27 of the Uniform Rules For Demand Guarantee, unless provided otherwise in the guarantee, the law governing the guarantee will be the law of

---

[11] The stated purpose of these rules, as set out in the foreword, is to "balance the interests of the different parties and to curb abuse in the calling guarantees".

the place of business of the guarantor. Whilst by art.28, unless the parties agree otherwise, disputes between the guarantor and the beneficiary shall be settled *exclusively* by a court in the country of the guarantor of the guarantee. It is important to remember that if the guarantor has more than one place of business then the country of the guarantor will be the country of the branch where the guarantee in question was issued.

Thus, if a guarantee under the Uniform Rules for Demand Guarantees (URDG) is issued by the Taipei branch of a French bank, the beneficiary will not be allowed to start proceedings against the bank in France and will have no other option but to bring its claim in the Taiwanese courts.   **4–026**

However under the Uniform Rules For Contract Bonds, Article 8(a) provides that the applicable law, if not otherwise chosen by the parties, shall be the law governing the contract. Also by art.8(c), unless the parties agree otherwise, disputes between the guarantor and the beneficiary shall be settled *exclusively* by a court in the country of the principle place of business of the guarantor. Unlike with the Demand Bonds, the guarantor has the option (not the obligation) to settle any disputes that may arise in the country of the branch where the guarantee in question was issued.

Thus, using the example set out above, the beneficiary would have the option to choose between the French and Taiwanese courts to hear the claim.

All the Annexes have been drafted in clear terms. They therefore do not fall foul of the condemnation by the House of Lords in *Trafalgar House Construction (Regions) Ltd v General Surety and Guarantee Co. Ltd*[12] in which Lord Jauncey said: **4–027**

> "I find great difficulty in understanding the desire of commercial men to embody so simple an obligation in a document which is quite unnecessarily lengthy, which obfuscates its true purpose and which is likely to give rise to unnecessary arguments and litigation as to its meaning."

The demand guarantees in favour of the Employer (Annexes B, C, E and F) are all similar in nature. Annex C, the Demand Guarantee provides that[13]:

> At the request of the Principal, we . . . hereby irrevocably undertake to pay you [Employer] any sum or sums not exceeding . . . upon receipt by us of your demand in writing and your written statement stating:

> That the Principal is in breach of his obligation(s) under the Contract, and
> The respect in which the Principal is in breach.

---

[12] 73 BLR 32.
[13] Thereby mirroring the requirement of art.20 of the ICC Rules No.458 for Demand Guarantees.

4-028   In other words, the Employer has to do more than simply state that there has been default. The fact that the Employer must set out details of the breach has probably been inserted to pressure an Employer to only seek to call the bond in genuine circumstances. Nevertheless, the Employer is under no obligation to provide any evidence of that default by the Contractor.

A similar requirement can be found in Annex G, which is the only demand guarantee in favour of the Contractor. However it has been drafted somewhat differently. As well as providing a demand in writing confirming that the Employer has failed to make payment in full, the Contractor must provide evidence as to that failure. This suggests that this means that the evidence supplied by the Contractor is liable to be checked and so Annexe G is not in the same category of guarantee as Annexes B, C, E and F.

Sub-clause 4.2 provides that the Employer shall not make a call on the Performance Security except in the circumstances listed out at items (a)–(d). The fact that these circumstances might not apply would not stop the Employer from making a call on the Demand Guarantee. The bond is an entirely separate document from the Contract. The guarantor must pay if the conditions of payment have been satisfied.

4-029   However, the prudent Employer would be advised to exercise caution before making a call on the Bond if the circumstances set out in sub-cll 4.2(a)–(d) did not apply, as the sub-clause also requires that the Employer shall indemnify the Contractor if he makes a call on the Performance Security to which he is not entitled under the Contract.

The same would apply in respect of the surety bonds. This is even though they are very different in nature to the demand guarantees. To take Annexe D as an example. This states:

"Upon Default, by the Principal to perform any Contractual Obligation or upon the occurrences of any of the events and circumstances listed in sub-clause 15.6 of the Conditions of Contract, the Guarantor shall satisfy and discharge the damages sustained by the beneficiary due to such default, event or circumstances . . ."

Therefore there must be default by the Contractor under the Contract. It is not enough for the Employer merely to provide a written notice that there has been default.

4-030   It can be seen that the demand guarantees tend to favour the Employer, whilst the surety bond is more favourable to the Contractor. However what form, if any, of Performance Security is chosen will be a matter for commercial negotiation[14] and whatever form is chosen, the Contractor and Employer

---

[14] The FIDIC Guide in fact advises against the on-demand form of security on the grounds that contractors may increase their tenders to reflect the risk of an unmeritorious call on the security.

will be both be aware of the restrictions imposed on the Employer by clause 4.2(a)–(d).

This is an important change from the Old Red Book 4th edn. Under cl.10 all the Employer had to do was notify the Contractor of the nature of the default prior to making a claim. Here, under sub-clause 4.2, the Employer can only make a call on the Performance Security in certain specific circumstances. Now, in respect of default by the Contractor, the Employer must give notice to the Contractor and allow the Contractor 42 days to remedy that default. However, the Contractor should be aware that there is no requirement on the Employer to, for example, wait for any decision which might be forthcoming from a DAB convened under the auspices of clause 20 before deciding to make a call on the performance guarantee.

There is no obligation to use any of the options suggested in the Annexe. However the Annexes have been drafted to ensure that when taken with the ICC rules, they conform with the requirements of the contract. Care should be taken if other forms of bond or performance security, which are in a different form, are used to ensure that they reflect the terms of the contract.

Of course, the obligation to maintain the Performance Security does come **4–031** to an end – eventually – although only when the Contractor has completed the Works and remedied any defects. The Employer should not forget to return the Performance Security to the Contractor within 21 days after receiving a copy of the Performance Certificate.

However, this does mean that the Performance Security is of no assistance to the Employer in respect of any defect that may arise once the defects notification period has come to an end.

MDB HARMONISED EDITION

There is one minor change, one more wide-ranging one and a couple of addi- **4–032** tions. In addition there are significant changes to the forms of Performance Security included as Annexes to the Contract.

First, the word "form" has been added to go before "amount" in the first sentence whilst, the reference to the "Appendix to Tender" has been changed to "Contract Data" in accordance with the change in the list of definitions.

Second, the four circumstances under which an Employer was entitled to make a call on the Performance Security have been replaced with the following:

The Employer shall not make a claim under the Performance Security, except **4–033** *for amounts to which the Employer is entitled under the Contract.*

Although the provision requiring the Employer to indemnify the Contractor in the event of a false claim remains, it does appear that this revised wording will make it easier for an Employer to make a call on any on-demand security.

In addition, the following additional paragraph has been added to the end of the clause:

4–034 *Without limitation to the provisions of the rest of this Sub-Clause, whenever the Engineer determines an addition or a reduction to the Contract Price as a result of a change in cost and/or legislation or as a result of a Variation amounting to more than 25 percent of the portion of the Contract Price payable in a specific currency, the Contractor shall at the Engineer's request promptly increase, or may decrease, as the case may be, the value of the Performance Security in that currency by an equal percentage.*

The first paragraph is self-explanatory. However, the second is perhaps more controversial. If the Contract Price goes up by more than 25 per cent, if the Engineer so requests, the Contractor must secure an increase in the Performance Security by an equivalent percentage. If the Contract Price falls by more than 25 per cent, then a similar reduction can be made to the Performance Security.

Perhaps the key factor here is that whether or not a reduction or increase is required is at the discretion of the Engineer. Therefore the Contractor cannot decrease the value of any security unless the Engineer agrees to such a reduction being made.

The Annexes

4–035 The MDB Harmonised Edition contains the following Annexes:

(i)   Annex F   Performance Security – Demand Guarantee
(ii)   Annex G   Performance Bond
(iii) Annex H   Advance Payment Security – Demand Guarantee
(iv)   Annex I   Retention Money Security – Demand Guarantee
(v)   Annex J   Parent Company Guarantee
(vi)   Annex K   Bid Security

It will be seen immediately that there is no form of payment guarantee in favour of the Contractor.

Only Annex G is a surety bond payable on default. Annexes F, H, I and K are all callable on demand. In addition the wording of the securities has changed.

4–036 Under Annex C (the on-demand form of security) of the standard Contract, any demand must be accompanied by a written statement stating that the principal is in breach of his obligations and in what respect the principal is in breach. Under Annex F of the MDB Harmonised Edition, the written statement need only state that the Contractor is in breach and suggested wording includes the phrase, *"without your needing to prove or to show grounds for your demand."* This is confirmed by the deletion of sub-paragraph (ii) of art.20 of the Uniform Rules For Demand Guarantee.

There are also changes to Annex G. Annex D of the standard FIDIC form provided that:

Upon Default by the Principal to perform any Contractual Obligation, or upon the occurrence of any of the events and circumstances listed in sub-clause 15.2 of the conditions of the Contract, the Guarantor shall satisfy and discharge the damages sustained by the Beneficiary due to such event, default or circumstance.

Annex G of the MDB Harmonised Edition is rather different and provides **4–037** that:

Now, therefore, the Condition of this Obligation is such that, if the Contractor shall promptly and faithfully perform the said Contract (including any amendments thereto), then this obligation shall be null and void; otherwise, it shall remain in full force and effect. Whenever the Contractor shall be, and declared by the Employer to be, in default under the Contract, the Employer having performed the Employer's obligations thereunder, the Surety may promptly remedy the default, or shall promptly:

(a)   complete the Contract in accordance with its terms and conditions; or

(b)   obtain a Bid or bids from qualified Bidders for submission to the Employer for completing the Contract in accordance with its terms and conditions, and upon determination by the Employer and the Surety of the lowest responsive Bidder, arrange for a Contract between such Bidder and Employer and make available as work progresses (even though there should be a default or a succession of defaults under the Contract or Contracts of completion arranged under this paragraph) sufficient funds to pay the cost of completion less the Balance of the Contract Price; but not exceeding, including other costs and damages for which the Surety may be liable hereunder, the amount set forth in the first paragraph hereof. The term "Balance of the Contract Price", as used in this paragraph, shall mean the total amount payable by Employer to Contractor under the Contract, less the amount properly paid by Employer to Contractor; or

(c)   pay the Employer the amount required by Employer to complete the Contract in accordance with its terms and conditions up to a total not exceeding the amount of this Bond.

There are three main differences between the two. First, the security under **4–038** the MDB Harmonised Edition is not subject to the Uniform Rules For Contract Bonds. Second, under the standard Contract Annex D, the Contractor must be in default under the contract. However, under Annex G of the MDB Harmonised Edition, the Contractor must be in default and declared to be so by the Employer. These are, of course, potentially two different things.

Finally, under the MDB Harmonised Edition, there are provisions for the Surety to become involved in performing the contract either by completing the Contract itself or becoming involved in any re-tendering process that may be necessary as a consequence of the Contactor default.

4.3 CONTRACTOR'S REPRESENTATIVE

4–039 *The Contractor shall appoint the Contractor's Representative and shall give him all authority necessary to act on the Contractor's behalf under the Contract.*

Unless the Contractor's Representative is named in the Contract, the Contractor shall, prior to the Commencement Date, submit to the Engineer for consent the name and particulars of the person the Contractor proposes to appoint as Contractor's Representative. If consent is withheld or subsequently revoked, or if the appointed person fails to act as Contractor's Representative, the Contractor shall similarly submit the name and particulars of another suitable person for such appointment.

The Contractor shall not, without the prior consent of the Engineer, revoke the appointment of the Contractor's Representative or appoint a replacement.

4–040 *The whole time of the Contractor's Representative shall be given to directing the Contractor's performance of the Contract. If the Contractor's Representative is to be temporarily absent from the Site during the execution of the works, a suitable replacement person shall be appointed, subject to the Engineer's prior consent, and the Engineer shall be notified accordingly.*

The Contractor's Representative shall, on behalf of the contractor, receive instructions under Sub-Clause 3.3 [Instructions of the Engineer].

The Contractor's Representative may delegate any powers, functions and authority to any competent person, and may at any time revoke the delegation. Any delegation or revocation shall not take effect until the engineer has received prior notice signed by the Contractor's Representative, naming the person and specifying the powers, functions and authority being delegated or revoked.

4–041 *The Contractor's Representative and all these persons shall be fluent in the language for communications defined in Sub-Clause 1.4 [Law and Language].*

OVERVIEW OF KEY FEATURES

4–042
- The Contractor must appoint a Contractor's Representative.
- The Contractor's Representative will either be named in the Contract or approved by the Engineer prior to the Commencement Date.
- If the Contractor wishes to replace the Contractor's Representative, the prior consent of the Engineer must be obtained.
- The Contractor's Representative *must* devote the whole of his time on site to directing the performance of the Contractor.

- It is the Contractor's Representative who receives instructions issued under sub-cl.3.3 by the Engineer.
- The Contractor's Representative must be fluent in whatever language is stated in the Appendix to Tender to be the language the parties should communicate in.
- The Contractor's Representative may delegate his powers provided the Engineer receives prior notice.

COMMENTARY

This sub-clause is concerned with the extensive obligations of the Contractor's Representative. It is a reworking of cl.15 of the Old Red Book, FIDIC 4th edn.   **4-043**

By sub-cl.1.1.25, the Contractor's Representative is defined as "the person named by the Contractor . . . who acts on behalf of the Contractor".

The Contractor's Representative is a key individual. If he is not named in the Contract, his appointment must have been approved by the Engineer. Although the Contractor's Representative can delegate his powers, the Engineer must receive prior notice of that delegation.

One of the more important requirements of this sub-clause is that the Contractor's Representative is either named in the Contract or approved before the Commencement Date. By sub-cl.8.1, the Commencement Date will be within 42 days of the Contractor receiving the Letter of Acceptance. Therefore it is important that this appointment is resolved at an early stage.   **4-044**

A typical problem on many sites is the failure by a contractor to man projects with the key personnel who lead the tender process and who are held out as going to lead the project. The clause is designed to combat this. In addition, it is always open to the Employer to suggest who the Contractor's Representative should be (or at least set out some minimum criteria), which is another way to resolve this problem.

The sub-clause also goes further by requiring that the Contractor's Representative spend his "whole time" directing the performance for the Contractor. It is clearly intended that the Contractor's Representative should be based on site. He must give notice of any temporary absence from site and someone must be appointed in his place during that absence.

On occasion, a fluency in more than one language might be required – for example if the ruling language of the contract is different from the language prevalent on site. In such circumstances, as suggested in the Particular Conditions, the Contractor might be required to ensure an interpreter is available.   **4-045**

The sub-clause is silent as to whether the Engineer can block the appointment of the Contractor's Representative. By cl.15 of the Old Red Book FIDIC 4th edn, the Engineer could withdraw his approval to the equivalent of the Contractor's Representative. That power no longer forms part of sub-clause 4.3 provided the Contractor's Representative is named in the Contract. If he is not, the Engineer can withhold or revoke his approval of the Contractor's Representative. This is not a clause where the Engineer is required to make a determination in accordance with sub-cl.3.5. However the Engineer cannot unreasonably withhold his consent: sub-cl.1.3 makes it clear that any approval shall not be unreasonably withheld or delayed.

MDB HARMONISED EDITION

4–046 First the words, "*in terms of Sub-clause 6.9 [Contractor's Personnel]*" have been added to the second paragraph. This should have the effect of preventing the unjustified removal of the Contractor's Personnel as sub-clause 6.9 lists the grounds under which the Engineer can require personnel to be removed. By implication those standards should be implied here.

Second, as anticipated in the Particular Conditions, the following straight-forward addition has been made to the final paragraph about the need for fluency in the Contractor's Representative:

The Contractor's Representative and all these persons shall be fluent in the language for communications defined in Sub-Clause 1.4 [Law and Language]. If the Contractor's Representative's delegates are not fluent in the said language, the Contractor shall make competent interpreters available during all working hours in a number deemed sufficient by the Engineer.

4.4 SUBCONTRACTORS

4–047 *The Contractor shall not subcontract the whole of the Works.*

The Contractor shall be responsible for the acts or defaults of any Subcontractor, his agents or employees, as if they were the acts or defaults of the Contractor. Unless otherwise stated in the Particular Conditions:

(a) *the Contractor shall not be required to obtain consent to suppliers of Materials, or to a subcontract for which the Subcontractor is named in the Contract;*

(b) *the prior consent of the Engineer shall be obtained to other proposed Subcontractors;*

(c) *the Contractor shall give the Engineer not less than 28 days' notice of the intended date of the commencement of each Subcontractor's work, and of the commencement of such work on the Site; and*

(d) *each subcontract shall include provisions which would entitle the Employer to require the subcontract to be assigned to the Employer under*

Sub-Clause 4.5 [Assignment of Benefit of Subcontract] (if or when applicable) or in the event of termination under Sub-Clause 15.2 [Termination by Employer].

OVERVIEW OF KEY FEATURES

- Unless otherwise agreed the Contractor shall *not* subcontract the whole of the Works.

  4–048

- If the Contractor is permitted to subcontract an element of the works, then the Contractor remains wholly responsible to the Employer for the acts of that Sub-contractor as if those acts had been carried out by him.
- The prior consent of the Engineer is required for all sub-contractors apart from suppliers and Sub-contractors named in the Contract.
- The Contractor must give the Engineer 28 days notice of both the intended and actual commencement date of any Sub-contractor's work.

COMMENTARY

This sub-clause, which is an extension of cl.4 of the Old Red Book, FIDIC 4th edn, deals with the appointment of Sub-contractors. It also makes it clear that the Contractor is wholly responsible for the performance of the Sub-contractors. This obligation extends not only to the Sub-contractors he chooses to appoint but also to the Sub-contractors nominated by the Employer in accordance with cl.5.

4–049

Although the Contractor cannot subcontract the whole of the Works, he can subcontract part of the Works. If the Sub-contractor is named in the Contract or if the Sub-contractor will merely be supplying materials, then the consent of the Engineer is not required. Otherwise, the consent of the Engineer will be required. The Particular Conditions include a sample paragraph suggesting that prior consent may not be required in all circumstances, perhaps if the value of the proposed sub-contract is very small compared with the overall contract value.

Similarly to sub-cl.4.3, this is not a clause where the Engineer is required to make a determination in accordance with sub-cl.3.5. However again in accordance with sub-cl.1.3 the Engineer's consent to any proposed subcontracting cannot be unreasonably withheld or delayed.

It is important that the Contractor appreciates how wide the obligations here are. The Contractor is responsible for all the acts and defaults of the Sub-contractors. This is in contrast to the position under English Law whereby a Contractor is not necessarily responsible for any design carried out

4–050

by a nominated Sub-contractor or compliance with performance specification and selection of goods and materials. In those circumstances the Contractor would only be liable for poor workmanship and poor quality of goods and materials.

Here the Contractor has strict contractual liability for any Sub-contractor. Therefore if a Sub-contractor commits a breach of contract, this might lead to the Contractor himself being considered to be in breach of contract. For this reason, the Contractor must be sure to consider whether to object to the nomination of a Sub-contractor in accordance with sub-cl.5.2. For example, one reasonable reason to object is if the provisions of the sub-contract do not include an indemnity in favour of the Contractor.

The sub-clause also requires the Contractor to give the Engineer not less than 28 days' notice of the intended start date for every sub-contractor. This is another clause which will enable the Engineer to keep an independent check on progress.

Finally, the sub-clause requires that every subcontract include provisions enabling the sub-contract to be assigned to the Employer in accordance with sub-cl.4.5 or in the event the Contractor's contract is terminated under sub-cl.15.2.

MDB HARMONISED EDITION

4–051   The word "solely" has been added to sub-paragraph (a), thereby confirming the restriction already set out in that sub-paragraph.

The following two paragraphs have been added to this sub-clause:

The Contractor shall ensure that the requirements imposed on the Contractor by Sub-Clause 1.12 [Confidential Details] apply equally to each Subcontractor.

Where practicable, the Contractor shall give fair and reasonable opportunity for contractors from the Country to be appointed as Subcontractors.

As set out above sub-cl.1.12 has been significantly extended in the MBD Harmonised Edition. Under sub-cl.1.12 of the Standard Contract, the Contractor must, if so required by the Engineer, disclose any such information (including any that may be confidential) that may be necessary to demonstrate compliance with the Contract. Under the MDB Harmonised Edition, sub-cl.1.12 extends on the Contractor a duty to treat any information about the contract, unless it is publicly available, as confidential. The purpose of the sub-clause here is to ensure that the Contractor extends that duty to any sub-contractors under the Contract.

4–052   The second additional paragraph amounts effectively to a request that the Contractor gives local contractors a fair opportunity to work on the project and is entirely in keeping with the aims of the World Bank to encourage local enterprise. This is also something suggested in the Particular Conditions of the 1999 Edition.

4.5 ASSIGNMENT OF BENEFIT OF SUBCONTRACT

If a Subcontractor's obligations extend beyond the expiry date of the relevant   **4–053**
Defects Notification Period and the Engineer, prior to this date, instructs the
Contractor to assign the benefit of such obligations to the Employer, then
the Contractor shall do so. Unless otherwise stated in the assignment, the
Contractor shall have no liability to the Employer for the work carried out by
the Subcontractor after the assignment takes effect.

OVERVIEW OF KEY FEATURES

- Where a sub-contractor's obligations last beyond the Defects Notification   **4–054**
  period, the Employer (through the Engineer) can choose to instruct the
  Contractor to assign to him the benefits of those obligations.
- The Employer, and not the Contractor, will be responsible for work
  carried out after any such assignment.

COMMENTARY

This clause deals with the situation where a sub-contractor's obligations   **4–055**
extend beyond the expiry of the Defects Notification period.

In these circumstances, as noted in sub-cl.4.4(d), the Employer can require
the Contractor to assign the benefits of the particular sub-contract.
Therefore the Employer will have the benefit of any guarantees provided by
the sub-contract. This may be important to an Employer, particularly in
circumstances when the Contactor may have no liability or when there are
concerns about the Contractor's solvency or attitude and general approach.

MDB HARMONISED EDITION

There is no change.   **4–056**

4.6 CO-OPERATION

The Contractor shall, as specified in the Contract or as instructed by the   **4–057**
Engineer, allow appropriate opportunities for carrying out work to:

(a)   the Employer's Personnel,
(b)   any other contractors employed by the Employer and
(c)   the personnel of any legally constituted public authorities,

who may be employed in the execution on or near the Site of any work not included in the Contract.

Any such instruction shall constitute a Variation if and to the extent that it causes the Contractor to incur Unforeseeable Cost. Services for these personnel and other contractors may include the use of Contractor's Equipment, Temporary Works or access arrangements which are the responsibility of the Contractor.

If, under the Contract, the Employer is required to give to the Contractor possession of any foundation, structure, plant or means of access in accordance with Contractor's Documents, the Contractor shall submit such documents to the Engineer in the time and manner stated in the Specification.

OVERVIEW OF KEY FEATURES

4–058   • The Contractor shall co-operate with the Employer's Personnel, other contractors and personnel from public authorities to enable them to carry out their work on or near to the site.
   • If the Contractor is instructed by the Engineer to co-operate in such a way, this shall constitute a Variation.
   • If the Contractor requires possession of any means of access, structure or foundation, or plant, this must be set out in the Contractor's Documents which must be submitted in accordance with the specification.

COMMENTARY

4–059   This is another sub-clause in two parts.

Co-operation

4–060   Despite being headed "co-operation", this sub-clause does not seem to refer to co-operation in the partnering sense of the word, or arguably in any real sense at all.

   The sub-clause sets out what the Contractor must do to assist others who may be carrying out work on or near the site. It is not immediately apparent as to whether there is an equivalent obligation on the Employer. Clause 2.3 provides that the Employer "shall be responsible for ensuring that" the Employer's personnel and the Contractor's "co-operate" with the

Contractor's efforts under cl.4.6. However it is the Contractor who must find the appropriate opportunities to enable the others to carry out their works.

Of course, these opportunities must be both appropriate to the Employer's Personnel and others named in the sub-clause as well as convenient to the Contractor's own programming requirements. The obligation required by sub-cl.4.6, specifically refers to both work near the site and work not included in the Contract.

If the Contract makes reference to the possibility of such work, the **4–061** prudent Contractor might be well advised to enquire of the Employer, at tender stage, as to the likely extent of any allowance he should make for the likelihood of such work.

Alternatively, the sub-clause does provide some comfort to the Contractor and recognition of the potentially disruptive and delaying effect of providing this "co-operation". This is through the reference to the Contractor being instructed by the Engineer to make allowance for this additional work. If such an instruction is issued, then it will be considered a variation and the Contractor can seek an extension of time in accordance with sub-clause 8.4(e).

The difficulty for the Contractor will be the recovery of Cost. He is only entitled to recover "Unforeseeable Cost". Sub-clause 1.1.6.8 defines "unforeseeable" as being "not reasonably foreseeable by an experienced Contractor by the date of submission of the Tender". Given the wide definition here, it is submitted that it may be difficult for the Contractor to demonstrate that such a situation was not foreseeable by an experienced contractor.[15] This is another reason why the prudent Contractor should make enquiries of the Employer during the tender stage.

Therefore it could be argued what this sub-clause is really saying is that: **4–062**

(i) if the Contractor is going to have to make appropriate allowances for those listed in sub-cll 4.6(a)–(c), then it will be noted in the Contract and the Contractor will be expected to have made the appropriate allowance for this in preparing his tender; alternatively

(ii) if it becomes apparent during the course of the Contract that the Contractor is going to have to make appropriate allowances for those listed in sub-cll 4.6(a)–(c), then the Engineer shall issue an instruction. The Contractor will be entitled to the Cost incurred by that instruction provided an experienced contractor would not have foreseen the need for those listed to carry out Works at or near the Site.

This is not really anything to do with co-operation. The final part of **4–063** sub-cause 4.6 is also not really anything to do with co-operation. It is also unrelated to the first half of the sub-clause.

---

[15] For further discussions of the definition of "experienced contractor" see sub-cll 4.7 and 1.1.6.8.

Other FIDIC contracts include here a clause requiring the Contractor to co-ordinate his own activities with those of other contractors to the extent specified. The FIDIC Guide notes that this paragraph has been omitted here because it was felt that it was usually impractical to require a Contractor here to so co-ordinate to an extent which could be defined in the contract and therefore allowed for in the tender.

Access

4-064   The final part of this sub-clause provides that if the Employer is to give the Contractor access to any foundation, structure, plant or means of access in accordance with the Contractor's Documents, then those documents must be submitted to the Engineer in accordance with the specification. It is unclear quite to what this is referring.

By sub-cl.2.1, the Employer is already required to give the Contractor right of access to, and possession of, all parts of the Site. Sub-clause 1.1.6.1 defines "Contractor's Documents" as the calculations, computer programs and other software, drawings, manuals models and other technical documents.

It is suggested that this clause must simply mean that, if the provision of access (or anything else listed here) is dependant on the Employer receiving information from the Contractor then that information shall be provided in the manner set out in the specification.[16]

MDB HARMONISED EDITION

4-065   The words *"suffer delays and/or to"* have been added to the first sentence of the second paragraph. Thus, an instruction shall constitute a Variation if it causes the Contractor to suffer delays and not just if it causes the Contractor to suffer Unforeseeable Cost.

4.7 SETTING OUT

4-066   *The Contractor shall set out the works in relation to original points, lines and levels of reference specified in the Contract or notified by the Engineer. The Contractor shall be responsible for the correct positioning of all parts of the Works, and shall rectify an error in the positions, levels, dimensions or alignment of the Works.*

---

[16] See sub-cl.2.3 and the discussion of *Ellis-Don Limited v the Parking Authority of Toronto.*

The Employer shall be responsible for any errors in these specified or notified items of reference, but the Contractor shall use reasonable efforts to verify their accuracy before they are used.

If the Contractor suffers delay and/or incurs Cost from executing work which was necessitated by an error in these items of reference, and an experienced contractor could not reasonably have discovered such error and avoided this delay and/or cost, the Contractor shall give notice to the Engineer and shall be entitled subject to Sub-Clause 20.1 [Contractor's Claims] to:

(a) *an extension of time for any such delay, if completion is or will be delayed, under Sub-Clause 8.4 [Extension of Time for Completion], and*   **4-067**

(b) *payment of any such Cost plus reasonable profit, which shall be included in the Contract Price.*

After receiving this notice, the Engineer shall proceed in accordance with Sub-Clause 3.5 [Determinations] to agree or determine (i) whether and (if so) to what extent the error could not reasonably have been discovered, and (ii) the matters described in sub-paragraphs (a) and (b) above related to this extent.

OVERVIEW OF KEY FEATURES

- It is the Employer's responsibility to set out the original points, lines and levels of reference specified in the Contract.   **4-068**
- Subject to this, the Contractor shall set out his works from this data.
- The Contractor has an obligation to check or use reasonable efforts to verify the accuracy of the Employer's setting out.
- If any error in the setting out causes delay or costs, then the Contractor should give notice to the Engineer within the time limits prescribed within sub-cl.20 and may be entitled to an extension of time and payment of any cost plus reasonable profit.
- It is for the Engineer to determine whether the error could have been found by "an experienced Contractor" using reasonable efforts to verify the data.

COMMENTARY

This sub-clause deals with who is responsible for any errors in setting out. It   **4-069** is a re-working of cl.17 of the Old Red Book FIDIC 4th edn.

The basic scheme of the clause is that the Employer is responsible for the original points, lines and levels. The Contractor is responsible for the setting out of his Works to those points.

This sub-clause, contrary to initial impressions, does not make the Employer responsible for every error in setting out. The Contractor is under an obligation to verify the accuracy of the data before it is used. The Contractor will only be entitled to an extension of time and payment for any increased costs if the Engineer considers that an "experienced contractor" carrying out reasonable checks would not have discovered the error.

4–070 The checking obligation is therefore an important one for the Contractor. It is also a stricter obligation than can be found under English law, where the usual criteria would be a requirement that the Contractor exercise reasonable skill and care. However as set out by McNair J.,[17] that is not the reasonable skill and care of an experienced contractor instead:

> "The test is the standard of the ordinary skilled man exercising and professing to have that special skill. A man need not possess the highest expert skill; it is well established law that it is sufficient if he exercises the skill of an ordinary competent man exercising that particular art . . . He is not guilty of negligence if he has acted in accordance with a practice accepted as proper by a responsible body of medical men skilled in that particular art . . . Putting it another way round, a man is not negligent if he is acting in accordance with such a practice merely because there is a body of opinion who would take a contrary view."

Here the standard is that of the "experienced contractor". If the Contractor considers that he has suffered delay, he needs to demonstrate why that delay was caused by a setting out error which an experienced Contractor would not have discovered? This will be a difficult hurdle to over-come as the experienced contractor, arguably, will be aware of the potential for delay if there are errors in the setting out and will accordingly make thorough checks.

4–071 If the Contractor considers that he has suffered delay, he must give notice in accordance with the provisions of cl.20, as soon as practicable but in any event not later than 28 days after he became aware of the problem.

A practical way to try and reduce the potential for conflict might be for the Contractor and Employer to agree the setting out points and levels upon (or before) commencement of the Works. Taking steps to do this would also at an early stage reveal whether the Contractor's tender was based on what might prove to be a fundamental misconception.

---

[17] *Bolam v Friern Hospital Management Committee* [1957] 1 W.L.R. 582.

MDB HARMONISED EDITION

"*Reasonable profit*" in sub-paragraph (b) has become "*profit*".[17A]   **4–072**

4.8 SAFETY PROCEDURES

The Contractor shall: **4–073**

(a)   *comply with all applicable safety regulations,*
(b)   *take care for the safety of all persons entitled to be on the Site,*
(c)   *use reasonable efforts to keep the Site and Works clear of unnecessary obstruction so as to avoid danger to these persons,*
(d)   *provide fencing, lighting, guarding and watching of the Works until completion and taking over under Clause 10 [Employer's Taking Over], and*
(e)   *provide any Temporary Works (including roadways, footways, guards and fences) which may be necessary, because of the execution of the Works, for the use and protection of the public and of owners and occupiers of adjacent land.*

OVERVIEW OF KEY FEATURES

- The Contractor must comply with the applicable safety regulations and   **4–074** take care for the safety of everyone entitled to be on site.
- The Contractor must keep the site clear of unnecessary rubbish.
- The Contractor must provide site security until the issue of the Taking Over Certificate.
- The Contractor must provide any Temporary Works which may be necessary as a consequence of the impact of the Works on the areas surrounding the site.

COMMENTARY

This sub-clause sets out the wide obligations imposed on the Contractor to   **4–075** ensure safety on Site and anywhere where contract works are carried out.

[17A] See discussion above in relation to sub-cl.1.2.

However contrary to the sub-clause heading, it does not solely deal with safety issues.

It is important that this sub-clause is read in conjunction with sub-cl.6.7, which sets out the obligations of the Contractor in relation to health and safety in more detail. For example, the Contractor must by virtue of sub-cl.6.7, appoint an accident prevention officer to deal with questions of safety and accident prevention.

Sub-clause 4.8(a) provides that the Contractor shall comply with all applicable safety regulations. An Employer would be well advised, particularly when dealing with foreign contractors, to include specific provision in the tender documentation in relation to the relevant health and safety legislation.

4–076   However, the Contractor will be aware that he is required by sub-cl.1.13 to comply with all applicable laws. The Contractor can ask the Employer, in accordance with sub-cl.2.2(a) for copies of any health and safety laws which may be relevant to the Contract. The Contractor must also check the requirements of the whole Contract carefully. It may not be necessarily enough for the Contractor simply to comply with the local safety regulations. The Contract may impose wider obligations.

The Contractor should also remember that by sub-cl.2.1, the Contractor is not necessarily given exclusive possession of the site. The fact that the Contractor's possession might not be exclusive does not free the Contractor from his obligations under this sub-clause or sub-cl.6.7.

By sub-cl.2.3(b) the Employer is responsible for ensuring that the Employer's Personnel and the Employer's other contractors on the Site take actions similar to those which the Contractor is required to comply with, namely the safety procedures of sub-paras 4.8(a), (b) and (c).

4–077   As sub-cl.2.3(b) notes, it is only the first three sub-clauses of 4.8 which specifically relate to safety procedures. The remaining two parts of this clause have a wider application. The first, the obligation imposed by sub-clause 4.8(d), relates to site security.

By sub-cl.4.8(d), the Contractor must provide fencing of the Works. The fencing must be adequate to keep intruders out. This is both to protect the Works but also for reasons of safety. Whilst sub-cl.4.8(b) only refers to the need to take care for people entitled to be on site, the Contractor might find himself under a duty to take care for the trespassers, or those not entitled to be on site.

It could be argued that as the Contractor must accept the responsibility for the security of the site, that responsibility must include the security of the site against trespassers (a term more easily understood by use of the phrase "non-visitor"). Accordingly, the Contractor should bear the consequences of trespassers' presence.

4–078   However, under English law that potential liability is encapsulated in the Occupiers Liability Act 1984. Section 1(3) of that Act provides that:

"(3) An occupier of premises owes a duty to another (not being his visitor) in respect of any such risk as is referred to in sub-section (1) above if –

(a)   he is aware of the danger or has reasonable grounds to believe that it exists;

(b)   he knows or has reasonable grounds to believe that the other is in the vicinity of the danger concerned or that he may come into the vicinity of the danger (in either case, whether the other has lawful authority for being in that vicinity or not); and

(c)   The risk is one against which, in all the circumstances of the case, he may reasonably be expected to offer the other some protection.

(4) Where, by virtue of this section, an occupier of premises owes a duty to another in respect of such risk, the duty is to take such care as is reasonable that in all the circumstances of the cases to see that he does not suffer injury on the premises by reason of the danger concerned."

The 1984 Act therefore imposes a duty on an occupier where: **4–079**

(i)   the state of the premises poses a danger;

(ii)   the danger is one that poses a risk of causing injury to a trespasser if he comes into the vicinity of the danger;

(iii) there are reasonable grounds for believing that the trespasser is or may come into the vicinity of the danger; and

(iv)   in all the circumstances of the case it is reasonable to afford the trespasser some protection against the risk.

The 1984 Act was designed to codify the words of Lord Diplock.[18] These are set out in detail below because they set out important considerations for any Contractor under any jurisdiction:

"*First*: The duty does not arise until the occupier has actual knowledge either of the presence of the trespasser upon his land or of facts which make it likely that the trespasser will come on to his land; and has also actual knowledge of facts as to the condition of his land or of activities carried out upon it which are likely to cause personal injury to a trespasser who is unaware of the danger. He is under no duty to the trespasser to make any inquiry or inspection to ascertain whether or not such facts do exist. His liability does not arise until he actually knows of them.

Secondly: Once the occupier has actual knowledge of such facts, his own   **4–080**
failure to appreciate the likelihood of the trespasser's presence or the risk to him involved, does not absolve the occupier from his duty to the trespasser if

[18] *Herrington v British Railways Board* [1972] A.C. 877.

a reasonable man possessed of the actual knowledge of the occupier would recognise that likelihood and that risk.

Thirdly: The duty when it arises is limited to taking reasonable steps to enable the trespasser to avoid the danger. Where the likely trespasser is a child too young to understand or heed a written or a previous oral warning, this may involve providing reasonable physical obstacles to keep the child away from the danger.

Fourthly: The relevant likelihood to be considered is of the trespasser's presence at the actual time and place of danger to him. The degree of likelihood needed to give rise to the duty cannot, I think, be more closely defined than as being such as would *impel a man of ordinary humane feelings to take some steps to mitigate the risk* of injury to the trespasser to which the particular danger exposes him. It will thus depend on all the circumstances of the case: the permanent or intermittent character of the danger; the severity of the injuries which it is likely to cause; in the case of children, the attractiveness to them of that which constitutes the dangerous object or condition of the land; the expense involved in giving effective warning of it to the kind of trespasser likely to be injured, in relation to the occupier's resources in money or in labour."

4–081 The obvious situation where a duty under the 1984 Act is likely to arise is where the occupier knows that a trespasser may come upon a danger that is latent. Occupiers of sites that contain equipment which may be regarded as allurements, (for example diggers or tractors) should be careful to guard against the risk of injury to trespassers should they be enticed onto the site as a result of such allurements. In such a case the trespasser may be exposed to the risk of injury without realising that the danger exists. It is these potential dangers that the Contractor should take care to minimise the risk of injuring not only the trespasser whose presence on the premises is envisaged but also those who are entitled to be on site. These are relevant considerations regardless of the jurisdiction. No-one wants anyone to suffer injury and in addition the Contractor must consider the potential consequences to progress of anyone suffering an injury.

The obligation conferred by sub-cl.4.8(d) to provide "guarding" and "watching" confirms that it is the Contractor who is responsible for site security of the Works.

Finally sub-cl.4.8(e) requires the Contractor to consider the impact of the Works on the surrounding area. If necessary, the Contractor must construct temporary roads, fences and bridges both to protect those who use adjacent land but also for the use of those who use that adjacent land. It is submitted that the use of the word "necessary" means that this obligation would extend to providing an alternative means of access if the Works blocked off a road or pathway which was previously in use.

MDB HARMONISED EDITION

There is no change.   **4–082**

4.9 QUALITY ASSURANCE

The Contractor shall institute a quality assurance system to demonstrate   **4–083**
compliance with the requirements of the Contract. The system shall be in accor-
dance with the details stated in the Contract. The Engineer shall be entitled to
audit any aspect of the system.

Details of all procedures and compliance documents shall be submitted to the
Engineer for information before each design and execution stage is commenced.
When any document of a technical nature is issued to the Engineer, evidence of
the prior approval by the Contractor himself shall be apparent on the document
itself.

Compliance with the quality assurance system shall not relieve the
Contractor of any of his duties, obligations or responsibilities under the
Contract.

OVERVIEW OF KEY FEATURES

If required by the Contract, the Contractor shall institute a quality assurance   **4–084**
system.

COMMENTARY

It may be that this sub-clause is unnecessary. If the Employer does not   **4–085**
require the Contractor to implement a quality assurance system then this
sub-clause will not be required. If this is the case then, as with any sub-clause
which is not used, both parties should ensure that the sub-clause is deleted.

If required to follow an approved Quantity Assurance system, the
Contractor cannot claim that complying with that system relieves him of any
of his duties under the Contract.

The requirement is that the Contractor provides details of procedures and
compliance documents before the commencement of each design and/or
execution stage of the project. If an Employer requires a thorough quality
assurance system, he might require that the Contractor prepare and submit

to the Engineer a full project quality assurance plan at an early stage in the project.

4–086 The benefits of a quality assurance plan for the Employer may include an increased level of checking and monitoring which should lead to a high quality project, with a reduced risk of latent defects. The downside will be increased cost. In deciding whether a quality assurance plan is necessary, and if so at what level of detail, the Employer should remember that often there will be an obligation to maintain records and this clause should be cross referenced against the obligations under sub-cl.4.21 to prepare regular detailed progress reports.

MDB HARMONISED EDITION

4–087 There is no change.

4.10 SITE DATA

4–088 *The Employer shall have made available to the Contractor for his information, prior to the Base Date, all relevant data in the Employer's possession on sub-surface and hydrological conditions at the Site, including environmental aspects. The Employer shall similarly make available to the Contractor all such data which comes into the Employee's possession after the Base Date. The Contractor shall be responsible for interpreting all such data.*

To the extent which was practicable (taking account of cost and time), the Contractor shall be deemed to have obtained all necessary information as to risks, contingencies and other circumstances which may influence or affect the Tender or Works. To the same extent, the Contractor shall be deemed to have inspected and examined the Site, its surroundings, the above data and other available information, and to have been satisfied before submitting the Tender as to all relevant matters, including (without limitation):

(a)   the form and nature of the site, including sub-surface conditions,

(b)   the hydrological and climatic conditions,

(c)   the extent and nature of the work and Goods necessary for the execution and completion of the works and the remedying of any defects,

(d)   the Laws, procedures and labour practices of the Country, and

(e)   the Contractor's requirements for access, accommodation, facilities, personnel, power, transport, water and other services.

OVERVIEW OF KEY FEATURES

- The Employer must make available, prior to the Base Date (i.e. 28 days prior to the latest date for submission of the Tender), all relevant data in his possession on sub-surface and hydrological conditions at the Site.   **4–089**
- This is an ongoing obligation.
- It is the responsibility of the Contractor to interpret the data.
- It will be assumed that the Contractor has both examined and inspected the site and all the available information provided to the Employer about the site and taken that information into account in compiling and submitting his Tender.

COMMENTARY

This is another sub-clause which is in two parts.   **4–090**

Provision of Data

The requirements of sub-cl.4.10 are similar to the requirements of sub-cl.11.1 of the Old Red Book FIDIC 4th edn. Sub-clauses 4.10–12 are some of the most important parts of clause 4. The question of the potential for problems surrounding different or unforeseeable site conditions is one of the largest risk areas in the type of civil engineering projects catered for under this contract.   **4–091**

By this sub-clause, the Employer confirms that he has made available to the Contractor before the tender is submitted, all relevant data in his possession on sub-surface and hydrological conditions at the Site, including environmental aspects. The Employer further warrants that he will continue to make this data available. It is likely that the Employer will have carried out certain investigations in establishing the viability of his project.

The obligation is not one to provide the data; it is to make the data available. The sub-clause does not say in what form the data is to be made available. Therefore an Employer might attempt to satisfy this obligation by providing the Contractor with details of the information he has in his possession. If this is how the Employer chooses to comply with this clause, then the Contractor must ask to see the information.

The FIDIC Guide says that the Employer does not have to make available expert's opinions or other non-factual interpretations as these are not "data". This is strictly correct. An expert compiles his report based on his interpretation of data. However a Contractor might find such a strict reading of the clause unhelpful and the FIDIC Guide does concede that the Employer "may" make such information available.   **4–092**

The Employer should bear in mind the extent of the definition of "Site" in sub-cl.1.1.6.7, when providing the relevant data. The definition is a wide one, including not just where the Permanent Works are to be carried out but also any other places specified in the Contract forming part of the site.

The question therefore arises as to whether the Employer makes any warranty as to the accuracy of the data provided. It is submitted that the answer must be that he does. It may well be for this reason that sub-cl.4.10 so carefully limits what the Employer has to provide. The data must be supplied prior to tenders being submitted. The Contractor will therefore rely on that data. For example, by sub-cl.4.11, the Contractor is, in part at least, deemed to have based his tender sum on the data provided.

4-093 In addition sub-cl.4.10 of the Silver Book where the Contractor, unlike the Red Book, *is* responsible for the design, specifically requires the Contractor not only to interpret the data but verify it. There is no specific obligation on the Contractor to verify the information in the Red Book.

It is considered that the Employer must act in good faith in complying with the requirements of this sub-clause. By good faith we mean the exercise of fair and open dealing with no concealed pitfalls or traps. The exact meaning of good faith depends on the jurisdiction the contract operates under. There is no recognition in English law of a general duty of good faith in contractual negotiations and it may be that in contrast to the more onerous requirements which are recognised in other jurisdictions, the concept of good faith in English law is largely restricted to the duty not to act fraudulently and the duty to co-operate.[19] However, the concept of good faith is recognised in many civil codes, including those operated under Sharia principles. These typically impose an over-riding obligation that contracts are performed both according to their terms *and* in a way compatible with good faith.

However a Contractor would be advised to keep an open mind as to the accuracy of the data provided. Under the Old Red Book FIDIC 4th edn, the Employer was required to supply all available information. Under sub-cl.4.10, the Employer is required to make available "all relevant data". Therefore it is for the Employer, at his own risk, to decide what is relevant.

4-094 The Employer only has to make the data available. It will most probably be provided in raw form. Sub-clause 4.10 is clear that it is the Contractor who must interpret that data. It might be that the Contractor when interpreting the data will have cause to question its accuracy. Obviously it would be far better to raise the issue prior to tenders being submitted and contracts being entered into.

Unlike other sub-clauses, see for example sub-cll 4.7 and 4.12, there is no suggestion that the Contractor will be deemed to have interpreted the data as an "experienced" contractor would. Therefore, under English jurisdictions,

---

[19] See *Ultraframe (UK) Limited v Tailored Roofing Systems Limited* [2004] BLR 341.

the standard by which the Contractor will be judged is that of the ordinary, competent contractor.[20]

Although this sub-clause is clearly primarily concerned with the provision of information prior to tenders being submitted, the obligation to provide the data is a continuing one, which it is suggested, presumably continues after the tender has been submitted. This may be of some importance.

In the case of *Stanmor Floors Ltd v Piper Construction Midlands Ltd*[21] the completion of flooring works was delayed. Stanmor argued that this delay was partly the responsibility of Piper who following a variation to the Contract, placed furniture in areas where flooring was to be laid and had increased the number and level of ducts in the floor. Piper relied on Stanmor's sub-contractor's knowledge under the contract which provided that the sub-contractor was deemed to:

4-095

"have made a thorough examination of the site and of drawings of documents which satisfied himself as to the nature of the works, general site conditions, means of access . . . as no claims through lack of knowledge in this respect will be considered"

Piper argued that when the contract was varied, Stanmor's knowledge was refreshed so that it also covered the varied works. The Court of Appeal agreed with the Trial Judge who found that the clause was to be read as at the date of the Contract and was not refreshed each time there was a variation. If it was intended that these obligations were to take effect in respect of any subsequent variations, then the clause needed to be expressly renewed.

Here is the Contractor is responsible for interpreting all site data and shall be deemed to have inspected the site prior to submitting its tender. This is the relevant date.

4-096

However the Contractor should not forget that the Employer is under a continuing obligation. Whilst, it is suggested that the Employer could claim that this sub-clause was refreshed if fresh data was provided to the Contractor, the Contractor should consider any new data carefully and immediately give notice if the data changes the position.

Tender Information

Indeed, it is important that the Contractor is not mislead by the title of the sub-clause and appreciates that it is not just about site data. Sub-clause 4.10 is in two distinct parts.

4-097

The second part provides that the Contractor is deemed to have obtained all necessary information to prepare his tender and that the Contractor shall

---

[20] See comments of McNair J. discussed under sub-cl.4.7.
[21] Unreported – Lawtel A.C. 9000449 – Court of Appeal May 4, 2000.

be deemed to have taken into account all the risks which might impact on his tender. Ignorance will be no defence.

As a consequence of this obligation, it will come as no surprise that by sub-cl.4.11, the Contractor will be deemed to accept the "correctness and sufficiency" of the Accepted Contract Sum. The Contractor must carry out his own investigations in order to ensure that his tender sum will an adequate one.

4–098   Further, the data and information referred to in the second part of sub-cl.4.10 are a much wider list than the data the Employer is required to make available in the first part of the sub-clause. Whilst subss. (a) and (b) talk about the nature of the site and hydrological and climatic conditions, subs.(c) talks about the nature of the work and the goods necessary to complete that Work, whilst subs.(d) refers to the laws, procedures and labour regulations of the Country where the project is based. Finally subs.(e) refers to the Contractor's own requirements.

Sub-clause 4.15 provides further clarification about the Contractor's access obligations, repeating that the Contractor shall be deemed to have satisfied himself as to the suitability and availability of access routes to the Site.

The Contractor should note that these items are listed "without limitation". They are really no more than examples. The Contractor's obligation is that he shall be deemed to have obtained *all* necessary information whatever it may be.

4–099   The Employer is required to provide limited assistance in relation to subs.(d). By sub-cl.2.2, the Employer shall, if the Contractor requests, provide copies of Laws of the Country which are relevant to the Contract but which are not readily available. The sub-clause also provides that the Employer will provide assistance to the Contractor in obtaining permits and licences.

The only limit on the Contractor's obligations, as provided by this clause, is that the Contractor will be deemed to have obtained all necessary information "*to the extent* " it was "*practicable (taking account of cost and time)*". The prudent Contractor should keep a record of the pre-tender period in order to be in a position, should it be necessary, to demonstrate that he acted reasonably in the obtaining all the necessary information relevant to the preparation of his tender as required by this sub-clause.

The question therefore arises as to whether a local contractor would be treated any differently to his foreign counterpart. It may be far easier and cheaper for those with local contacts to carry out a more thorough investigation. Given the qualifications imposed here, it is likely that different contractors would be treated differently and this is why the keeping of records is to be encouraged. What type of investigations could be carried out given the timescale of the tender period or the locality of the project? This is very much an area which will depend on the particular facts of each individual situation.

4–100   The difficulties a Contractor might have in making a claim under sub-cause 4.10 are demonstrated below in discussion of sub-clause 4.12.

MDB HARMONISED EDITION

There is no change.   **4–101**

4.11 SUFFICIENCY OF THE ACCEPTED CONTRACT AMOUNT

The Contractor shall be deemed to: **4–102**

(a)   have satisfied himself as to the correctness and sufficiency of the Accepted Contract Amount, and
(b)   have based the Accepted Contract Amount on the date, interpretations, necessary information, inspections, examinations and satisfaction as to all relevant matters referred to in Sub-Clause 4.10 [Site Data].

Unless otherwise stated in the Contract, the Accepted Contract Amount covers all of the contractor's obligations under the Contract (including those under Provisional Sums, if any) and all things necessary for the proper execution and completion of the Works and the remedying of any defects.

OVERVIEW OF KEY FEATURES

- The Contractor will be deemed to have based his tender sum on the basis   **4–103** of the information provided by the Employer under sub-cl.4.10 and on the basis of the Contractor's own checks on that information as required by sub-cl.4.10
- It is for the Contractor to ensure that his tender sum covers everything the Contractor needs to be able to carry out the Works as required by sub-cl.4.1.

COMMENTARY

Sub-clause 4.11 provides that it is for the Contractor to satisfy himself as to   **4–104** the adequacy of his tender. Although it must be read in close conjunction with sub-cl.4.10, the basis of the Contractor's tender is not limited to the information supplied or obtained in accordance with sub-cl. 4.10.

This is because whilst under sub-cl.4.11(b) the Contractor is deemed to have based the Contract Amount on the information provided under cl.4.10

by the Employer, the list of items referred to at sub-cll 4.10 (a)–(e) is specifically stated to be a non exclusive one. If the Contractor discovers any potential discrepancies in the information provided by the Employer, these should be raised prior to the tender being submitted.

It is only if there is an error in the site data provided by the Employer, which is found out at a later stage and which error it is held the Contractor could not have found out through his own enquiries, that a Contractor may be able to make a claim for additional payment. This is, of course, provided the Contractor has interpreted the data correctly.

4–105   The question of whether, as with sub-cl.4.10, (and therefore unlike sub-cll 4.7 and 4.12), the relevant standard by which the Contractor will be judged is the "ordinary" or the "experienced" contractor"[22] is a difficult one. There is nothing within the wording of the sub-clause to suggest that the Contractor will be judged according to the standards of the "experienced" Contractor. Therefore, it is submitted that the Contractor will, following the "Bolam" test, be judged (in English jurisdictions) as an "ordinary" contractor.

That said, some care is necessary. Under sub-cl.8.4(d), the Contractor is entitled to an extension of time in relation to unforeseeable shortages in the availability of personnel or goods caused by epidemic or governmental actions. Such shortages, if deemed foreseeable, would have an obvious impact on the Contractor's tender price. By sub-cl.1.1.6.8, the forseeability test is judged according to the standards of the experienced Contractor.

However, this sub-clause also provides a shield for the Employer to potential claims made by the Contractor for additional payment or time. By the final paragraph, the Contractor warrants that the Accepted Contract Amount covers everything necessary for the proper execution and completion of the Works and the remedying of any defects. When making any claim, a Contractor must bear in mind that his claim may be met with the defence that the Contractor should have taken the subject matter of that claim into account when preparing its tender. This is not necessarily that straightforward.

4–106   It is therefore important that the Contractor keeps adequate records[23] of everything that lead him to submit his tender in the form he did. For example, by sub-cl.4.12, the Engineer may take account of evidence provided by the Contractor in relation to the physical conditions he foresaw when submitting the Contract Sum. The Contractor should take this suggestion as a starting point of the type of information to keep.

---

[22] See sub-cl.4.7 for a discussion on what "experienced" contractor means.
[23] See discussion under sub-cl.6.10 and cl.20 as to what type of records may be admissible.

MDB HARMONISED EDITION

There is no change.   **4–107**

4.12 UNFORESEEABLE PHYSICAL CONDITIONS

In this Sub-Clause, "physical conditions" means natural physical conditions and **4–108**
man-made and other physical obstructions and pollutants, which the Contractor
encounters at the Site when executing the Works, including sub-surface and
hydrological conditions but excluding climatic conditions.

If the Contractor encounters adverse physical conditions which he considers
to have been Unforeseeable, the Contractor shall give notice to the Engineer as
soon as practicable.

The notice shall describe the physical conditions, so that they can be
inspected by the Engineer, and shall set out the reasons why the Contractor
considers them to be Unforeseeable. The Contractor shall continue executing
the Works, using such proper and reasonable measures as are appropriate for
the physical conditions, and shall comply with any instructions which the
Engineer may give. If an instruction constitutes a Variation, Clause 13
[Variations and Adjustments] shall apply.

If and to the extent that the Contractor encounters physical conditions which **4–109**
are Unforeseeable, gives such a notice, and suffers delay and/or incurs Cost due
to these conditions, the Contractor shall be entitled subject to Sub-Clause 20.1
[Contractor's Claims] to:

(a) an extension of time for any such delay, if completion is or will be
delayed, under Sub-Clause 8.4 [Extension of Time for Completion], and
(b) payment of any such Cost, which shall be included in the Contract Price.

After receiving such notice and inspecting and/or investigating these physical
conditions, the Engineer shall proceed in accordance with Sub-Clause 3.5
[Determinations] to agree or determine (i) whether and (if so) to what extent
these physical conditions were Unforeseeable, and (ii) the matters described in
sub-paragraphs (a) and (b) above related to this extent.

However, before additional Cost is finally agreed or determined under sub- **4–110**
paragraph (ii), the Engineer may also review whether other physical conditions
in similar parts of the works (if any) were more favourable than could reason-
ably have been foreseen when the Contractor submitted the Tender. If and to the
extent that these more favourable conditions were encountered, the Engineer
may proceed in accordance with Sub-Clause 3.5 [Determinations] to agree or
determine the reductions in Cost which were due to these conditions, which may
be included (as deductions) in the Contract Price and Payment Certificates.

However, the net effect of all adjustments under sub-paragraph (b) and all these reductions, for all the physical conditions encountered in similar parts of the works, shall not result in a net reduction in the Contract Price.

The Engineer may take account of any evidence of the physical conditions foreseen by the Contractor when submitting the Tender, which may be made available by the Contractor, but shall not be bound by any such evidence.

OVERVIEW OF KEY FEATURES

4–111
- The Contractor must give notice to the Engineer if he encounters adverse physical conditions which he considers to have been unforeseeable.
- "Physical conditions" means both natural phenomena and man-made obstructions.
- Unforeseeable is defined by cl.1.1.6.8 as being "not reasonably foreseeable by an experienced Contractor by the date of submission of the Tender".
- The notice must describe the problem and demonstrate why it was unforeseeable.
- Upon receipt of such a Notice, the Engineer shall proceed to make a determination in accordance with cl.3.5.
- If the Engineer so determines, the Contractor may be entitled to an extension of time under cl.8.4 and to Payment of Cost.
- Before the Engineer finally agrees any cost that may be due, the Engineer must take account of whether any physical conditions were more favourable to the Contractor than had been anticipated.

COMMENTARY

4–112   The question of who is responsible for unforeseen physical conditions is unsurprisingly a contentious one. Here sub-cl.4.12 provides that a Contractor might be entitled to an extension of time for delay and the recovery of Cost incurred as a consequence of that delay, if he encounters physical conditions which were unforeseeable at the date of submission of the tender. It is therefore a particularly important sub-clause and one which inevitably will be the focus of many Contractor claims.

The sub-clause begins with a definition of "physical conditions" something which was missing from the Old Red Book FIDIC 4th edn. The definition is a wide-ranging one and includes reference to "natural" and "man-made" obstructions. This therefore presumably includes fossils and archaeological artefacts.[24]

---

[24] See discussion under sub-cl.4.24 below.

The immediate difficulty for the Contractor if he is delayed by such a condition will be to demonstrate that the physical conditions encountered were unforeseeable. We discussed in relation to sub-cl.4.7, some of the potential hurdles the Contractor must overcome. The same obstacles can be found here.

Sub-clause 1.1.6.8 defines "unforeseeable" as being "not reasonably foreseeable by an experienced Contractor by the date of submission of the Tender". In addition to this, sub-cl.4.10 details the very wide scope of information the Contractor will be deemed to have taken into account when compiling the Tender. Taking these two sub-clauses together, on one view, it is difficult to see from an Employer's point of view, what physical conditions might be not foreseeable by an experienced contractor. **4–113**

This, it is submitted, is not the intended result of this sub-clause. Presumably the idea behind the sub-clause is to impart to the Employer, the risk of those physical conditions which could not have reasonably been foreseen by an experienced contractor by the date for submission of the Tender. Otherwise why have the clause at all?

As the clause stands, the burden of proof lies with the Contractor who must demonstrate successfully not only that the reasonable contractor would not have foreseen the "physical condition" which caused the delay, but also that the experienced contractor would not have included the risks of encountering such a physical condition in preparing his programme and tender. Does this mean that an event will only be regarded as unforeseeable when adequate precautions could not have been reasonably taken by an experienced contractor? If something is foreseeable then the experienced contractor would be expected to take the appropriate precautions and to price for them in his tender.

Although this appears to be a difficult issue, the answer according to FIDIC itself would appear to be no. We say this because the May 2005 amendment to sub-clause 1.1.6.8 of the MDB Harmonised Edition specifically introduced the phrase "*and against which adequate preventative precautions could not be taken*" thereby suggesting that this could not be implied into the original clause as drafted. **4–114**

Although the definition contained within the sub-clause of "physical conditions"[25] is a wide one, it does not include climatic conditions. At first glance this seems surprising, as in many areas of the world one needs to be aware of the risks caused by hurricanes in the Caribbean or tropical rains in Asia. However these tend to be seasonal and so foreseeable. In addition, exceptionally adverse climatic conditions is a ground for claiming an extension of time under sub-cl.8.4.

The notice given by the Contractor must explain why the physical conditions encountered were unforeseeable. This is one reason why we suggested in

---

[25] No such definition was provided by sub-cl.12.2 of the Old Red Book FIDIC 4th edn.

relation to sub-cl.4.10, that a prudent Contractor should keep records of the pre-tender investigations.

4–115   Helpfully for the Contractor, the sub-clause ends by noting that the Contractor may provide evidence of the physical conditions foreseen when compiling the tender. Whether a Contractor should do this is a difficult question to answer. In compiling such a list the Contractor is nailing his colours to the mast. On balance it might be better to do so, as any failure is likely to be seized on as evidence that no such physical condition was foreseen. Whatever course is chosen, less helpfully for the Contractor, the sub-clause spells out that the Engineer is not bound by this when making his decision.

What the Contractor in his notice must do, is establish that based on the information that was available at the relevant time, a reasonably thorough investigation would not have brought to light the particular adverse physical condition that caused the delay.

In addition, the notice must describe the unforeseeable physical conditions. This will give the Engineer an idea of what he needs to do to investigate the situation.

4–116   The sub-clause also provides for the Contractor to give notice "as soon as practicable". However, no definition is provided as to what this means. The reason for this is that what is practicable will depend on the circumstances of the discovery of the problem. However, although no strict time limit is set, a Contractor would be advised not to delay issuing such a notice.

What the Contractor cannot do on encountering unforeseeable physical conditions is stop working. The sub-clause specifically notes that the Contractor shall continue executing the Works and take such measures as are reasonable in the circumstances to take account of the problem. This might mean the Contractor will need to change his programme or method of working. The sub-clause seems to envisage that the Contractor and Engineer work together to resolve the difficulties caused by the unforeseeable physical conditions. Even if that is not what was intended, it is submitted that the prudent Contractor should work with the Engineer to resolve the problem. If this leads to the issuing of a variation instruction under clause 13, then the Contractor would not need to consider making a claim in accordance with cl.20.

If the Engineer agrees that the physical conditions encountered were unforeseeable, the Contractor will be entitled to an extension of time and payment of Cost. However, before any additional cost is agreed, the Engineer may review other physical conditions in similar parts of the Works to carry out a balancing exercise to see if these are more favourable to the Contractor than could have been reasonably foreseen when the tender was submitted. If the Engineer considers they are more favourable, he may determine a reduction in Cost, although any such reduction must not lead to a reduction in the Contract Price. A typical example given of this might be a tunnel where conditions at both ends have changed, one to the detriment of the Contractor, the other to his advantage.

This provision is new to this edition of the Red Book. The fact that it is a **4–117** new introduction gives the impression that it is unfair to the Contractor. However, in reality, it provides for a balancing exercise which suggests a degree of fairness to all parties to the contract. However it is important for the Contractor that the Engineer understands, if it is appropriate, that physical conditions which appear to be more favourable to the Contractor might not in fact be so. Any conditions which are different to those envisaged will not have been taken into account when the tender was prepared.

Where the Contractor recovers additional costs, these costs do not include an element of profit, even though it is likely to have been necessary to carry out additional works and possibly to employ additional plant and labour.

By way of contrast, under the Silver Book all the risk is transferred to the Contractor who is:

". . . deemed to have obtained all necessary information as to risks, contingencies, and other circumstances which may influence or affect the Works. By signing the Contract, the Contractor accepts total responsibility for having foreseen all difficulties and costs of successfully completing the Works."

The Contractor also, under the Silver Book, accepts that the Contract Price **4–118** will not be adjusted to take account of any unforeseen difficulties or costs.[26]

The Particular Conditions provide a sample clause for use where the allocation of the risk of sub-surface conditions is shared.

MDB HARMONISED EDITION

There are some minor changes to this sub-clause. The words *"notice under"* **4–119** have been inserted before *"Sub-Clause 20.1"* in the fourth paragraph whilst, the word *"After"* at the beginning of the fifth paragraph, has been replaced by *"Upon"*.

The final paragraph now reads:

The Engineer <u>shall</u> take account of any evidence of the physical conditions foreseen by the Contractor when submitting the Tender, which <u>shall</u> be made available by the Contractor, but shall not be bound by <u>the Contractor's interpretation of any</u> such evidence.

However, whilst the word "may" has been replaced by "shall", which means that the Engineer must take account of any evidence put forward by the

---

[26] A situation that will be familiar to those with experience of Hong Kong.

Contractor, there is no change to the ultimate meaning of this paragraph as the Engineer is not bound by that evidence

4–120   Under the May 2005 Version of the MDB Harmonised Edition there was another change and sub-cl.1.1.6.8 defined unforeseeable as meaning:

> *Not reasonably foreseeable <u>and against which adequate preventative precautions could not be taken</u> by an experienced contractor by the date for the submission of the Tender.*

This was a controversial amendment. The European International Contractors group criticised the addition, referring to it as a *"twist of Catch-22 proportions"*[27] and the editors of the International Construction Law Review indicated that the change called for *"rational justification and explanation of its practical application."*[28]

4–121   It could be argued that the addition to the clause served to add clarity to the original definition, no more. For example, in demonstrating that the physical conditions would have been unforeseeable to an experienced contractor, the Contractor would already have to show that there were no adequate precautions which could have reasonably been taken. However, the EIC raised three questions of the amendment:

(i)   Is it the intention that contractors should make allowances for precautionary steps for unforeseeable events and circumstances?

(ii)   Is it the intention of the MDB harmonised edition to shift the balance of risk under the contract for unforeseeable events to the contractor?

(iii)   How can you take reasonable precautions against an event which is not reasonably foreseeable?

The answer to the first question does appear to be yes, which may well have an impact on the tender returns. If the answer to the second question is yes then this would suggest that the contract drafters are moving towards the risk profile adopted under the Silver Book. A simple comparison between the two forms of wording seem to make it clear that this is not the intention behind the new clause.

4–122   It was the third question which demonstrated the real difficulty with the new wording and this may have been one factor which lead to the revision being dropped. However, as there have been two versions, care should be taken to check which definition has been adopted.

---

[27] See "The Contractor's View on the MDB Harmonised Version of the New Red Book" – Richard Appuhn and Eric Eggink [2006] ICLR 4.
[28] Introduction to Vol.23 of the ICLR January 2006.

4.13 RIGHTS OF WAY AND FACILITIES

The Contractor shall bear all costs and charges for special and/or temporary   **4–123**
rights-of-way which he may require, including those for access to the Site. The
Contractor shall also obtain, at his risk and cost, any additional facilities
outside the Site which he may require for the purposes of the Works.

OVERVIEW OF KEY FEATURES

- The Contractor is responsible for site access.   **4–124**
- The Contractor is responsible for special and/or temporary rights of way.
  This includes those routes outside the Site but which are necessary for
  Access to the Site.
- The Contractor is responsible to obtain any facilities outside of the Site
  which he needs to execute his Works at his own cost and risk.

COMMENTARY

This is a relatively straightforward sub-clause.   **4–125**
   The Contractor is responsible for (in terms of risk and cost) providing and
maintaining site access. These risks and costs should be taken into account in
arriving at the tender sum. The Contractor should remember that he must
not only organise and secure the access routes but maintain them during the
currency of the Works. Sub-clause 4.13 must be read in conjunction with sub-
cl.4.15 which confirms that it is for the Contractor to satisfy himself as to the
suitability and availability of access.
   The Contractor should also remember that in accordance with sub-cl.2.2,
the Employer must only provide reasonable assistance (nothing more) in
obtaining any permits, licences or approvals which might be necessary to
obtain rights of way or access. This assistance will only be provided if
actually requested by the Contractor.
   Finally, sub-cl.4.13 confirms that the Contractor is responsible both in   **4–126**
terms of cost and risk for any additional facilities (for example design or
manufacturing) which may be needed to carry out the Works.

MDB HARMONISED EDITION

This clause now reads: **4–127**

Unless otherwise specified in the Contract the Employer shall provider access to and possession of the Site including special and/or temporary rights-of-way which are necessary for the Works. The Contractor shall obtain, at his own risk and cost, any additional rights of way or facilities outside the Site which he may requite for the purpose of the Works.

This is quite important for the Contractor as it confirms that it is the Employer who must provide access to and possession of the Site. This is all in accordance with sub-cl.2.1.

In addition, it is the Employer and no longer the Contractor who is responsible for the costs of any ordinary, special or temporary rights of way that are necessary for the Works – the words "The contractor shall bear all costs and charges" having been deleted in the MDB Harmonised Version.

4.14 AVOIDANCE OF INTERFERENCE

4–128 *The Contractor shall not interfere unnecessarily or improperly with:*

(a)   the convenience of the public, or
(b)   the access to and use and occupation of all roads and footpaths, irrespective of whether they are public or in the possession of the Employer or of others.

The Contractor shall indemnify and hold the Employer harmless against and from all damages, losses and expenses (including legal fees and expenses) resulting from any such unnecessary or improper interference.

OVERVIEW OF KEY FEATURES

4–129   • The Contractor shall not "unnecessarily or improperly" interfere with the convenience of the public or access of all roads and footpaths.
   • The Contractor shall indemnify the Employer against any claims resulting from any such unnecessary or improper interference.

COMMENTARY

4–130   By sub-cl.4.13, the Contractor is responsible for the costs of maintaining any access routes and temporary rights of way. By sub-cl.4.14 here, the

Contractor is responsible for costs arising out of unnecessary or improper interference with the public and roads and footpaths.

No definition is provided as to what "unnecessary or improper" means. This meaning will vary according to local laws. However, it is submitted that the words relate to the public and not the Contractor. For example, it would be no defence for the Contractor to say that the interference with the public came about because it was necessary for the carrying out of his Works.

An example of what might constitute "unnecessary or improper interference" is provided by sub-cl.4.23, which provides that the Contractor shall confine his operations to the Site and that the Contractor shall take all necessary precautions to keep his equipment and personnel within the Site and to keep them off adjacent land.

The third party indemnity in respect of third party claims to be provided to the Employer is all-embracing, including damages and legal expenses. The Contractor would be well advised to take account of the possibility of public access to the site and site access roads and routes when preparing his tender.   **4–131**

The sub-clause itself will also have a wide implication, as it refers to the "convenience of the public". It does not differentiate between the public on-and-off site. Therefore the Contractor should read this clause together with the health and safety obligations of sub-cl.4.8, the requirement under sub-cl.4.18 to protect the environment and the need to maintain site security required by sub-cl.4.22. All these sub-clauses might lead to the public being "inconvenienced". The Contractor would then be liable to indemnify the Employer in respect of any claims that may be made.

MDB HARMONISED EDITION

There is no change.   **4–132**

4.15 ACCESS ROUTE

The Contractor shall be deemed to have been satisfied as to the suitability and   **4–133**
availability of access routes to the Site. The Contractor shall use reasonable efforts to prevent any road or bridge from being damaged by the contractor's traffic or by the contractor's Personnel. These efforts shall include the proper use of appropriate vehicles and routes.

Except as otherwise stated in these Conditions:

(a)   the Contractor shall (as between the Parties) be responsible for any maintenance which may be required for his use of access routes;

117

(b)   the Contractor shall provide all necessary signs or directions along access routes, and shall obtain any permission which may be required from the relevant authorities for his use of routes, signs and directions;

(c)   the Employer shall not be responsible for any claims which may arise from the use or otherwise of any access route;

(d)   the Employer does not guarantee the suitability or availability of particular access routes, and

(e)   Costs due to non-suitability or non-availability, for the use required by the Contractor, of access routes shall be borne by the Contractor.

OVERVIEW OF KEY FEATURES

4–134 • It is for the Contractor to satisfy himself as to the suitability and availability of access to the Site.
• The Contractor is responsible for obtaining permissions to use the access route and all necessary signage.
• The clause expressly provides that the Employer does not guarantee the suitability or availability of particular access routes.
• The Contractor bears all the costs and risks associated with access.

COMMENTARY

4–135 Sub-clause 4.13 confirms that the Contractor is responsible for access to the site. Sub-clause 4.15 sets out the Contractor's obligations in relation to access.

This clause might be viewed as a change from the Old Red Book FIDIC 4th edn, whereby by sub-cl.30.3, the Employer in certain circumstances had to pay (and indemnify the Contractor) for damage to roads and bridges caused by the transportation of Goods albeit that that obligation was subject to the Engineer considering that the Contractor was not at fault. By sub-cl.4.16(c) that liability is firmly placed on the Contractor.

The sub-clause is clear. Whilst the Employer by sub-cl.2.1 must provide the Contractor with access to the site, this is entirely different to the obligation here on the Contractor to ensure that access routes to the site are adequate and suitable to enable him to carry out his Work in accordance with the contract.

4–136 This sub-clause cross-refers to cl.4.10(e) which notes that the Contractor will be deemed to have taken into account any requirements for access before submitting the tender. The reference to potential damage to roads and bridges should serve as a reminder to the Contractor to consider the likely effect of any heavy loads that need to be brought to site. If a Contractor's lorries cause damage to roads and bridges, this sub-clause provides that the

Contractor cannot make any claim against the Employer arising from the use or otherwise of any access route. Therefore the Contractor might not only suffer delay to his Works, for which he will be responsible, but he might also find himself having to make recompense under local laws for any damage caused.

MDB HARMONISED EDITION

The words *"at Base Date"* have been added at the end of the first sentence. **4–137**
By sub-cl.1.1.3.1, the base date, means the date 28 days prior to the latest date for submission of the Tender. This addition therefore adds some certainty should a dispute arise about the suitability of the access routes since the relevant date of the Contractor's knowledge should be fixed in time.

4.16 TRANSPORT OF GOODS

Unless otherwise stated in the Particular Conditions: **4–138**

(a) the Contractor shall give the Engineer not less than 21 days' notice of the date on which any Plant or a major item of other Goods will be delivered to the Site;

(b) the Contractor shall be responsible for packing, loading, transporting, receiving, unloading, storing and protecting all Goods and other things required for the works; and

(c) the Contractor shall indemnify and hold the Employer harmless against and from all damages, losses and expenses (including legal fees and expenses) resulting from the transport of goods, and shall negotiate and pay all claims arising from their transport.

OVERVIEW OF KEY FEATURES

- The Contractor must give the Engineer not less than 21 days' notice of **4–139** delivery of Plant or major items of other Goods.
- The Contractor is responsible for the transportation, storage and protection of all Goods necessary to carry out the Works.
- The Contractor is responsible for all costs arising out of the transport of goods.
- The Contractor shall indemnify the Employer against any claims arising from the transport of goods.

COMMENTARY

4–140   This sub-clause confirms that the Contractor is responsible for the costs and risk associated with the transfer of Goods to site. Goods are defined in sub-cl.1.1.5.2 as being the "Contractors' Equipment, Materials, Plant and Temporary Works", in short everything the Contractor needs to carry out his work on site. That responsibility extends to indemnifying the Employer as a result of any claims that might arise. Presumably this might include claims following accidents whilst Goods are being transported. In addition a Contractor cannot seek to hold the Employer responsible for delays or additional expenses caused by customs hold-ups or by Goods being damaged or deteriorating whilst being stored.

The obligations under sub-cl.4.16 should be read in conjunction with a number of other sub-clauses particularly sub-cll 4.13 and 4.15 which relate to access. This is an important obligation. If the Contractor cannot transport his Goods, his progress might become delayed which will impact on his programme. Where Goods are stored will equally impact on the programme.

4–141   Notwithstanding that the Contractor is responsible for the delivery and storage of Goods, he must give 21 days' notice to the Engineer of the delivery of Plant or other major items to the site. The giving of these notices required by this clause will provide a mechanism whereby the Engineer can keep an eye on the extent to which the Contractor is keeping up with his programme, since delivery is one of the items which must be recorded in the monthly progress reports required by sub-cl.4.21. Therefore a Contractor would be well advised to keep detailed records of the progress of any shipments or planned deliveries including planned and actual dates.

MDB HARMONISED EDITION

4–142   There is no change.

4.17 CONTRACTOR'S EQUIPMENT

4–143   *The Contractor shall be responsible for all Contractor's Equipment. When brought on to the Site, Contractor's Equipment shall be deemed to be exclusively intended for the execution of the works. The Contractor shall not remove from the Site any major items of Contractor's Equipment without the consent of the Engineer. However, consent shall not be required for vehicles transporting Goods or Contractor's Personnel off Site.*

OVERVIEW OF KEY FEATURES

- The Contractor is responsible for his own equipment.   **4–144**
- All the Contractor's Equipment on site is for the exclusive use of the Works on site.
- Major items of Contractor's Equipment may only be moved off site with the consent of the Engineer.

COMMENTARY

By sub-cl.1.1.5.1, "Contractor's Equipment" means all apparatus, machinery,   **4–145** vehicles and other things required for the execution and completion of the works and the remedying of any defects. The definition, however, specifically excludes Temporary Works, Employer's Equipment (if any), Plant, Materials and anything else intended to form part of the Permanent Works. For clarity, if there is to be any Employer's Equipment, the nature and extent of this, should be specified.

Unsurprisingly, the Contractor is responsible for his own equipment. In addition, any equipment which is brought on site must be used exclusively to carry out the Works which are the subject of the Contract. Therefore a Contractor cannot make use of equipment (for example by manufacturing items) on site for the benefit of other projects.

It is unclear why the Contractor is prohibited from removing any major items of equipment from site without the consent of the Engineer. Presumably this requirement can be used as a check on progress as the Engineer will be aware whether the equipment which the Contractor wants to remove from the site is no longer needed.

In addition, if the Engineer is aware that particular tasks will need to be   **4–146** carried out imminently, he will be able to prevent the Contractor from removing the equipment, thereby potentially ensuring that progress is kept to programme. Therefore, if a Contractor needs particular equipment on two different projects, he should take care to ensure that the programme demands of the two projects are such that he is able to switch equipment between projects.

The Particular Conditions provide the following sub-clause, if any vesting of the Contractor's Equipment is required:

The Contractor's Equipment which is owned by the Contractor (either directly or indirectly) shall be deemed to be the property of the Employer with effect from its arrival on the Site. This vesting of property shall not:

(a)   Affect the responsibility or liability of the Employer,

(b) *Prejudice the right of the Contractor to the sole use of the vested Contractor's Equipment for the Purpose of the Works, or*

(c) *Affect the Contractor's responsibility to operate and maintain Contractor's Equipment.*

The property in each item shall be deemed to revest in the Contractor when he is entitled either to remove it from the site or to receive the Taking-over Certificate for the Works, whichever occurs first.

MDB HARMONISED EDITION

4–147   There is no change.

4.18 PROTECTION OF THE ENVIRONMENT

4–148   *The Contractor shall take all reasonable steps to protect the environment (both on and off the Site) and to limit damage and nuisance to people and property resulting from pollution, noise and other results of his operations.*

The Contractor shall ensure that emissions, surface discharges and effluent from the Contractor's activities shall not exceed the values indicated in the Specification, and shall not exceed the values prescribed by applicable Laws.

OVERVIEW OF KEY FEATURES

4–149
- The Contractor shall take all reasonable steps to protect the environment.
- That obligation includes the environment both on and off site.
- Emissions, surface discharges and effluent caused by the Contractor's activities must not exceed the greater of the values set out in the Specification or that allowed for by the applicable laws.

COMMENTARY

4–150   The obligation provided by this sub-cause is a simple yet wide-ranging one. The Contractor is responsible for taking all reasonable steps to protect the environment whilst carrying out his Works.

In fact, there are a number of potential difficulties with this clause. The first part of the sub-clause is perhaps too wide-ranging – no definition of "reasonable steps" is provided. Equally there is considerable scope for disagreement about both what protecting the environment means and what it is necessary to do to ensure that the environment is protected.

However, it is possible to gain an understanding of what the requirements of this sub-clause mean.

This sub-clause should be read in conjunction with sub-clause 4.10(d) **4–151** which provides that the Contractor shall be deemed to have inspected and examined the Laws and procedures of the Country where the Work is to be carried out before submitting the tender. The reason for this is two-fold. First, the Contractor needs to take account of the effect of the environmental legislation on the programme and likely costs. Secondly, by sub-cl.1.13, the Contractor must comply with all local laws when executing his work. See, for example the discussion under sub-cl.4.23 of the 2005 Hazardous Waste Regulations.

This sub-clause should also be read in conjunction with sub-cl.4.14(a) which provides that the Contractor shall not "unnecessarily or improperly" interfere with the convenience of the public and that the Contractor shall indemnify the Employer as a consequence of any such interference. It is entirely feasible that a claim might be made by the public as a consequence of pollution or noise. It might well be said that such a claim had arisen as a result of interference with the convenience of the public. Hence, if a claim was made against the Employer as owner of the site, the Contractor would have to indemnify him.

The sub-clause is in two halves. The second half (which also relates to the environment) is far more specific. The Contractor should check the specification and other contractual requirements carefully as these should set out limits for emissions, surface discharges and effluent. If they are not, the Contractor should make enquiry before submitting the tender as to what these levels are. By the final paragraph of this sub-clause, the Contractor must also ensure that these levels exceed neither those required by the Specification nor the applicable laws. It may be that the applicable laws will provide a more stringent standard than that required in the Specification.

By sub-clause 4.22, the Contractor is responsible for site security and **4–152** ensuring that unauthorised personnel are kept away from site. The prudent Contractor would therefore need to consider the likely environmental impact of the project and the extent to which it might lead to protests. Not only might the impact of such a protest on the programme be significant, the Contractor should also bear in mind his health and safety responsibilities under sub-cl.4.8.

MDB HARMONISED EDITION

The final paragraph has been amended and reads as follows: **4–153**

> *The Contractor shall ensure that emissions, surface discharges and effluent from the Contractor's activities shall not exceed the values stated in the Specification or prescribed by applicable Laws.*

The first change, namely the substitution of the word "stated" in place of "indicated", has little practical effect. The second, effectively the change of "and" with "or" serves to strengthen the requirement that the Contractor must ensure that any emissions, discharge or effluent do not exceed the levels stated in the Contract or by local law.

4.19 ELECTRICITY, WATER AND GAS

4–154   *The Contractor shall, except as stated below, be responsible for the provision of all power, water and other services he may require.*

The Contractor shall be entitled to use for the purposes of the works such supplies of electricity, water, gas and other services as may be available on the Site and of which details and prices are given in the Specification. The contractor shall, at his risk and cost, provide any apparatus necessary for his use of these services and for measuring the quantities consumed.

The quantities consumed and the amounts due (at these prices) for such services shall be agreed or determined by the Engineer in accordance with Sub-Clause 2.5 [Employer's Claims] and Sub-Clause 3.5 [Determinations]. The Contractor shall pay these amounts to the Employer.

OVERVIEW OF KEY FEATURES

4–155   • The Contractor is responsible for the provision of all services needed to carry out the Works.
   • The Contractor is entitled, at his own cost, to use such supplies as are available on site.
   • The cost of these services shall be agreed or determined by the Engineer.

COMMENTARY

4–156   The Contractor must check the Specification and the other data provided to him during the tender stage to ensure that adequate supplies of electricity, water, gas and other services will be available. If not, the Contractor must at his own risk and cost provide such facilities.

Sometimes, an Employer will specifically provide that no electricity or gas supplies will be available on site for the Works. If the Employer does this, there should be no scope whatsoever for any subsequent disagreement.

Alternatively, if such services are available, it is important that the location of these services is made clear. Equally importantly, the payment for these sums must be agreed. If they are not, then the Engineer will determine the costs in accordance with sub-cl.3.5. However pursuant to sub-cl.2.5, there is no requirement on the Employer to give the Contractor any prior notice that he has referred his claim in the event of non-payment to the Engineer for determination.

MDB HARMONISED EDITION

The first paragraph has been extended and qualified to read: **4–157**

The Contractor shall, except as stated below, be responsible for the provision of all power, water and other services he may require <u>for his construction activities and to the extent defined in the Specifications, for the tests</u>.

4.20 EMPLOYER'S EQUIPMENT AND FREE-ISSUE MATERIAL

The Employer shall make the Employer's Equipment (if any) available for the **4–158** *use of the Contractor in the execution of the works in accordance with the details, arrangements and prices stated in the Specification. Unless otherwise stated in the Specification:*

(a) the Employer shall be responsible for the Employer's Equipment except that

(b) the Contractor shall be responsible for each item of Employer's Equipment whilst any of the Contractor's personnel is operating it, driving it, directing it or in possession or control of it.

The appropriate quantities and the amounts due (as such stated prices for the use of Employer's Equipment shall be agreed or determined by the Engineer in accordance with Sub-Clause 2.5 [Employer's Claims] and Sub-Clause 3.5 [Determinations]. The contractor shall pay these amounts to the Employer.

The Employer shall supply, free of charge, the "free-issue materials" (if any) **4–159** *in accordance with the details stated in the Specification. The Employer shall, at his risk and cost, provide these materials at the time and place specified in the Contract. The Contractor shall then visually inspect them, and shall promptly* *give notice to the Engineer of any shortage, defect or default in these materials. Unless otherwise agreed by both Parties, the Employer shall immediately rectify the notified shortage, defect or default.*

After this visual inspection, the free-issue materials shall come under the care, custody and control of the Contractor. The Contractor's obligations of inspection, care, custody and control shall not relieve the Employer of liability for any shortage, defect or default not apparent from a visual inspection.

OVERVIEW OF KEY FEATURES

4–160
- To the extent set out in the specification, the Employer shall provide and the Contractor may use the Employer's Equipment.
- The Contractor will be responsible for the cost of using such equipment at rates to be agreed or if not agreed, determined by the Engineer.
- The Employer is responsible for the Employer's Equipment except whilst the Contractor is using or controlling it.
- To the extent set out in the Specification, the Employer shall provide and the Contractor may use "free-issue materials".
- "Free issue materials" shall be provided free of charge
- After the Contractor has inspected the "free issue materials", he becomes responsible for their use, custody and control.

COMMENTARY

4–161 By sub-cl.1.1.6.3, "Employer's Equipment" means the apparatus, machinery and vehicles (if any) made available by the Employer for the use of the Contractor in the execution of the Works. Any such equipment must be clearly set out in the Specification. Therefore the Contractor will know what equipment the Employer has available for use when preparing his programme and tender. Employer's Equipment does not include Plant, which has not been taken over by the Employer.

"Free-use materials" is not defined elsewhere in the Contract. However, again, the Contractor will know from the Specification what (if any) free-use materials the Employer intends to provide during the currency of the project.

An Employer should give careful consideration before deciding to provide the Contractor with either equipment or free-use materials. If they are to be provided, they must be provided in accordance with the "details, arrangements" ("Employers Equipment") and "at the time and place specified in the Contract" ("Free-use materials"). Therefore they are provided at the Employer's risk. Any delay (or other difficulty) in providing them may well provide the Contractor with grounds to claim an extension of time and associated cost particularly if the Contractor was relying on their provision at a certain time.

If the Employer states in the specification that either "Employer's Equipment or "free-use materials" will be made available, a Contractor does not have to make use of them, unless it is a condition of the contract. For example, the equipment might be too expensive. **4–162**

If there is a dispute about the amount due to the Employer for the use of the Employer's Equipment, this shall be determined by the Engineer in accordance with sub-cl.3.5. Again, as with sub-cl.4.19 in accordance with sub-cl.2.5, the Employer does not need to give the Contractor notice that he has referred the matter to the Engineer for determination.

The Employer will be responsible for the Employer's Equipment unless the Contractor is using it. Equally, the Employer is responsible for the free-use materials until they come into the control of the Contractor. Before they do come into the Contractor's control, the Contractor must visually inspect them for any defect. However the Employer remains responsible for any defect not apparent from a visual inspection. As the Contract does not specifically say anything about the standard of the inspection, it is submitted that it will be that of a reasonable and not the experienced contractor.

Sub-sections (a) and (b) are important because they clearly define who is responsible for, or who has control of, the Employer's Equipment and when. This clarity is important if any accident should occur and care should be taken if this clause is amended. **4–163**

For example, in the 1947 case of *Mersey Docks Harbour Board v Coggins and Griffith* (Liverpool)[29] a harbour authority let a mobile crane to a firm of stevedores for loading a ship. It also provided a craneman who was employed and paid and liable to be dismissed by it, although the general hiring conditions stipulated that cranemen so provided should be the servants of the hirers. The craneman injured a third person. The stevedores had the immediate direction and control of the operation in respect of how the crane was used but had no power to direct how the crane should be worked or the controls manipulated. The House of Lords held that the Harbour Authority as the general permanent employer was liable. The test turned on where the authority lay to direct, or to delegate to the workmen, the manner in which the vehicle was driven. It would have been different if the hirers had intervened to give directions as to how to drive which they have no authority to give, and the driver complied with them.

Here the FIDIC Conditions provide that it is the Contractor who is responsible for the Employer's Equipment when he is using it, directing or controlling it.

---

[29] [1946] 2 ALL E.R. 345.

MDB HARMONISED EDITION

4–164   There is no change.

4.21 PROGRESS REPORTS

4–165   *Unless otherwise stated in the particular Conditions, monthly progress reports shall be prepared by the Contractor and submitted to the Engineer in six copies. The first report shall cover the period up to the end of the first calendar month following the Commencement Date. Reports shall be submitted monthly there-after, each within 7 days after the last day of the period to which it relates.*

Reporting shall continue until the Contractor has completed all work which is known to be outstanding at the Completion Date dated in the Taking-Over Certificate for the Works.

Each report shall include:

(a)   charts and detailed descriptions of progress, including each stage of design (if any), Contractor's Documents, procurement, manufacture, delivery to Site, construction, erection and testing; and including these stages for work by each nominated Subcontractor (as defined in Clause 5 [Nominated Subcontractors],

(b)   photographs showing the status of manufacture and of progress on the Site;

(c)   or the manufacture of each main item of Plant and Materials, the name of the manufacturer, manufacture location, percentage progress, and the actual or expected dates of:

   (i)   commencement of manufacture,
   (ii)   Contractor's inspections,
   (iii)   tests, and
   (iv)   shipment and arrival at the Site;

(d)   the details described in Sub-Clause 6.10 [Records of Contractor's Personnel and Equipment];

(e)   copies of quality assurance documents, test results and certificates of Materials;

(f)   list of notices given under Sub-Clause 2.5 [Employer's Claims] and notices given under Sub-Clause 20.1 [Contractor's Claims];

(g)   safety statistics, including details of any hazardous incidents and activities relating to environmental aspects and public relations; and

(h)   comparisons of actual and planned progress, with details of any events or circumstances which may jeopardise the completion in accordance with

the Contract, and the measures being (or to be) adopted to overcome delays.

OVERVIEW OF KEY FEATURES

- If required by the Contract, the Contractor shall provide regular progress reports.   **4–166**
- The provision of the progress report amounts to a condition of payment.
- These reports shall be in the detail outlined in this sub-clause.
- Six copies of the report shall be provided to the Engineer.
- The first report shall be produced at the end of the calendar month following the Commencement Date.
- Thereafter the reports shall be provided on a monthly basis until the Contractor has completed all the work listed as outstanding at the Completion Date.

COMMENTARY

The requirement to produce detailed monthly progress reports provided by   **4–167**
sub-cl.4.21 is a new feature of the Red Book. As set out below, the provision of this report appears to be a condition of payment.

A Contractor must ensure that he is aware of the very detailed requirements of this clause. It is essential that adequate procedures are put in place from the beginning of the Contract to ensure that the information can be collected and collated in good time. The reports are to be provided monthly. Given the nature of the information required by the sub-clause, it is likely that the Contractor will need to make use not just of his own internal records but also of information which can only be provided by others.

The required format is very detailed. The EIC has described the format as being "unnecessarily detailed"[30] and over prescriptive for most types of contract. If it is thought that this is the case, then the Contractor would be advised to try and agree an accepted format for reporting with the Employer or Engineer. However, there is no requirement that the Engineer agree the format.

The prime purpose of the reports required by sub-cl.4.21 must be to   **4–168**
monitor progress – hence in particular sub-clauses (c) and (h). By sub-cl.8.3 the Contactor shall submit a detailed time programme to the Engineer within

---

[30] The EIC Contractor's Guide to the FIDIC Conditions of Contract for Construction, which provides a useful and informative view from a Contractor's viewpoint of this contract. See also, for example, sub-cl.4.12, below.

28 days after receiving the Notice of Commencement. The Contractor is likely to be able to make use of this base programme, as adapted, to demonstrate progress to fulfil the requirement of this sub-clause, including that of sub-clause (h) to compare actual and planned progress.

Although the requirements of sub-cl.4.21 are very detailed, they do not include that the Contractor provide any form of Contract Price forecast to include, for example, the information under Clause 12 (Measurement and Evaluation) and Clause 13 (Variations and Adjustments). The reason for this is presumably that the Contractor must in accordance with sub-cl.14.3 submit the progress report as part of the documentation supporting his application for interim payment. Although there is no specific link between sub-cll 4.21 and 14.3, the fact that six copies of the progress report and six copies of the interim application are required suggests that they are closely linked.

Indeed under sub-cl.14.3, payment will only be made within 28-days of receipt of the application for payment *and* the supporting documents. In other words the provision of the progress reports is a condition of payment.

4–169   Whilst the importance of ensuring that the progress reports are accurate might seem obvious, His Honour Judge Wilcox in a recent case[31] involving a construction manager highlighted some of the potential difficulties where that reporting is not accurate.

Under the terms of the particular contract, the construction manager was described as being the only person on the project with access to all of the information and the various programmes. He was the only available person who could make an accurate report to the client at any one time, of both the current status of the Project and the likely effects both on timing and on costs. He was at the "the centre of the information hub" of the Project.

It is only with knowledge of the exact status of the Project on a regular basis that the construction manager can deal with problems that have arisen, and therefore anticipate potential problems that may arise, and make provisions to deal with these work fronts. That is not dissimilar from the status of the Contractor under the FIDIC conditions.

4–170   An Employer will need accurate information of the likely completion date, as well as the costs, because this would effect his pre-commencement preparation and financing costs. Any change to the likely completion date would give an Employer the chance to adjust its operational dates. Judge Wilcox concluded:

> "Where a completion date was subject to change the competent Construction Manager had a clear obligation to accurately report any change from the original projected completion date, and the effect on costs."

---

[31] *Great Eastern Hotel Company Limited v John Laing Construction Ltd and Laing Construction Plc* – 99 Con L.R. 45.

Sub-clause (h) confirms that the Contractor has a similar obligation here. If the Employer and the Engineer are aware of the true position they might be able to bring their combined resources to improve the position on site and so assist the Contractor.

This is recognised by sub-cl.8.3, which requires the Contractor to give prompt notice of events or circumstances which may adversely delay the work or increase the contract price. Indeed sub-cl.(f) requires that the Progress Reports also list out any notices served by either the Employer or Contractor pursuant to sub-cll 2.5 and 20.1.

There is no contractual requirement to hold regular site meetings.[32] Some **4–171** Employers have been known to require a clause imposing a duty on the Contractor to participate in site meetings. This should not really be necessary as regular meetings are of obvious benefit to all parties concerned. Here they might serve as an obvious adjunct to the progress reports to check on progress.

If meetings are held, then all parties who attend should ensure that any minutes are an accurate record of matters discussed. This might be particularly important if any claim is made. In addition, in some jurisdictions it is not unusual to find clauses imposed which make any agreements recorded in meeting minutes binding on all concerned if no objection is raised.

MDB HARMONISED EDITION

There is no change. **4–172**

4.22 SECURITY OF THE SITE

Unless otherwise stated in the Particular Conditions: **4–173**

(a)   the Contractor shall be responsible for keeping unauthorised persons off the Site, and

(b)   authorised persons shall be limited to the Contractor's Personnel and the Employer's Personnel; and to any other personnel notified to the Contractor, by the Employer or the Engineer, as authorised personnel of the Employer's other contractors on the Site.

---

[32] Particular condition 3.6 provides an option whereby either the Engineer or the Contractor's Representative can call for a meeting to review the arrangements for future work.

OVERVIEW OF KEY FEATURES

4–174
- The Contractor is responsible for keeping unauthorised personnel off site.
- Authorised personnel are limited to the Contractor's Personnel and the Employer's Personnel unless otherwise notified to the Contractor by the Employer or Engineer.

COMMENTARY

4–175 By virtue of sub-cl.4.1, the Contractor is, unless otherwise stated, responsible for site security. Sub-clause 4.22 is one of the sub-clauses which relate to that obligation. It should, in particular, be considered together with sub-cl.4.8 which deals with safety procedures and confirms that the Contractor must provide, where necessary, fencing and guards. If the site is shared with other contractors, then presumably this obligation will also be shared. If so, the contract must make this clear.

The definition of personnel is a wide one and the Contractor should ensure that he knows who is in fact authorised to be on site.

By sub-cl.1.1.2.7, "Employer's Personnel" means the Engineer, his assistants as referred to in sub-cl.3.2, all other staff, labour and other employees of the Engineer and of the Employer, and any other personnel notified to the Contractor, by the Employer or the Engineer, as Employer's Personnel.

4–176 By sub-cl.1.1.2.6, "Contractor's Personnel" means the Contractor's Representative and all personnel who may include the staff, labour and other employees of the Contractor and of each Subcontractor assisting the Contractor in the execution of the Works.

Whilst in theory there should be little difficulty initially for the Contractor in establishing both who are his own personnel and who are the Employer's personnel, the Contractor should ensure that an adequate procedure is in place so that he is kept informed of any additional personnel who might be brought on to the site.

Sub-clause 4.6 provides that the Contractor must co-operate with any other contractors employed by the Employer who may be working on or near the Site. This raises one of the potential problems where there is more than one contractor working on the site. The employees of other contractors do not automatically fall within the definition of Employer's Personnel unless they have been notified as such by the Employer or Engineer. Unless they have been so notified, strictly, they should be treated as unauthorised.

4–177 In addition, if there is more than one contractor on site, it is suggested that the responsibility for keeping unauthorised persons off the Site should be a joint and several one as between all the contractors.

MDB HARMONISED EDITION

4–178

There is no change.

4.23 CONTRACTOR'S OPERATIONS ON SITE

The Contractor shall confine his operations to the Site, and to any additional **4–179**
areas which may be obtained by the Contractor and agreed by the Engineer as
working areas. The Contractor shall take all necessary precautions to keep
Contractor's Equipment and Contractor's Personnel within the Site and these
additional areas, and to keep them off adjacent land.

During the execution of the works, the Contractor shall keep the Site free from
all unnecessary obstruction, and shall store or dispose of any contractor's
Equipment or surplus materials. The Contractor shall clear away and remove from
the Site any wreckage, rubbish and Temporary Works which are no longer required.

Upon the issue of a Taking-Over Certificate, the Contractor shall clear away
and remove from that part of the Site and Works to which the Taking-Over
Certificate refers, all Contractor's Equipment, surplus material, wreckage,
rubbish and Temporary Works. The Contractor shall leave that part of the Site
and the Works in a clean and safe condition. However, the Contractor may
retain on Site, during the Defects Notification Period, such Goods as are
required for the Contractor to fulfil obligations under the Contract.

OVERVIEW OF KEY FEATURES

- The Contractor's operatives are confined to the site and any additional **4–180**
  areas agreed by the Engineer.
- The Contractor's Equipment and Personnel should keep to these areas.
- The Contractor is responsible for keeping the site tidy and otherwise free
  from obstruction.
- On issue of the Taking-Over Certificate, the Contractor should remove its
  equipment off site.
- When the Contractor removes his equipment he should leave the site in a
  clean and safe condition.
- The Contractor is permitted to retain on site such goods and equipment
  as are required to enable him to fulfil his obligations under the Defects
  Notification period.

COMMENTARY

4–181   Although this sub-clause requires that the Contractor confine his operations to the Site, by sub-cl.4.13, the Contractor is responsible for obtaining any temporary rights of way that might be required in order to obtain access to the site.

As part of sub-cl.4.22, a Contractor must keep unauthorized personnel off site. By this sub-clause, the Contractor must take steps to keep his own personnel within the site. Sub-clause 4.23 is also closely related to the obligation under sub-cl.4.14 that the Contractor must not unnecessarily or improperly interfere with the convenience of the public and access of public roads.

This sub-clause also requires that the Contractor keep the site tidy and free from debris and obstructions. This has obvious close links with the health & safety requirements of sub-cl.4.8.

4–182   Once the Contractor has completed the execution of his works and the Taking Over Certificate has been issued, the Contractor must remove all his equipment (including rubbish and Temporary Works) from the Site, again taking care to ensure that the site is left in a clean and tidy condition.

The removal of rubbish might also have environmental considerations and the Contractor should remember the obligations imposed by sub-cl.4.18 to protect the environment. For example, in the UK new regulations were introduced in the summer of 2005[33] to deal with the management of hazardous waste. This is part of a worldwide trend of increasing regulation of waste management and environment protection. The Contractor will be expected, through sub-cl.2.2, to have obtained details of all local legislation dealing with the disposal of rubbish. As a consequence the Contractor will also have been expected to take account of the availability of landfill sites necessary to dispose of that rubbish when compiling his tender and programme.

MDB HARMONISED EDITION

4–183   The words "*as additional*" have been added to go before the word "*Working*" at the end of the first sentence. This change merely serves to reinforce the original meaning.

---

[33] Hazardous Waste Regulations July 2005.

4.24 FOSSILS

All fossils, coins, articles of value or antiquity, and structures and other remains   **4–184**
or items of geological or archaeological interest found on the Site shall be
placed under the care and authority of the Employer. The contractor shall take
reasonable precautions to prevent Contractor's Personnel or other persons from
removing or damaging any of these findings.

The contractor shall, upon discovery of any such finding, promptly give notice
to the Engineer, who shall issue instructions for dealing with it. If the
Contractor suffers delay and/or incurs Cost from complying with the instruc-
tions, the Contractor shall give a further notice to the Engineer and shall be
entitled subject to Sub-Clause 20.1 [Contractor's Claims] to:

(a)   an extension of time for any such delay, if completion is or will be
delayed, under Sub-Clause 8.4 [Extension of Time for Completion], and
(b)   payment of any such Cost, which shall be included in the Contract Price.

After receiving this further notice, the Engineer shall proceed in accordance
with Sub-Clause 3.5 [Determinations] to agree or determine these matters.

OVERVIEW OF KEY FEATURES

- All fossils and other antiquities are to be placed under the control of the   **4–185**
  Employer.
- The Contractor shall promptly give notice to the Engineer of the
  discovery of any such antiquity.
- If the discovery of any such antiquity leads to delay or additional cost,
  upon the determination of the Engineer, the Contractor may be entitled
  to an extension of time or payment of any cost.

COMMENTARY

Under the Old Red Book FIDIC 4th edn, cl.27 provided that all fossils and   **4–186**
other antiquities belonged to the Employer. Here, under sub-cl.4.24 they are
to be placed under the care and authority of the Employer. This slight refine-
ment is designed to take account of local conditions and traditions. The
purpose of the clause remains the same.

Leaving aside the potential value of any such antiquities, the discovery of
fossils or items of archaeological interest might have a significant impact on
the Contractor's programme. This is the reason why the Contractor is

135

required to give prompt notice to the Engineer if any such objects are discovered.

The Contractor, subject to the claims procedures required by cl.20, may be entitled to an extension of time and payment of subsequent cost as a consequence of the impact of the finding of fossils or other antiquities. However, can the Employer suggest that the Contractor should have been aware of the possibility of finding fossils or other such artefacts?

4–187   By sub-cl.4.10, the Contractor shall be deemed to have inspected and examined the Site, its surroundings, any data supplied by the Employer and other available information, and to have been satisfied before submitting his Tender as to all relevant matters, including at subsection (a) the form and nature of the site, including sub-surface conditions.

Further, "physical conditions" are defined at sub-cl.4.12 as:

natural physical conditions and man-made and other physical obstructions and pollutants, which the Contractor encounters at the Site when executing the Works, including sub-surface and hydrological conditions but excluding climatic conditions.

This would seem to include quite possibly fossils (a natural physical condition) and archaeological artefacts (man-made obstructions). By sub-cl.4.12, if the physical condition was foreseeable, then the Contractor is not entitled to any relief. Under sub-cl.4.12 the standard by which the Contractor is judged when considering whether any physical condition was unforeseeable or not is that of the experienced Contractor.

4–188   Clearly the Employer might think that the experienced contractor, if the site and position of the site demanded, should make enquiries of the nature of any archaeological investigations that may have been carried out or contemplated by the Employer as part of his tender investigations.

This is why sub-cl.4.24 clearly sets out that, provided the Contractor gives prompt notice to the Engineer he will be entitled to an extension of time and/or cost in respect of any delay that might arise as a consequence of finding any fossils. The risk lies with the Employer.

MDB HARMONISED EDITION

4–189   There is no change.

CLAUSE 5 – NOMINATED SUBCONTRACTORS

5.1 DEFINITION OF "NOMINATED SUBCONTRACTOR"

In the Contract, "nominated Subcontractor" means a Subcontractor: **5–001**

(a)   who is stated in the Contract as being a nominated Subcontractor, or
(b)   whom the Engineer, under Clause 13 [Variations and Adjustments], instructs the Contractor to employ as a Subcontractor.

OVERVIEW OF KEY FEATURES

A nominated sub-contractor is someone who is either so-named in the   **5–002**
Contract or who the Engineer has instructed the Contractor to employ as a
sub-contractor.

COMMENTARY

This straightforward sub-clause merely provides a definition of the "nomi-   **5–003**
nated sub-contractor". The reason for the inclusion of this clause is not
because the use of nominated sub-contractors is recommended but because
they are sometimes necessary.[1]

   The FIDIC Guide sets out three potential advantages to the Employer or
Engineer:

(i)   involvement in the choice of a specialist sub-contractor;
(ii)   involvement in the choice of plant;
(iii)   the avoidance of participation in the co-ordination of the interface
        between the nominated sub-contactors and the Contractors' Works.[2]

If there are to be nominated sub-contractors, it is preferable for the Employer
to make this clear in the tender documents. Then the Contractor will know
exactly where it stands when pricing for the project.

---

[1] P.L. Booen – The Four FIDIC 1999 Contract Conditions: Their Principles, Scope & Details, *www.fidic.org/resources.*
[2] Although the Guide concedes that this is frequently not achieved.

137

5–004   If there are to be no nominated sub-contractors, this clause should be deleted.

The Contractor should be aware that at the time he enters into the Contract, he might not know of the identity of all the sub-contractors. Sub-clause 5.1(b) provides that the Engineer can instruct, in accordance with the variation procedure supplied by cl.13, the Contractor to employ a sub-contractor.

However, an Engineer cannot unilaterally impose a sub-contractor as sub-cl.5.2 provides that a Contractor can raise a reasonable objection by notice to the Engineer about any proposed appointment.

MDB HARMONISED EDITION

5–005   The words *"subject to Sub-Clause 5.2 [Objection to Notification]"* have been added to the end of sub-paragraph (b).

5.2 OBJECTION TO NOMINATION

5–006   *The Contractor shall not be under any obligation to employ a nominated Subcontractor against whom the Contractor raises reasonable objection by notice to the Engineer as soon as practicable, with supporting particulars. An objection shall be deemed reasonable if it arises from (among other things) any of the following matters, unless the Employer agrees to indemnify the Contractor against and from the consequences of the matter:*

(a)   there are reasons to believe that the Subcontractor does not have sufficient competence, resources or financial strength;

(b)   the subcontract does not specify that the nominated Subcontractor shall indemnify the Contractor against and from any negligence or misuse of Goods by the nominated Subcontractor, his agents and employees; or

(c)   the subcontract does not specify that, for the subcontracted work (including design, if any), the nominated Subcontractor shall:

(i)   undertake to the Contractor such obligations and liabilities as will enable the Contractor to discharge his obligations and liabilities under the Contract; and

(ii)   indemnify the Contractor against and from all obligations and liabilities arising under or in connection with the Contract and from the consequences of any failure by the Subcontractor to perform these obligations or to fulfil these liabilities.

OVERVIEW OF KEY FEATURES

- If the Contractor objects to the employment of a nominated sub- **5–007**
  contractor, he must do so by notice to the Engineer as soon as possible.
- An objection will be deemed to be reasonable if:

  (a) There are reasons to believe the sub-contractor lacks competence or
       sufficient finances or resources.
  (b) The sub-contract does not specify that the sub-contractor shall
       provide the Contractor with an appropriate indemnity in relation to
       any negligence.
  (c) The sub-contract does not specify that the sub-contractor will
       undertake to perform its work in such a way so as to ensure that the
       Contractor will discharge his obligations under the Contract.
  (d) The sub-contract does not say that the sub-contractor will indemnify
       the Contractor in respect of any failure to perform its obligations.

- An objection will be deemed reasonable if the Employer does not agree to
  indemnify the Contractor in respect of a reasonable objection.

COMMENTARY

This clause sets out what a Contractor must do, if he objects to the attempted **5–008**
nomination of a sub-contractor by the Employer. The Contractor should
remember that in accordance with sub-cl.4.4, he is responsible for the acts
and defaults of all sub-contractors.

Any objection must be made as soon as practicable. A bare objection will
not suffice. It must be reasoned and detailed. The objection will be deemed to
be reasonable if it falls within the items listed within sub-cll (a)-(d).

The Employer can, if he chooses, overcome any objection to a sub-
contractor of his choosing by agreeing to provide an indemnity to the
Contractor in respect of the specific objection made by the Contractor.

No specific provision is made for the resolution of any dispute which might **5–009**
arise over the issue. It is possible that a dispute may arise if the reasons the
Contractor puts forward as to why he believes that the Subcontractor does
not have sufficient competence, resources or financial strength, are
challenged.

MDB HARMONISED EDITION

5–010 There are a number of small changes:

First, the Employer agreement in paragraph one, must be "*in writing.*" This is sensible for reasons of certainty. The definition of what constitutes "in writing" can be found at sub-cl.1.2(d).

Second, sub-para.(b) has been changed so that a reasonable objection can be made if the "*nominated Subcontractor does not accept to*" indemnify the Contractor. This is a change in wording rather than meaning.

There is a similar change to sub-para.(c). The relevant objection is the Subcontractor not accepting "*to enter into a subcontract which specifies*" the requirements of the further sub-clauses.

5–011 A third sub-para. has been added to (c) which reads:

be paid only if and when the Contractor has received from the Employer payments for sums due under the Subcontract referred to under Sub-Clause 5.3 [Payment to nominated Subcontractors].

In other words, the Nominated Subcontractor has to accept this "pay when paid" clause, which gives potentially significant protection to the Contractor. In putting this new sub-clause in, FIDIC and the Banks are moving against the trend at least in the UK, where increasingly pay-when-paid clauses are prohibited, expressly so by s.113 of the Housing Grants Construction & Regeneration Act 1996.

5.3 PAYMENTS TO NOMINATED SUBCONTRACTORS

5–012 *The Contractor shall pay to the nominated Subcontractor the amounts which the Engineer certifies to be due in accordance with the subcontract. These amounts plus other charges shall be included in the Contract Price in accordance with sub-paragraph (b) of Sub-Clause 13.5 [Provisional Sums], except as stated in Sub-Clause 5.4 [Evidence of Payments].*

OVERVIEW OF KEY FEATURES

5–013
- The Engineer will certify sums due under any sub-contract.
- The Contractor shall pay the Sub-Contractor the sums certified.
- Any sums certified will form part of the Contract Price.

COMMENTARY

This sub-clause deals with the payment of Sub-Contractors by the Contractor. **5–014**

The requirement to pay the sums certified by the Employer is a mandatory one. It is also a requirement which is policed by the requirements of sub-cl.5.4 which enable the Engineer to require evidence of payment by the Contractor of sums previously certified.

MDB HARMONISED EDITION

There has been a small change for reasons of clarity. The words, "*shown on* **5–015** *the nominated Subcontractor's invoices approved by the Contractor*" have been added after the word "*amounts*". This sub-clause therefore appears to contradict the addition made at sub-cl.5.2(c)(iii) which says that the Contractor only has to pay the Sub-contractor when it actually receives its money from its Employer.

5.4 EVIDENCE OF PAYMENTS

Before issuing a Payment Certificate which includes an amount payable to a **5–016** *nominated Subcontractor, the Engineer may request the Contractor to supply reasonable evidence that the nominated Subcontractor has received all amounts due in accordance with previous Payment Certificates, less applicable deductions for retention or otherwise. Unless the Contractor:*

(a)   Submits this reasonable evidence to the Engineer, or

   (i)   satisfies the Engineer in writing that the Contractor is reasonably entitled to withhold or refuse to pay these amounts, and
   (ii)   submits to the Engineer reasonable evidence that the nominated Subcontractor has been notified of the Contractor's entitlement,

then the Employer may (at his sole discretion) pay, direct to the nominated Subcontractor, part or all of such amounts previously certified (less applicable deductions) as are due to the nominated Subcontractor and for which the Contractor has failed to submit the evidence described in sub-paragraphs (a) or (b) above. The Contractor shall then repay, to the Employer, the amount which the nominated Sub-contractor was directly paid by the Employer.

OVERVIEW OF KEY FEATURES

5–017
- The Engineer may request from the Contractor evidence that it has paid the Sub-contractor all sums due under previous Payment Certificates.
- Unless the Contractor can provide evidence that the Sub-Contractor has been paid or satisfies the Engineer that he is reasonably entitled to withhold money and has notified the Sub-Contractor of this then the Employer will pay the Sub-Contractor direct.
- If the Employer is forced to pay the Sub-Contractor direct then the Contractor must repay that sum to the Employer.

COMMENTARY

5–018   The purpose of this clause is to enable the Employer to keep a check on whether the Contractor is paying the nominated sub-contractor.

If the Engineer chooses, he can require the Contractor to provide evidence that he has paid the Sub-contractor the sums due under the payment certificates or if he has not paid the sums due, that he has served the appropriate notices on the Sub-contractor entitling the Contractor to withhold payment.

If the Contractor has failed to do this the Employer may, at his own discretion, choose to pay the Sub-contractor direct. If the Employer does that, he can recover the money from the Contractor. The requirement that the Contractor repay the Employer in these circumstances is a mandatory one.

MDB HARMONISED EDITION

5–019   There is no change.

CLAUSE 6 – STAFF AND LABOUR[1]

6.1 ENGAGEMENT OF STAFF AND LABOUR

Except as otherwise stated in the Specification, the Contractor shall make **6–001**
arrangements for the engagement of all staff and labour, local or otherwise, and
for their payment, housing, feeding and transport.

OVERVIEW OF KEY FEATURES

The Contractor shall engage all staff and labour and make arrangements for **6–002**
their payment, housing, feeding and transport unless stated otherwise in the
Specification.

COMMENTARY

The Contractor should remember that sub-cl.4.1 makes him responsible for **6–003**
providing all the necessary personnel to carry out the design, execution and
completion of his works as well as to remedy any defects.

This sub-clause places a wider responsibility onto the shoulders of the
Contractor to ensure not only that he engages the labour and staff but
also that he makes welfare arrangements for them. It should be noted,
however, that the obligation in this sub-clause relates to "all staff and
labour" as opposed to "Contractor's Personnel", the latter being an
arguably wider category of people as it is defined as including all staff and
labour.

This obligation may reflect the international flavour of the projects and the
type of projects for which FIDIC contracts are often used. In particular, with
overseas projects where the Contractor may employ staff from his own
country and also where those projects are in remote locations (for example
process plants), then this provision of welfare is key. The Contractor will
therefore need to allow for such provision in his tender and take into account
in doing so the particular location of the works and the difficulties that might
be encountered in making arrangements for payment (e.g. local currency

[1] The authors are grateful for the assistance of Karen Gidwani, a partner at Fenwick Elliott
LLC with this section.

conversion), housing (e.g. location), feeding (e.g. local laws/religious customs) and transport (e.g. local infrastructure).

6–004   As to the cost of the labour itself, we note that sub-clause 13.8 provides the adjustment of the amount paid to the Contractor due to the rise or fall in the cost of labour, goods or other inputs to the works.

Although the Contractor's prima facie responsibility extends to local staff and labour, in engaging local staff and labour, the Contractor should also be aware of local labour laws (which may not be the same as the law governing the Contract) and, in particular, aspects of that law which might conflict with the obligations of the Contract (see below). Where problems such as this arise, the Contractor will have to ascertain whether he can "contract out" of local, conflicting labour laws or whether he is bound by them. In either case, the ensuing risk must be built into the Contractor's price for the Works. It is noted that the Contractor has, in any event, an obligation to comply with applicable Laws (which include local laws) pursuant to sub-cl.1.13 of the Contract. The Contractor should also consider whether insurance should or must be taken out in relation to his obligations to engage all staff and labour and provide for their welfare. (This also applies to some of the later sub-clauses of cl.6).

The Guidance provided in the 1999 Edition for the preparation of Particular Conditions, sets out some examples of sub-clauses that can be added to cl.6 to take account particular circumstances and the locality of the Site. These, for example, cover matters such as the provision and importation of foreign staff and labour, measures against insect and pest nuisance, alcoholic liquor and drugs, arms and ammunition and festivals and religious customs. These provisions will come as no surprise to those familiar with the optional Particular Conditions which also made similar, although more wide-ranging,[1A] provision for particular circumstances and locality.

6–005   Indeed as set out below, the new Multilateral Development Bank Harmonised Edition includes approximately 10 of these "particular locality" clauses as part of the General Conditions covering, for example, supply of foodstuffs, supply of water, alcoholic liquor and drugs and funeral arrangements.

If a party is planning to include particular locality clauses as Particular Conditions, then they should note that the Particular Conditions have priority over the General Conditions in the case of conflict.

There may be situations where the Employer wishes to engage certain staff or labour and/or provide facilities for them. In such cases, as noted in the FIDIC Guide, the obligations that the Employer undertakes should be specified precisely. Any confusion over the shared responsibility of engaging labour could not only lead to a dispute between the Employer and the Contractor but may also have the undesirable affect of disaffecting staff and

---

[1A] See for example in relation to epidemics, burial of the dead and repatriation.

144

labour who may not get paid as a result of the dispute.

It is noted that, by stating that this clause is subject to the Specification, this reverses the order of priority of documents at sub-cl.1.5 of the Contract where the General Conditions are stated to take precedence over the Specification. **6–006**

MDB HARMONISED EDITION

There have been two changes.

In the first, the words *"feeding, transport and, when appropriate, housing"* have been added to the end of the first paragraph. This serves to increase the obligation of the Contractor. It should be noted that the addition does not make the Contactor responsible for the cost of making arrangements for the feeding, transport and, if appropriate, housing of the staff and labour. **6–007**

The following additional paragraph has also been included:

The Contractor is encouraged, to the extent practicable and reasonable, to employ staff and labor with appropriate qualifications and experience from sources within the Country.

This is self-explanatory and, like a number of the other amendments, reflects a desire on the part of the World Bank to encourage local enterprise.[2] **6–008**

6.2 RATES OF WAGES AND CONDITIONS OF LABOUR

The Contractor shall pay rates of wages, and observe conditions of labour, which are not lower than those established for the trade or industry where the work is carried out. If no established rates or conditions are applicable, the Contractor shall pay rates of wages and observe conditions which are not lower than the general level of wages and conditions observed locally by employers whose trade or industry is similar to that of the Contractor. **6–009**

OVERVIEW OF KEY FEATURES

- The Contractor must pay wages at a rate comparable with local labour rates.
- The Contractor must ensure that labour conditions on site are no lower than local labour conditions. **6–010**
- If the Contractor has no comparators for wages and conditions then he

---

[2] See for example sub-cl.4.4.

shall maintain wages and conditions which are not lower than those observed locally by employers in a similar trade or industry.

COMMENTARY

6–011
The first point to note about this sub-clause is that the level that the Contractor must meet appears to be subjective and therefore open to much interpretation. However the sprit of this clause is in keeping with the obligation on the Contractor to comply with local laws under sub-cll 1.13 and 6.4 (see below). On that basis, there may well be legislation in the country where the Works are being carried out that restricts the minimum wage that the Contractor can pay and which specifies particular work conditions, for example in relation to health and safety.[3]

There are also likely to be limits on the age of potential local labour. For example, over 140 countries are signed up to Art.182 of the International Labour Organisation[4] which provides for minimum labour ages.

This sub-clause appears to be designed to ensure compatibility with other employers so that local labour and trade markets are not adversely affected by the Contract Works. However, the clause does not refer to the Contractor paying higher wages than those locally and, where this happens, this too could adversely affect local trade conditions.

6–012
Whilst this clause gives guidance to the Contractor if there is no comparable local labour; there is no fall-back if there is no comparable industry locally. It is submitted that if this were to occur then the Contractor should look at the payment and conditions prevalent in that industry in that particular country.

If the Contractor were to breach this obligation, as the Guide points out, the Employer may have difficulty proving any loss. However, this would not stop the Employer from using the breach to resist a claim from the Contractor for payment due to the consequences of the non-compliance.

MDB HARMONISED EDITION

6–013
The following additional paragraph has been included:

---

[3] For example see the Construction (Design & Management) Regulations 1994 and the National Minimum Wages Act 1998 in England and Wales.

[4] *www.ilo.org* – a number of countries require compliance with the various conventions which include rights of freedom of association, freedom from forced labour, freedom from discrimination on the grounds of race, colour, sex, religion, political opinion and social origin.

The Contractor shall inform the Contractor's Personnel about their liability to pay personal income taxes in the Country in respect of such of their salaries, wages and allowances as are chargeable under the Laws of the Country for the time being in force, and the Contractor shall perform such duties in regard to such deductions thereof as may be imposed on him by such Laws.

This addition serves to take the obligation on the Contractor one step further by imposing upon him an obligation to inform the Contractor's Personnel about income tax liability in the country where the works are being carried out and further states that the Contractor shall perform his duties in relation to the deduction of taxes as may be imposed by the law of that country.

In addition, as set out below, the MDB Harmonised Edition includes at sub-cll 6.12–6.22 a number of additional sub-clauses which relate to local labour conditions.

6.3 PERSONS IN THE SERVICE OF THE EMPLOYER

The Contractor shall not recruit, or attempt to recruit, staff and labour from   **6–014**
amongst the Employer's Personnel.

OVERVIEW OF KEY ISSUES

The Contractor shall not recruit or try to recruit the Employer's Personnel as   **6–015**
part of his own workforce.

COMMENTARY

Employer's Personnel is defined at sub-cl.1.1.2.6 of the Contract as "the   **6–016**
Engineer, the assistants referred to in sub-cl.3.2 [*Delegation by the Engineer*] and all other staff, labour and other employees of the Engineer and of the Employer; and any other personnel notified to the Contractor, by the Employer or the Engineer, as Employer's Personnel". Sub-clause 6.3 is therefore a very wide clause encompassing more than just those directly employed by the Employer.

It is easy to see the purpose of the clause. It guards against conflicts of interest and the misappropriation of ideas and resources. It also assists in maintaining commercial confidentiality. However, the weight of this clause in practice is another matter altogether. The construction and engineering industry may be vast but it is split into a number of specialised sectors. Those using this type of contract are likely to be based in the process plant sectors and in oil and gas. Movement between key contractors, consultants and employers is inevitable.

Accordingly, it may well be that this clause is not really intended to militate against commercial movement within the industry but instead it is set up to protect the Employer from the "poaching" of its key staff by the Contractor. Typically an Employer will be based in the locality where the project is being undertaken, whereas a Contractor is more likely to be based elsewhere. Local knowledge can be key.

6–017   The parties might want to consider a short amendment to the effect that the parties shall agree a procedure in the event that an employee expresses a desire to be employed by the Contractor. However, we do query the extent to which the Employer could prove and quantify damage for a breach of this sub-clause in any event.

MDB HARMONISED EDITION

6–018   There is no change.

6.4 LABOUR LAWS

6–019   *The Contractor shall comply with all relevant labour laws applicable to the Contractor's Personnel, including Laws relating to their employment, health, safety, welfare, immigration and emigration, and shall allow them all their legal rights.*

The Contractor shall require his employees to obey all applicable Laws, including those concerning safety at work.

OVERVIEW OF KEY FEATURES

6–020   • The Contractor shall comply with all relevant labour laws which apply to the Contractor's Personnel.
   • The Contractor shall not do anything which might prevent the Contractor's personnel from exercising their legal rights.
   • The Contractor must take steps to require his employees to obey all applicable Laws, in particular those concerning safety at work.

COMMENTARY

"Contractor's Personnel" is defined at sub-cl.1.1.2.7 as "the Contractor's **6–021** Representative and all personnel whom the Contractor utilises on Site, who may include the staff, labour and other employees of the Contractor and of each Subcontractor; and any other personnel assisting the Contractor in the execution of the Works". This is very wide. The last phrase "any other personnel assisting the Contractor" could almost be read to include the Employer's employees.

The sub-clause here also appears to be very broad in scope referring to all applicable labour laws. In fact this is doing no more than confirming the requirement of sub-cl.1.13 whereby the Contractor must comply with all local laws.

However, it is sensible to define the type of legislation that typically affects staff. Accordingly the Contractor must ensure that he is acting in accordance with the applicable Laws in relation to employment, health, safety, welfare, immigration and emigration. This is followed by a catch-all provision whereby the Contractor must allow his "personnel" their "legal rights". This is quite nebulous and would almost certainly encompass the Human Rights legislation that has been put in place in the European community.

The flip side to this obligation is that the Contractor must also keep in **6–022** mind, when preparing his tender, the extent to which the local laws may not afford the protection to his employees that they might otherwise expect. This is particularly so if the Contractor is a company foreign to the place of works and is employing staff from his own country. To an extent this should be able to be overcome by ensuring that the relevant contracts of employment are subject to the laws of the Contractor's country. However, if this is the case, then the Contractor must also ensure that those contracts of employment do not in any way contravene local labour laws.

The law of the Contract may not be the same as the law of the local area. The Contractor will have to research his obligations before tendering in order to price the risk that this obligation will impose upon him. Whilst the Guide queries the extent to which the Employer would be able to claim for loss arising out of a breach of this clause, it is clear that a breach could have serious consequences for the Contractor. If the Contractor does not allow the Contractor's Personnel their rights and there is then a strike, causing delay to the project and causing cost to be incurred by the Employer in relation to the effects of that strike, then, in English law at least, the Contractor would be liable for those costs as damages.

Therefore this is another risk that the Contractor must carefully consider when tendering as the variables will change from country to country and knowledge of local labour laws, including health and safety, will be paramount.

6–023   It is noted that the obligation relates to laws "applicable" to the Contractor's Personnel. This sub-clause does not make the Contractor liable for breaches of the applicable Laws by the Contractor's Personnel although it does state that the Contractor must "require" its employees to comply with the applicable laws.

Finally, sub-cl.13.7 provides for the adjustment of the Contract Price if the applicable laws change (including by way of judicial or governmental interpretation) after the Base Date and the Contractor suffers (or will suffer) delay or incurs (or will incur) additional cost as a result of the change.

MDB HARMONISED EDITION

6–024   There is no change.

6.5 WORKING HOURS

6–025   *No work shall be carried out on the Site on locally recognised days of rest, or outside the normal working hours stated in the Appendix to Tender, unless:*

(a)   otherwise stated in the Contract,

(b)   the Engineer gives consent, or

(c)   the work is unavoidable, or necessary for the protection of life or property or for the safety of the works, in which case the contractor shall immediately advise the Engineer.

OVERVIEW OF KEY FEATURES

6–026   No work is to be carried out on Site on:

(i)   locally recognised days of rest; or

(ii)   outside the normal working hours stated in the Appendix to Tender;

unless:

(i)   the Contract states otherwise;

(ii)   the Engineer gives consent; or

(iii)   the work is unavoidable or necessary for the protection of life or property or for the safety of the works.

COMMENTARY

This sub-clause is designed to protect the rights of local workers, particularly **6–027** in areas of strong religious traditions. The Contractor therefore should have due regard to all recognised festivals, days of rest and religious/other customs.

The sub-clause is, however, subject to three exceptions, set out at 6.4(a), 6.4(b) and 6.4(c) and is also subject to what the Appendix to Tender states. Under sub-cl.1.5, the Appendix to Tender takes priority over the General Conditions in any event (as it forms part of the Letter of Tender).

Therefore whilst there is some scope for conflict between sub-cl.6.5 and the Appendix to Tender, the Appendix to Tender takes precedence. It is submitted that sub-cl.6.4 could be read as either (a) subject to/qualified by the Appendix to Tender; or (b) amended by the Appendix to Tender by virtue of sub-cll 6.4(a) and 1.5. It will depend upon the extent to which the sub-clause and the Appendix to Tender are inconsistent. The most practical way to avoid inconsistency, however, is to ensure that the parties to the Contract are aware of local conditions in relation to hours to be worked and that these are correctly reflected in the Appendix to Tender.

The FIDIC Guide states that when the tender documents are being **6–028** prepared, consideration should be given as to whether this clause is in fact needed, otherwise it may be deleted. This is also reflected in the guidance in the 1999 Edition of the Red Book in relation to drafting the Particular Conditions. However, in our view, specifying working hours in advance in the Appendix to Tender can be very helpful for the management and planning of activities. Either the Contractor or the Employer can fill in the working hours part of the Appendix to Tender if it suits their purposes.

If hours are to be worked on locally recognised days of rest outside normal working hours, then sub-cl.6.5(b) allows this to be achieved by the Contractor obtaining consent from the Engineer. Under cl.1.3, such consent should be given in writing and should not be unreasonably withheld or delayed. It is also noted that under sub-cl.8.6 the Contractor may be instructed to adopt measures to expedite the progress of the Works, which may include working outside of normal working hours.

The parties will need to bear in mind local legislation which might effect the number of hours that can be worked. For example, members of the European Community will have adopted variants of the Working Time Directive.

As to the Contractor's programming obligations, it is clear from sub-cl.8.3 **6–029** that the Contractor must take this sub-clause into account in drafting the programme that it will submit to the Engineer.

Sub-clause 6.4(c), the last exception, states that work can be carried outside of the defined hours if it is unavoidable and necessary because of danger to life, property or the works. The Contractor must advise the

Engineer when such circumstances arise. This is important otherwise this part of the sub-clause could be open to abuse as an excuse to work longer hours.

MDB HARMONISED EDITION

6–030   The words "Appendix to Tender" have been replaced by "Contract Data."

6.6 FACILITIES FOR STAFF AND LABOUR

6–031   *Except as otherwise stated in the Specification, the Contractor shall provide and maintain all necessary accommodation and welfare facilities for the Contractor's Personnel. The Contractor shall also provide facilities for the Employer's Personnel as stated in the Specification.*

The Contractor shall not permit any of the Contractor's Personnel to maintain any temporary or permanent living quarters within the structures forming part of the Permanent Works.

OVERVIEW OF KEY FEATURES

6–032
- The Contractor shall provide all necessary accommodation and welfare facilities for his personnel.
- The Contractor's personnel are not allowed to live within the permanent works.
- If required by the Specification, the Contractor shall provide similar facilities for the Employer's personnel.

COMMENTARY

6–033   This sub-clause should be read in conjunction with sub-cl.6.1 which provides for the Contractor to make arrangements for, inter alia, the housing of all staff and labour. This sub-clause relates to the "Contractor's Personnel" and, in part, the "Employer's Personnel" which, as noted previously in this chapter, is a very wide range of people.

It is submitted that in order to ensure that this obligation is not unduly onerous on the Contractor then the Specification[5] should be used to specify exactly what the Contractor will provide as accommodation and welfare facilities and what, if anything, the Employer will provide. This will ensure a measure of certainty which, in the absence of any definition of "necessary", is welcome. Otherwise we envisage that whether or not a particular facility is necessary could easily become a matter of dispute.

In any event, the sub-clause is clear that no-one is allowed to camp out on site.

MDB HARMONISED EDITION

There is no change.   6–034

6.7 HEALTH AND SAFETY

The Contractor shall at all times take all reasonable precautions to maintain the   6–035
health and safety of the Contractor's Personnel. In collaboration with local
health authorities, the Contractor shall ensure that medical staff, first aid facil-
ities, sick bay and ambulance services are available at all times at the Site and
at any accommodation for Contractor's and Employer's Personnel. The
Contractor shall also ensure that suitable arrangements are made for all
necessary welfare and hygiene requirements and for the prevention of epidemics.

The Contractor shall appoint an accident prevention officer at the Site,
responsible for maintaining safety and protection against accidents. This person
shall be qualified for this responsibility, and shall have the authority to issue
instructions and take protective measures to prevent accidents. Throughout the
execution of the Works, the Contractor shall provide whatever is required by
this person to exercise this responsibility and authority.

The Contractor shall send, to the Engineer, details of any accident as soon as
practicable after its occurrence. The Contractor shall maintain records and
make reports concerning health, safety and welfare of persons, and damage to
property, as the Engineer may reasonably require.

---

[5] It is noted that the order of priority of documents in this sub-clause, as in sub-cl.6.1 is reversed with the Specification taking precedence over the General Conditions.

OVERVIEW OF KEY FEATURES

6–036
- The Contractor is *at all times* to take all reasonable precautions to look after the health and safety of his personnel.
- The Contractor shall ensure that necessary medical and first aid facilities are available at all times on site and any accommodation for both his own and the Employer's Personnel.
- The Contractor shall appoint an Accident Prevention Officer at site who must be suitably qualified and will be responsible for maintaining safety and protection.
- The Accident Prevention Officer will have authority to issue instructions and take protective measures to prevent accidents.
- The Contractor shall send to the Engineer details of any accident as soon as practical after it happens.
- The Contractor shall maintain records and reports about health and safety issues such as the Engineer may reasonably require.

COMMENTARY

6–037   This sub-clause is an amalgam of a number of Old Red Book FIDIC 4th edn Particular Conditions.

Sub-clauses 6.4 and 6.7 mirror the obligations of sub-cl.4.8 whereby the Contractor shall comply with all applicable safety regulations and take care for the safety of all persons entitled to be on the Site.

By sub-cl.6.4, the Contractor shall comply with all relevant labour laws. These specifically include laws in relation to Health & Safety. The Contractor is also required to ensure that his employees obey all applicable laws including those concerning the Safety at Work. Sub-clause 6.7 spells out those duties in more detail. The sub-clause is very different to clauses in typical UK domestic building contracts. However, whilst, for example, in the UK there is a body of health and safety legislation which will either be incorporated into the contract or which will govern what happens on site, this is not always the case.

6–038   The Contractor must check the relevant health and safety legislation that could be binding whether by virtue on the governing law of the contract or by virtue of the fact that the parties are carrying out work in a particular country.

Of particular importance is the Accident Prevention Officer. The Accident Prevention Officer, who must be suitably qualified, has wide-ranging responsibilities and will have the authority to issue instructions and/or take protective measures to prevent accidents. The Contractor is required to ensure that the Accident Prevention Officer is not prevented from carrying out his responsibilities.

By sub-cl.4.21(g), the regular Progress Reports which the Contractor is required to prepare must include details of safety statistics including details of any hazardous incidents or activities relating to environmental aspects or public relations. This obligation is separate from, but obviously linked to, the obligation under sub-cl.6.7 that the Contractor shall following any accident, maintain records *and* reports in relation to Health & Safety and welfare issues.

The obligation also requires the Contractor to maintain records in relation   **6–039**
to damage to property. The keeping of these records is subject to what is reasonably required by the Engineer.

Whatever is required by sub-cl.6.7 must be carried out at the Contractor's cost. They could be quite considerable. The Guide goes so far as to note that if local facilities seem likely to be insufficient, the Contractor must overcome the shortfall and even, if necessary, provide a properly equipped hospital.

This sub-clause is not just confined to health and safety in terms of accident prevention. For example, the Contractor should give consideration to the possibility of an outbreak of illness or epidemic and the prevention thereof.[6] In such circumstances the Contractor will need to comply with any regulations or orders made by the government or local health authorities. The Contractor will however, in accordance with sub-cl.8.3(d) be entitled to an extension of time in respect of unforeseeable shortages in the availability of goods or personnel by reason of epidemic.

Similarly, other health and safety considerations might include insect infes-   **6–040**
tation, rats or other wild animals. Stagnant pools or water might increase the threat from malaria.

The fact that the Contractor must provide suitable arrangements for all welfare and hygiene requirements will include the necessity to provide an adequate supply of drinking water – a simple sounding but not always straightforward task.

Further "particular locality" clauses (see commentary below on sub-cll 6.12–6.22 of the MDB Harmonised Edition) can also be added to assist this sub-clause.[7] The Contractor must tailor this form to the particular conditions that he expects to encounter at the site in question in order to manage his economic and contractual risks effectively.

MDB HARMONISED EDITION

The additional paragraphs here make specific provision for the prevention of   **6–041**
HIV and AIDS transfer. The Contractor should note that this provision

---

[6] Avian flu is one such scenario, also (and perhaps more common) cholera and typhoid.
[7] For example, see the sub-clauses in relation to the supply of food and the supply of drinking water.

needs to be incorporated into the programme which must be provided in accordance with sub-cl.8.3:

HIV-AIDS Prevention. The Contractor shall conduct an HIV-AIDS awareness programme via an approved service provider, and shall undertake such other measures as are specified in this Contract to reduce the risk of the transfer of the HIV virus between and among the Contractor's Personnel and the local community, to promote early diagnosis and to assist affected individuals.

The Contractor shall throughout the contract (including the Defects Notification Period): (i) conduct Information, Education and Consultation Communication (IEC) campaigns, at least every other month, addressed to all the Site staff and labour (including all the Contractor's employees, all Subcontractors and Consultants' employees, and all truck drivers and crew making deliveries to Site for construction activities) and to the immediate local communities, concerning the risks, dangers and impact of, and appropriate avoidance behaviour with respect to, Sexually Transmitted Diseases (STD)-or Sexually Transmitted Infections (STI) in general and HIV/AIDS in particular; (ii) provide male or female condoms for all Site staff and labour as appropriate; and (iii) provide for STI and HIV/AIDS screening, diagnosis, counseling and referral to a dedicated national STI and HIV/AIDS programme, (unless otherwise agreed) of all Site staff and labour.

6–042   *The Contractor shall include in the programme to be submitted for the execution of the Works under Sub-Clause 8.3 an alleviation programme for Site staff and labour and their families in respect of Sexually Transmitted Infections (STI) and Sexually Transmitted Diseases (STD) including HIV/AIDS. The STI, STD and HIV/AIDS alleviation programme shall indicate when, how and at what cost the Contractor plans to satisfy the requirements of this Sub-Clause and the related specification. For each component, the programme shall detail the resources to be provided or utilized and any related sub-contracting proposed. The programme shall also include provision of a detailed cost estimate with supporting documentation. Payment to the Contractor for preparation and implementation this programme shall not exceed the Provisional Sum dedicated for this purpose.*

6.8 CONTRACTOR'S SUPERINTENDENCE

6–043   *Throughout the execution of the Works, and as long thereafter as is necessary to fulfil the Contractor's obligations, the Contractor shall provide all necessary superintendence to plan, arrange, direct, manage, inspect and test the work.*

Superintendence shall be given by a sufficient number of persons having adequate knowledge of the language for communications (defined in Sub-

Clause 1.4 [Law and Language]) and of the operations to be carried out (including the methods and techniques required, the hazards likely to be encountered and methods of preventing accidents), for the satisfactory and safe execution of the Works.

OVERVIEW OF KEY FEATURES

- For as long as is necessary to undertake his obligations under the Contract, the Contractor shall provide all necessary supervision to plan, direct and manage the work.
- The supervision must be provided by enough number of people to ensure the satisfactory and safe execution of the works.
- Those supervising must have adequate knowledge both of the language for communications, as defined in sub-cl.1.4 and of the likely work that will be carried out.

6–044

COMMENTARY

The prime purpose of this sub-clause is to help ensure the "satisfactory and safe" execution of the works. Therefore, unlike the preceding sub-clause, it is not solely related to health and safety.

6–045

This superintendence obligation is a wide one. It might also be felt to be an obvious one. However, by spelling the obligations out in the Contract, the Employer, through the Engineer, is provided with a means to keep a check on the Contractor's performance.

Unfortunately, no definition is provided of "superintendence", "necessary" or "sufficient". In relation to "superintendence", those providing the superintendence will be in addition to the Contractor's Representative who by sub-cl.4.3 shall direct performance of works. The lack of a definition of "necessary" and "sufficient" is more problematical and this is likely to cause difficulties if a breach of this sub-clause is ever alleged. It is submitted that what is "necessary" and "sufficient" will very much depend on the individual circumstances of each project.

By sub-cl.1.4, the language for communication shall be that stated in the Appendix to Tender. If no language is stated, then the language for communications shall be the language in which the Contract (or most of it) is written. There is a potential here for problems if the language for communication is not specified in the Appendix to Tender. It is possible, if not highly likely in certain circumstances, that the language spoken by the majority of personnel on site may differ from the language in which the Contract is written. An experienced Contractor should be alive to the potential problems

6–046

and ensure that those supervising have an adequate level of communication skills.

The guidance in the 1999 Edition of the Red Book for the drafting of the Particular Conditions suggests that if the ruling language is not the same as the language for day to day communications (under sub-cl.1.4) or if for any other reason it is necessary to stipulate that the superintending staff shall be fluent in a particular language then a sentence to this effect (set out in the guidance) can be added to this sub-clause.

MDB HARMONISED EDITION

6–047   There is no change.

6.9 CONTRACTOR'S PERSONNEL

6–048   *The Contractor's Personnel shall be appropriately qualified, skilled and experienced in their respective trades or occupations. The Engineer may require the Contractor to remove (or cause to be removed) any person employed on the Site or Works, including the Contractor's Representative if applicable, who:*

(a)   persists in any misconduct or lack of care,
(b)   carries out duties incompetently or negligently,
(c)   fails to conform with any provisions of the contract, or
(d)   persists in any conduct which is prejudicial to safety, health, or the protection of the environment.

If appropriate, the Contractor shall then appoint (or cause to be appointed) a suitable replacement person.

OVERVIEW OF KEY FEATURES

6–049   • The Contractor's Personnel shall be of the appropriate qualifications and experience in their profession.
   • The Engineer may require the Contractor to remove anyone employed on the site or works who persists in misconduct, is incompetent, fails to conform with the provisions of the Contract or persists in conduct which is prejudicial to Health & Safety or protection of the environment.
   • If someone is removed, the Contractor shall then appoint a suitable replacement.

COMMENTARY

This sub-clause is self-explanatory. It provides that the Engineer may require **6–050** the removal from site of any person employed on the site who acts in a way prejudicial to the carrying out of the Contract. As this sub-clause applies to "any person employed on the site", it also is likely that the clause will relate to not only the Contractor's Personnel, but any of the Subcontractor's personnel as well. This right of the Engineer will be in addition to the obvious rights of the Contractor to remove or deal with his employees in the manner he sees fit. However it is important that there is a procedure in place for the removal of personnel in order to prevent abuse and also possible non-compliance with any local legislation.

The sub-clause provides 4 categories of behaviour which might constitute grounds for removal. However, there is no definition of, for example, the term "misconduct" and the parties may have to have recourse to the relevant legislation under the law of the Contract in order to properly interpret this sub-clause.

Further, the sub-clause does not provide for what might happen if the Contractor disagrees with the Engineer and does not want to lose the personnel in question. This is particularly important where the Contractor's Representative is concerned. Presumably the Contractor could invoke the DAB procedure required by cl.20. However this could cause delay and slow the project down. The pragmatic approach in such circumstances might be to comply with any such instructions to remove personnel and then seek to demonstrate that such removal was unjustified and seek appropriate recompense. The only difficulty with this would be ascertaining what the recompense would be.

It will not always be the case that the Contractor will incur great financial **6–051** loss in such a situation. Whilst an Arbitrator may not have the power to reinstate the person who was removed from site, he or she should be able to make an order that the particular Engineer's instruction is invalid. However, even if this happens, the Contractor may not want to reinstate the employee as doing so may adversely affect cordial relations with the Employer. Therefore once an employee has been removed from site, in all likelihood, they will not return.

MDB HARMONISED EDITION

There is no change. **6–052**

6.10 RECORDS OF CONTRACTOR'S PERSONNEL AND EQUIPMENT

6–053   *The Contractor shall submit, to the Engineer, details showing the number of each class of Contractor's Personnel and of each type of Contractor's Equipment on the Site. Details shall be submitted each calendar month, in a form approved by the Engineer, until the Contractor has completed all work which is known to be outstanding at the completion date stated in the Taking-Over Certificate for the Works.*

OVERVIEW OF KEY FEATURES

6–054   • The Contractor shall submit to the Engineer details showing the number of personnel and type of equipment on site.
         • These details shall be submitted on a monthly basis until the Contractor has completed all work known to be outstanding at the Completion Date stated in the Taking Over Certificate.

COMMENTARY

6–055   This clause is particularly important as the information required by the sub-clause will form part of the regular Progress Reports required by sub-clause 4.21. As discussed above in relation to sub-cl.4.21, the submission of this Progress Report appears to be a condition of payment

         The Contractor should ensure that he takes the appropriate steps to establish the book-keeping necessary to comply with the requirement of this sub-clause at an early stage in the contract. In addition the data required here may form the basis of information necessary to evaluate any claims and variations made pursuant to sub-cll 13 and 20.1. Further, as pointed out by the FIDIC Guide, sub-clause 6.10 may well become an "other sub-clause which may apply to a claim" mentioned in the last paragraph of sub-cl.20.1.

         It is also important that the records are contemporary. In *Attorney General for the Falklands Island v Gordon Forbes Construction(Falklands) Limited*,[8] Forbes and the Government entered into a FIDIC 4th edn contract. A dispute arose that was referred to arbitration. Clause 53.4 of the FIDIC conditions required claims to be verified by contemporary records. Forbes wanted to introduce witness statements to cover those parts of the

---

[8] (2003) 6 BLR 280.

claim where no contemporary records existed. The Arbitrator refused an application by the Attorney General inviting him to rule on the meaning of "contemporary records", and also the extent to which statements could cover the absence of such records.

The issue for determination was whether, on the true meaning of cl.53, witness statements could be introduced into evidence to supplement contemporary records. Acting Judge Sanders held that "contemporary records" meant original or primary documents or copies produced or prepared on or about the time giving rise to the claim. These documents could be produced by the parties. However, contemporary records did not mean witness statements that were produced long after the event. Thus, where there were no such contemporary records in support of a claim, that claim must fail. Witness statements could only be used to identify or clarify contemporary records but not substitute them.

6–056

Clause 53 therefore required the contractor to keep contemporary records in order to support his claim. Any failure to keep those contemporary records may mean that the contractor is unable to support his claim and that the claim will fail. It is submitted that the same considerations will apply here to claims in accordance with the provisions of the contract.

MDB HARMONISED EDITION

There is no change.

6–057

6.11 DISORDERLY CONDUCT

The Contractor shall at all times take all reasonable precautions to prevent any unlawful, riotous or disorderly conduct by or amongst the Contractor's Personnel, and to preserve peace and protection of persons and property on and near the Site.

6–058

OVERVIEW OF KEY FEATURES

- The Contractor shall at all times take reasonable precautions to ensure there is no unlawful, riotous or disorderly conduct by his personnel on or near the site.
- The failure by the Contractor to ensure the good behaviour of his personnel could lead to a claim in relation to sub-cll 4.14, 17.1 and 18.3.

6–059

COMMENTARY

The Contractor is already by virtue of clause 4 responsible for the general conduct of his staff. This sub-clause attempts to go one step further and make the Contractor responsible for any unlawful, riotous or disorderly conduct that might occur. It is difficult to say whether the sub-clause achieves this. The fact that the Contractor shall take "reasonable precautions" means that, provided the Contractor can demonstrate that he has taken those reasonable precautions, the Contractor will escape any liability should any riots or other such disturbances break out.

It is submitted that what are "reasonable precautions" will depend on the circumstances of the individual case as the likelihood or unlawful, riotous and disorderly conduct is higher is some countries than in others (compare the likelihood of unlawful or riotous behaviour in say Afghanistan as opposed to say Canada). So, when allowing for and planning reasonable precautions, the Contract must take this into account.

MDB HARMONISED EDITION

6–060   There is no change.

6.12–6.22 MDB HARMONISED EDITON – THE PARTICULAR LOCALITY CLAUSES

6–061   As set out above,[9] the Guidance to the Standard Contract included a number of sub-clauses that could be added to clause 6 primarily to take account of particular local circumstances. These were however optional. Under the MDB Harmonised Edition the following 10 sub-clauses are no longer included as optional sub-clauses at the end of the Contract. As can be seen, the provisions set out below are largely self-explanatory. It is also unlikely that they add any significant obligation to the Contractor as they will be reflected in the requirements of local labour and health and safety legislation – the Contractor being required by the provisions of sub-cl.1.13 to comply with the applicable laws.

---

[9] See discussion on sub-cl. 6.1.

6.12 FOREIGN PERSONNEL

The Contractor shall be responsible for the return of these personnel to the place   **6–062**
where they were recruited or to their domicile. In the event of the death in the
Country of any of these personnel or members of their families, the Contractor
shall similarly be responsible for making the appropriate arrangements for their
return or burial.

6.13 SUPPLY OF FOODSTUFFS

The Contractor shall arrange for the provision of a sufficient supply of suitable   **6–063**
food as may be stated in the Specification at reasonable prices for the
Contractor's Personnel for the purposes of or in connection with the Contract.

6.14 SUPPLY OF WATER

The Contractor shall, having regard to local conditions, provide on the Site an   **6–064**
adequate supply of drinking and other water for the use of the Contractor's
Personnel.

6.15 MEASURES AGAINST INSECT AND PEST NUISANCE

The Contractor shall at all times take the necessary precautions to protect the   **6–065**
Contractor's Personnel employed on the Site from insect and pest nuisance, and to
reduce their danger to health. The Contractor shall comply with all the regulations
of the local health authorities, including use of appropriate insecticide.

6.16 ALCOHOLIC LIQUOR OR DRUGS

The Contractor shall not, otherwise than in accordance with the Laws of the   **6–066**
Country, import, sell, give, barter or otherwise dispose of any alcoholic liquor
or drugs, or permit or allow importation, sale, gift, barter or disposal thereto by
Contractor's Personnel.

6.17 ARMS AND AMMUNITION

6–067 *The Contractor shall not give, barter, or otherwise dispose of, to any person, any arms or ammunition of any kind, or allow Contractor's Personnel to do so.*

6.18 FESTIVALS AND RELIGIOUS CUSTOMS

6–068 *The Contractor shall respect the Country's recognized festivals, days of rest and religious or other customs.*

6.19 FUNERAL ARRANGEMENTS

6–069 *The Contractor shall be responsible, to the extent required by local regulations, for making any funeral arrangements for any of his local employees who may die while engaged upon the Works.*

6.20 PROHIBITION OF HARMFUL CHILD LABOUR

6–070 *The contractor shall not employ "forced or compulsory labour" in any form. "Forced or compulsory labour" consists of all work or service, not voluntarily performed, that is extracted from an individual under threat of force or penalty.*

6.21 PROHIBITION OF HARMFUL CHILD LABOUR

6–071 *The Contractor shall not employ any child to perform any work that is economically exploitative, or is likely to be hazardous to, or to interfere with, the child's education, or to be harmful to the child's health or physical, mental, spiritual, moral, or social development.*

6.22 EMPLOYMENT RECORDS OF WORKERS

6–072 *The Contractor shall keep complete and accurate records of the employment of labour at the Site. The records shall include the names, ages, genders, hours worked and wages paid to all workers. These records shall be summarized on a* *monthly basis and shall be available for inspection by the Engineer during normal working hours. These records shall be included in the details to be submitted by the Contractor under Sub-Clause 6.10 [Records of Contractor's Personnel and Equipment].*

CLAUSE 7 – PLANT, MATERIALS AND WORKMANSHIP

7.1 MANNER OF EXECUTION

The Contractor shall carry out the manufacture of Plant, the production and manufacture of Materials, and all other execution of the Works: **7–001**

(a) in the manner (if any) specified in the Contract,
(b) in a proper workmanlike and careful manner, in accordance with recognised good practice, and
(c) with properly equipped facilities and non-hazardous Materials, except as otherwise specified in the Contract.

OVERVIEW OF KEY FEATURES

- The Contractor shall carry out his work in the manner specified in the **7–002**
  Contract.
- The Contractor shall carry out his work in a workmanlike and careful manner.
- The Contractor shall carry out his work in accordance with recognised good practice.
- The Contractor must ensure his facilities are properly equipped.
- The Contractor shall carry out his work using non-hazardous materials, unless otherwise specified in the Contract.

COMMENTARY

Materials are defined under sub-cl.1.1.5.3 as "things of all kinds (other than **7–003**
Plant) intended to form or forming part of the Permanent Works, including the supply-only materials (if any) to be supplied by the Contractor under the Contract".

Plant is defined under sub-cl.1.1.5.5 as "the apparatus, machinery and vehicles intended to form or forming part of the Permanent Works".

Sub-clause 7.1 deals with the way in which the Contractor carries out his work. It should be read in conjunction with other clauses such as sub-cl.7.8.2 with regard to the disposal of surplus material.

7–004   Sub-clause 7.1(b) imposes upon the Contractor the obligation to demonstrate good workmanship ("proper workmanlike and careful manner") in executing the Works. The standards imposed by this clause are general in nature and their concrete meaning might vary from country to country. It is probable, therefore, that an Employer will further define the standard of skills required from the Contractor within the Specifications.

In England, the extent of a contractor's duty to demonstrate proper workmanship will depend on all the circumstances of a particular case, including the degree of skill advertised by the contractor to the employer. This duty may also be found to include certain obligations with regard to the design of the works. In cases where the design clearly appears to be flawed, the contractor will be under an obligation to warn the employer of his findings.[1]

Indeed in recent years, performance specifications have been developed for many construction operations. Rather than specifying the required construction process, these specifications refer to the required performance or quality of the finished facility. The exact method by which this performance is obtained is left to the construction contractor. Here, sub-cl.4.1 provides that the Contractor must complete the Works in accordance with the Contract and that the Contractor shall, when required by the Employer, submit details of the methods which it proposes to adopt. The Contractor is not to make any significant alternations to this without notifying the Engineer.

7–005   This is important. Clause 8(3) of the ICE form of Contract provides that the Contractor shall take full responsibility for the methods of construction. However the English Courts[2] have held that, notwithstanding this apparently clear provision, if a method statement was incorporated into the Contract, then the Contractor would be obliged to work in accordance with that statement. Thus entitling the Contractor to payment for works deemed to be over and above that required by the statement.

No definition is provided of either "properly equipped facilities" or "non-hazardous". Whilst the meaning of "properly equipped" is something which will become clear if the Contractor is in breach of the obligation (i.e. falling behind because he did not have sufficient plant), the meaning of "non hazardous" is less clear-cut.

A typical definition of a hazardous material would be:

---

[1] See discussion in relation to sub-cl.4.1 and *Plant v Adams.*
[2] *Yorkshire Waste Authority v Sir Alfred McAlpine & Son (Northern) Ltd* 32 BLR 114 and *Holland Dredging (UK) Ltd v Dredging and Construction Courts Limited* 37 BLR 1. Indeed following the *Yorkshire Water Authority* case, parties to the contract might want to confirm expressly that any pre-tender programme or other documents do not form part of the contract itself.

"Any material or artificial substance (with a solid, liquid, gas, ion, vapour, electro magnetic or radiation) which whether alone or in combination with others is capable of causing harm or having a significant deleterious effect on human health or the Environment."

It is usually the case that a contract will forbid the use of prohibited or dele-   **7–006**
terious materials or as here require that materials be "non-hazardous". This is clearly the intention here. However in the interests of clarity both parties might want to include a definition within the contract. Traditionally such a definition would have been provided by use of a list. However the more recent approach is to adopt a much more general approach.[3] For example:

"The Contractor shall ensure that the Works shall not, when completed, incorporate any goods, materials, substances or equipment which contravene any relevant British and/or European Union Standards, Codes of Practice or good building practice or techniques or which are stated in the Employer's Requirements to be prohibited or which are generally known at the time of use to be deleterious to health and safety or the durability of the completed works or any part thereof in the particular circumstances in which they are to be used having regard to the guidance note "Good Practice in Selection of Construction Materials" dated May 16, 1997 sponsored by the British Property Federation and British Council for Offices or any new edition, revision or amendment of the same, current at the date the works."[4]

The FIDIC Guide notes that the Engineer is not empowered to relax the requirements of this sub-clause and that the Contractor will be in breach of this sub-clause even if the Engineer has consented to the use of Materials which are subsequently found to be hazardous. Liability is strict and the Contractor, following sub-cl.3.1(c), will have to replace the material in question.

MDB HARMONISED EDITION

There is no change.   **7–007**

---

[3] Although, of course, care must be taken to ensure the definition is project specific. A chemical plant might have different requirements to standard civil engineering works.
[4] Obviously the guidance may vary from jurisdiction to jurisdiction.

7.2 SAMPLES

7–008   *The Contractor shall submit the following samples of Materials, and relevant information, to the Engineer for consent prior to using the Materials in or for the works:*

    (a)   manufacturer's standard samples of Materials and samples specified in the Contract, all at the Contractor's cost, and

    (b)   additional samples instructed by the Engineer as a Variation.

OVERVIEW OF KEY FEATURES

7–009   • The Contractor must submit:

    (i)   manufacturer's standard samples of Materials; and

    (ii)   samples specified in the Contract, to the Engineer, for approval, prior to using the Materials.

    • The Contractor will supply these samples at his own cost.

    • The Contractor must supply any other additional sample, which may be requested by the Engineer.

    • The Contractor will supply these other samples at the Employer's cost.

COMMENTARY

7–010   Clause 7.2 introduces a system of "quasi-total quality control" with regard to the Materials. In this system, no defective items are allowed anywhere in the construction process.

As the Contractor is responsible for the cost of testing both the manufacturer's standard samples of Materials and the samples specified in the Contract, the costs of these tests must be included in the contract sum.

By sub-cl.1.3, the Engineer cannot unreasonably withhold his consent to the use of Materials. Any decision by the Engineer to reject certain materials will thus need to be supported by scientific evidence (e.g laboratory test reports).

MDB HARMONISED EDITION

7–011   There is no change.

7.3 INSPECTION

The Employer's Personnel shall at all reasonable times: **7–012**

(a) have full access to all parts of the Site and to all places from which natural Materials are being obtained, and

(b) during production, manufacture and construction (at the Site and elsewhere), be entitled to examine, inspect, measure and test the materials and workmanship, and to check the progress of manufacture of Plant and production and manufacture of Materials.

The Contractor shall give the Employer's Personnel full opportunity to carry out these activities, including providing access, facilities, permissions and safety equipment. No such activity shall relieve the Contractor from any obligation or responsibility.

The Contractor shall give notice to the Engineer whenever any work is ready and before it is covered up, put out of sight, or packaged for storage or transport. The Engineer shall then either carry out the examination, inspection, measurement or testing without unreasonable delay, or promptly give notice to the Contractor that the Engineer does not require to do so. If the Contractor fails to give the notice, he shall, if and when required by the Engineer, uncover the work and thereafter reinstate and make good, all at the Contractor's cost.

OVERVIEW OF KEY FEATURES

- The Employer is entitled to inspect the entire Site and anywhere where   **7–013** work is being carried out, even off-site.
- The purpose of the inspection is to examine materials and workmanship and check progress of the manufacture of Plant or Materials.
- The Contractor cannot reasonably refuse a request to inspect and must give the Employer's Personnel including the Engineer full opportunity to carry out the inspection.
- The fact that an inspection has been carried out will not relieve the Contractor of any liability.
- The Contractor shall notify the Engineer when work is ready in order that the Engineer, if he chooses, may carry out an inspection.
- If the Contractor fails to give notice, he must uncover the work to enable an inspection to take place, and then reinstate and make it good at its own cost.

COMMENTARY

7–014 This is a wide-ranging sub-clause giving the Employer's Personnel the right to inspect (upon reasonable notice) not only any work achieved on site but also Materials or Plant produced or manufactured off-site before these are actually used by the Contractor and incorporated in the works. It should be remembered that by sub-cl.1.1.2.7, Employer's Personnel has a wide-ranging definition including the Engineer and those employed by the Engineer.

In addition, the Contractor is under a strict obligation to give notice when any work is completed and ready for inspection prior to it being covered up or put out of sight. No definition is provided of "any work" and there is scope for disagreement and potential disruption at site if the Engineer wants to inspect work which may already have been covered up.

The purpose of this clause is to prevent defects before they actually occur. It reflects the current trend of total quality control management adopted by the construction industry.

7–015 In practical terms, it will be sensible for the parties to agree on a specific notice procedure for informing the Engineer whenever any work is ready for inspection. This, for example, can be done by means of weekly or monthly progress reports and/or any regular site meetings which may take place.

MDB HARMONISED EDITION

7–016 There is no change.

7.4 TESTING

7–017 *This Sub-Clause shall apply to all tests specified in the Contract, other than the Tests after Completion (if any).*

The Contractor shall provide all apparatus, assistance, documents and other information, electricity, equipment, fuel, consumables, instruments, labour, materials, and suitably qualified and experienced staff, as are necessary to carry out the specified tests efficiently. The Contractor shall agree, with the Engineer, the time and place for the specified testing of any Plant, Materials and other parts of the Works.

The Engineer may, under Clause 13 [Variations and Adjustments], vary the location or details of specified tests, or instruct the Contractor to carry out additional tests. If these varied or additional tests show that the tested Plant, materials or workmanship is not in accordance with the Contract, the cost of carrying out this Variation shall be borne by the Contractor, notwithstanding other provisions of the Contract.

The Engineer shall give the Contractor not less than 24-hours' notice of the Engineer's intention to attend the tests. If the Engineer does not attend at the time and place agreed, the Contractor may proceed with the tests, unless otherwise instructed by the Engineer, and the tests shall then be deemed to have been made in the Engineer's presence.   **7–018**

If the Contractor suffers delay and/or incurs Cost from complying with these instructions or as a result of a delay for which the Employer is responsible, the Contractor shall give notice to the engineer and shall be entitled subject to Sub-Clause 3.5 [Determinations] to agree or determine these matters.

The Contractor shall promptly forward to the Engineer duly certified reports of the tests. When the specified tests have been passed, the Engineer shall endorse the Contractor's test certificate, or issue a certificate to him, to that effect. If the Engineer has not attended the tests, he shall be deemed to have accepted the readings as accurate.

OVERVIEW OF KEY FEATURES

- The Contractor and Engineer will agree the time and place for any specified tests.   **7–019**
- The Contractor will provide everything necessary to carry out the tests at his own cost, including qualified personnel.
- If, pursuant to cl.13, the Engineer varies a specified test or orders an additional test, the cost of this test will be borne by the Employer, unless the varied or additional test shows that work is not in accordance with the Contract, in which case the Contactor will bear the cost of this varied or additional test.
- The Engineer must give not less than 24 hours' notice of his intention to attend any test.
- If the Engineer does not attend an agreed test, the test can nevertheless proceed and will be deemed to have been carried out in the Engineer's presence.
- If the Contractor suffers delay because of the tests, he must give notice to the Engineer who will then determine whether or not to grant time and/or money to the Contractor in accordance with Cl.3.5 [Determinations].
- The Contractor must forward certified test reports to the Engineer.

COMMENTARY

7–020 This sub-clause sets out how specified tests including Tests On completion and tests instructed pursuant to cl.13 should be carried out. It does not deal with any tests that may be carried out after completion. This is covered by cl.9.

The way this clause is intended to work is that the Contractor will give notice under cl.7.3 that an item of work is ready to be tested. The time and date of the test will be agreed and the Engineer will give not less than 24 hours' notice of his intention to attend. In practice, it is likely that the parties will know well in advance when an item is due to be completed and so tests will often be arranged ahead of completion.

Sub-clause 7.5 deals with items which fail the tests and are rejected by the Engineer.

7–021 The costs of a test instructed pursuant to cl.13 will normally be borne by the Employer, unless this test is unsuccessful. In reality, however, most tests will be borne by the Contractor as they will fall within the wide ambit of sub-cl.7.3 or otherwise will be specifically required elsewhere in the Contract. It is therefore of the utmost importance that the Contractor take the costs of all possible tests into account when pricing the works in the tender.

MDB HARMONISED EDITION

7–022 The words "*Except as otherwise specified in the contract*" have been added to the beginning of the second paragraph.

7.5 REJECTION

7–023 *If, as a result of an examination, inspection, measurement or testing, any Plant Materials or workmanship is found to be defective or otherwise not in accordance with the Contract, the Engineer may reject the Plant, Materials or workmanship by giving notice to the Contractor, with reasons. The Contractor shall then promptly make good the defect and ensure that the rejected item complies with the Contract.*

If the Engineer requires this Plant, Materials or workmanship to be retested, the tests shall be repeated under the same terms and conditions. If the rejection and retesting cause the Employer to incur additional costs, the Contractor shall subject to Sub-Clause 2.5 [Employer's Claims] pay these costs to the Employer.

OVERVIEW OF KEY FEATURES

- If an item is defective or not in accordance with the Contract it may be rejected by notice from the Engineer.   **7–024**
- The Engineer must give reasons for the rejection.
- The Contractor must remedy the rejected item so that it complies with the Contract.
- The Engineer may require a retest.
- If any rejection or retesting causes the Employer to incur additional costs, these will be borne by the Contractor.

COMMENTARY

If any item of plant or workmanship fails or is found to be defective, the   **7–025**
Contractor will bear the cost associated with this failure. This is because the
risk of such items failing will have been included in the Contract Sum and by
virtue of sub-cll 4.1 and 7.1 the Contractor must carry out his work in accordance with the Contract.

The Contractor will also logically be responsible for the cost of retesting a
defective item insofar as the Contractor is responsible for the failure.

Although there is apparently nothing controversial in this clause, the
Engineer must give reasons for any rejection and it is possible that these may
be challenged. In addition, the Contractor might dispute whether the
Employer has any entitlements to additional costs. If a Contractor successfully challenges a rejection, then the question must be asked as to who is
responsible for the costs of any re-testing that may have taken place.

Further, the Contractor should remember that in accordance with sub-   **7–026**
cl.15.2(c)(ii), the Employer is entitled to terminate the Contract if the
Contractor fails, without reasonable excuse, to comply with any notice issued
by the Engineer under this clause.

MDB HARMONISED EDITION

There is no change.   **7–027**

7.6 REMEDIAL WORK

7–028 *Notwithstanding any previous test or certification, the Engineer may instruct the Contractor to:*

(a) remove from the Site and replace any Plant or materials which is not in accordance with the Contract,
(b) remove and re-execute any other work which is not in accordance with the Contract, and
(c) execute any work which is urgently required for the safety of the works, whether because of an accident, unforeseeable event or otherwise.

The Contractor shall comply with the instruction within a reasonable time, which shall be the time (if any) specified in the instruction, or immediately if urgency is specified under sub-paragraph (c).

If the Contractor fails to comply with the instruction, the Employer shall be entitled to employ and pay other persons to carry out the work. Except to the extent that the Contractor would have been entitled to payment for the work, the Contractor shall subject to Sub-Clause 2.5 [Employer's Claims] pay to the Employer all costs arising from this failure.

OVERVIEW OF KEY FEATURES

7–029 • The Contractor must, within a reasonable time, comply with any Engineer's instruction to:

   (i)   remove and replace any Plant or materials which do not conform with the Contract; or
   (ii) remove and re-execute any works which do not conform with the Contract.

   • The Contractor must comply immediately with any Engineer's instruction to execute works, which are urgently required for the safety of the works.
   • If the Contractor fails to comply with the Engineer's instruction, the Employer will be entitled to employ other persons to carry out the instruction.
   • In the event the Employer employs other persons to carry out the instruction, then, unless the works instructed by the Engineer would have entitled the Contractor to payment under the Contract, the Contractor will pay the cost of these works.

COMMENTARY

This sub-clause, uncontroversially, commences by confirming that the 7–030 Engineer can instruct the Contractor to remove or re-execute non-compliant or dangerous plant or materials. However, the sub-clause continues to give the Engineer the option of instructing others, at the Contractor's cost, to remedy the work in question if the Contractor fails within a reasonable time to carry out the remedial work.

In addition, as with sub-cl.7.5, the Contractor should remember that in accordance with sub-cl.15.2(c)(ii), the Employer is entitled to terminate the Contract if the Contractor fails, without reasonable excuse, to comply with any notice issued by the Engineer under this clause.

These options available to the Engineer pursuant to cl.7.6 are radical ones and should only be used in situations where it would be unreasonable to repair the defective material or works.

Under most, if not all, applicable laws, if the Contractor is able to estab- 7–031 lish that an Engineer's instruction under cl.7.6 was unreasonable as, for example, repairs could have successfully been undertaken, then the Employer will be responsible for the costs arising from the replacement less the cost of what a reasonable repair would have cost.

The exact meaning of what constitutes a reasonable time for the Contractor to comply with the Engineer's instruction will depend on the particular circumstances of the works and on the interpretation under the applicable law.

MDB HARMONISED EDITION

There is no change. 7–032

7.7 OWNERSHIP OF PLANT AND MATERIALS

Each item of Plant and Materials shall, to the extent consistent with the Laws 7–033 *of the Country, become the property of the Employer at whichever is the earlier of the following times, free from liens and other encumbrances:*

(a) when it is delivered to the Site;
(b) when the Contractor is entitled to payment of the value of the Plant and Materials under Sub-Clause 8.10 [Payment for Plant and Materials in Event of Suspension].

OVERVIEW OF KEY FEATURES

7-034 • Provided the Laws of the Country do not provide otherwise, items of Plant and Materials will become the property of the Employer either:

(i) when they are delivered to the site; or,

(ii) when the Contractor is entitled to payment for them, whichever takes place first.

COMMENTARY

7-035 The legal possession as to the ownership of Plant and Materials on the cause of insolvency will vary form jurisdiction to jurisdiction.

A property-vesting clause

7-036 The primary goal of this property-vesting sub-clause is to provide some form of security to the Employer for the sums it has had to disburse in advance of completion. It also enables the immediate replacement of the Contractor by another to complete the works, if this ever becomes necessary.

It is a property-vesting clause, and as such it can be distinguished from intermediate protective measures such as cl.65.2 of the 7th edn of the ICE Standard Form of Contract. Clause 65.2 does not grant any proprietary rights to the Employer but simply provides it with the option to:

"... at any time sell any of the ... Contractor's Equipment temporary works goods and materials on any part of the Site and apply the proceeds of the sale in or towards the satisfaction of any sums due or which may become due to him from the Contractor under the Contract."

In England & Wales, in the case of *Smith v Bridgend County Borough Council*,[5] the House of Lords held that this arrangement amounted to the placing of a floating charge over the contractor's assets. In the absence of registration in accordance with Part XII of the Companies Act 1985, such floating charge will be void against the contractor's administrator. In contrast, under sub-clause 7.7, the Employer is granted full ownership of the items of plant and materials. No registration is therefore necessary.

---

[5] [2002] 1 A.C. 336 HL; [2002] 1 All E.R. 292.

Subject to the provisions of the law of the site

Sub-clause 7.7 grants property-vesting rights to the Employer, "to the extent   **7–037**
consistent with the "Laws of the country". The Contractor by virtue of sub-
cl.2.1 can ask the Employer for copies of relevant laws which may effect the
Contract and by sub-cl.4.10 will be deemed to have taken into account the
effect of any such laws when compiling his tender.

This means that, regardless of the law governing the Contract, if the legislation of the country where the site is located, grants security rights to the
Contractor or its Sub-contractors over the items of Plant and material, these
provisions will take precedence over the rights granted to the Employer under
this sub-clause. There is thus no potential for conflict between local legislation and sub-cl.7.7 as the former will always prevail over the latter. This issue
is highly relevant, as several jurisdictions possess a type of contractor's lien
legislation, which provides statutory security in the form of a lien for
payment of money owing to contractors.[6]

In the situation where a subcontractor relies on such statutory provisions
to regain control of the materials he supplied to the site, the Contractor will
not be in breach of sub-cl.7.7 in respect of the Employer, as sub-cl.7.7
specifically grants priority to "the Laws of the country".

MDB HARMONISED EDITION

The words "*Except otherwise provided in the contract*" have been added to the   **7–038**
beginning of this sub-clause.

In addition there have been two changes to sub-cll (a) and (b). These now
read:

(a)   when it is <u>*incorporated in the Works*</u>;
(b)   when the Contractor is <u>*paid the corresponding value*</u> of the *Plant and
        Materials under Sub-Clause 8.10 [Payment for Plant and Materials in
        Event of Suspension]*.

These changes can be considered to favour the Contractor, as previously the
Plant became the property of the Employer when it was delivered to site or
when the Contractor was entitled to be paid. Now the Plant has to be incorporated in the Works and/or the Contractor actually has to be paid for that
Plant before ownership changes hands.

---

[6] See, for example, British Columbia Builders Lien Act 1997 – Florida Statute 713.13.

7.8 ROYALTIES

7–039   *Unless otherwise stated in the Specification, the Contractor shall pay all royalties, rents and other payments for:*

    (a)   natural Materials obtained from outside the Site, and
    (b)   the disposal of material from demolitions and excavations and of other surplus material (whether natural or man-made), except to the extent that disposal areas within the Site are specified in the Contract.

OVERVIEW OF KEY FEATURES

7–040   • The Contractor shall pay for natural Materials obtained from outside the Site.
    • The Contractor shall pay for the removal of surplus material from the site, except where the Contract provides for disposal areas within the site.

COMMENTARY

7–041   This straightforward clause confirms that the Contractor must allow for the payment of any Royalties for any natural materials together with the disposal costs of debris and surplus materials in compiling his tender.

    In addition, if the Contract provides for disposal areas, then the Contractor is also obliged to dispose of all surplus material, be it demolition debris or unearthed soil, into the designated disposal area(s).

MDB HARMONISED EDITION

7–042   There is no change.

7.9 ORIGINS OF GOODS

7–043   *All Goods shall have their origin in eligible source countries as defined in: [insert name of published guidelines for procurement]*

    Goods shall be transported by carriers from these eligible source countries, unless exempted by the Employer in writing on the basis of potential excessive costs or delays. Surety, insurance and banking services shall be provided by insurers and bankers from the eligible source countries.

OVERVIEW OF KEY FEATURES

This is a Particular Condition only. Therefore it will only appear in the **7–044** Contract if the Parties chose to incorporate it.

This is the type of sub-clause that might be required where the institution funding the project, or part of it, has particular rules (or is governed by particular rules) which impose a restriction on the use of its funds. [7]

---

[7] See also comments about sub-cl.4.1 above.

CLAUSE 8 – COMMENCEMENT, DELAYS AND SUSPENSION

8.1 COMMENCEMENT OF WORKS

The Engineer shall give the Contractor not less than 7 days' notice of the **8–001**
Commencement Date. Unless otherwise stated in the Particular Conditions, the
Commencement Date shall be within 42 days after the Contractor receives
the Letter of Acceptance.

The Contractor shall commence the execution of the works as soon as is
reasonably practicable after the Commencement Date, and shall then proceed
with the Works with due expedition and without delay.

OVERVIEW OF KEY FEATURES

- The Engineer shall give the Contractor not less than seven days' notice of **8–002**
  the Commencement Date.
- That Commencement Date, unless otherwise stated, shall be within 42
  days after the Contractor receives the Letter of Acceptance.
- The Contractor shall commence his works as soon as reasonably
  practicable after the Commencement Date.
- The Contractor shall proceed with his works with due expedition and
  without delay.

COMMENTARY

This sub-clause deals with the date for commencement of the Contractor's **8–003**
work.

The Contract does not actually set out a date as to when the Contractor
may commence work. Instead a seven-day notice period is provided for.
Although the notice period of the Commencement Date provided for by
sub-cl.8.1 may seem tight, the Contractor knows that the Commencement
Date will be within a 42-day window from the date he receives the Letter of
Acceptance. In addition, the Contractor does not have to physically start
work on that date. He is required to start as soon as is "reasonably
practical".

The Letter of Acceptance is defined by sub-cl.1.1.1.3 as being a formal letter signed by the Employer signifying acceptance of the Contractor's Letter of Tender.

8–004   If the formalities required by sub-cl.1.1.1.3 are not observed, the parties are thrown back on the Contract Agreement itself.[1] The Contract procedure provided for by the FIDIC terms and conditions does not envisage there being a letter of intent and both parties to the Contract should understand the potential pitfalls if a letter of tender is issued which could be interpreted as a letter of intent.

Those potential pitfalls arise because the Commencement Date is the precursor of a number of events. The two most important of these are the Time for Completion and the time for access to the site:

(i)   The Time for Completion, set out in more detail at sub-cl.8.2, is calculated from the Commencement Date. By the Appendix to Tender this period should be calculated in days.

(ii)   The time for access to the site, set out in more detail in sub-cl.2.1, is also calculated in the number of days from the Commencement Date.

One potential problem is although the Time for Completion might have started to run on the Commencement Date, the date the Contractor is due to obtain access might be a significant number of days after the Commencement Date. Whilst, lead-in times can be used profitably, the Contractor should check that there is no discrepancy with his intended programme, as he might have difficulties in arguing that he is entitled to an extension of time in accordance with sub-cl.8.4(e).

8–005   By sub-cl.1.1.3.9, "day" means calendar and not working day. This therefore includes weekends. With both these dates, it is suggested that the parties should try and agree the actual calendar date in order to avoid any potential disagreement over the day on which these two key events are to take place. It is critical that the commencement and end dates are clearly identified and established to avoid uncertainty.

In addition, by sub-cl.4.2 the Contractor must, if required by the Contact, provide the Performance Security within 28 days of receiving the Letter of Acceptance. Sub-clause 4.3 requires that the Contractor must have submitted the name of the Contractor's Representative for approval prior to the Commencement Date. Whilst by sub-cl.8.3, the Contractor shall submit a detailed time programme to the Engineer within 28 days after receiving the sub-cl.8.1 notice.

The Contract does not say what will happen if the Engineer is not able to give at least seven days' notice of the Commencement Date such that it falls

---

[1] Sub-cl.1.1.1.3 also provides that if there is no such Letter of Acceptance, the expression "Letter of Acceptance" means the Contract Agreement and the date of issuing or receiving the Letter of Acceptance means the date of signing the Contract Agreement.

within the 42-day period defined by sub-clause 8.1 or if the Employer is not able to give access to the Contractor within the defined period. Obviously it would be open for the Contractor and Employer to agree the changed periods. However, given the mandatory words of this sub-clause, it is submitted that the giving of a late notice would be a breach of contract such that the Contractor might well be entitled to consider the Contract to be at an end. Sub-clause 16.2(d) provides that if the Employer substantially fails to perform his obligations under the Contract then the Contractor may, having given 14 days notice, terminate the Contract. The failure to give notice of the Commencement Date is clearly a substantial failure.

As noted above, the final paragraph of sub-cl.8.1 requires the Contractor **8–006** to commence execution of the Works "as soon as reasonably practicable" and to carry out those works with "due expedition and without delay". These phrases are not defined in the Contract, but the over-riding obligation on the Contractor is to complete his Works within the Time for Completion of the Works. Therefore strictly it might be considered that these terms are not required. Where a Contract includes an express obligation for a contractor to complete the works by a specified date then, in English jurisdictions, a term will not be implied that the Contractor is to proceed regularly and diligently or with due expedition with those works. Without these words, all the Contractor has to do is plan the works as he sees fit, provided that he completes the Works as required by the Contract.[2]

However the fact that these express terms have been included within sub-cl.8.1 means that in theory even if the Contractor does complete in time, if it can be shown that he is in breach of Contract because he has not proceeded with due expedition, then the Employer will:

(i)   potentially be able to terminate the Contract under clause 15; and
(ii)   provided he can establish a loss, have available a potential remedy of damages.[3]

To take one example, in the case of *Hounslow v Twickenham Garden Developments*[4] it was said that even where the Contractor is well ahead with the Works, he was not to be allowed to slow down so that the work is completed on time. Instead the Contractor remained under an obligation to continue to proceed regularly and diligently.

The more usual phrase to be found in construction contracts is the obliga- **8–007** tion on the Contractor to proceed "regularly and diligently". It is suggested

---

[2] *GLC v Cleveland Bridge & Eng Co. Ltd* (1984) 34 BLR 50.
[3] If, for example, the Employer has made arrangements, of which the Contractor is aware, which are dependent on the regular progress of the Works, then the Employer might well suffer a loss if the Contractor's poor performance means that he has to re-organise these arrangements.
[4] [1970] 7 BLR 89.

that the obligation to proceed with "due expedition" is likely to have a similar meaning.

Some guidance on the obligation to proceed regularly and diligently was provided in the case of *West Faulkner Associates v London Borough of Newham*[5] where Justice Brown commented:

"Taken together the obligation upon the contractor is essentially to proceed continuously, industriously and efficiently with appropriate phys-ical resources so as to progress the works steadily towards completion substantially in accordance with the contract requirements as to time, sequence and quality of work."

Justice Brown then conceded that:

"Beyond that I think it impossible to give useful guidance. These are after all plain English words and in reality the failure of which clause 25(1)(b) speaks is, like the elephant, far easier to recognise than to describe."

8–008   If no time period for completion is contained within the Appendix to Tender then under English Law a term will be implied by s.14 of the Supply of Goods and Services Act 1982 that the Contractor's obligation is to complete the Works within a reasonable time.

In addition if no time period for completion is contained then time will be said to be at large and the Employer will not be entitled to deduct Delay Damages as set out in sub-cll 8.7 and 14.15(b). Time will also be at large if the Employer prevents the Contractor from completing the Works by the time required under the Contract. In the House of Lords decision in *Percy Bilton v GLC*,[6] Lord Fraser of Tullybelton said:

1.   The general rule is that the main contractor is bound to complete the work by the date of completion stated in the contract. If he fails to do so, he will be liable for liquidated damages to the employer.
2.   That is subject to the exception that the employer is not entitled to liquidated damages if by his acts or omissions he has prevented the main contractor from completing his work by the completion date.

There then remains the question as to what constitutes a reasonable time. Where no time period is specified and a contractor has been selected through the competitive tender process, then a court would be likely to take an objec-tive view based on how the reasonable contractor in the actual circumstances would have carried out the works. If time is at large following an act of prevention or breach by the Employer, then the original completion date

[5] [1992] 71 BLR 6.
[6] (1982) 20 BLR 1 See also *Peak Construction v McKinney* (1970) 1 BLR 114.

provides good but not conclusive evidence. It is not conclusive because one needs to take account of not only the fact that a contractor would be expected to have planned the works in order to achieve the original completion date but also of the delay and/or disruption caused by the breach or act of prevention.

Thus, in *British Steel Corporation v Cleveland Bridge & Eng Co*[7] Lord Goff said:

8–009

> "I have first to consider what would, in ordinary circumstances, be reasonable time for the performance of the relevant services; and I have then to consider to what extent the time for performance by BSC [the Contractor] was in fact extended by extraordinary circumstances outside their Control."

Alternatively, the Court of Appeal[8] quoted with approval the following definition given by HHJ Seymour Q.C. in a situation where he said that the question as to whether a reasonable time has been exceeded is:

> "a broad consideration, with the benefit of hindsight, and viewed from the time at which one party contends that a reasonable time for performance has been exceeded, of what would, in all the circumstances which are by then known to have happened, have been a reasonable time for performance. That broad consideration is likely to include taking into account any estimate given by the performing party of how long it would take him to perform; whether that estimate has been exceeded and, if so, in what circumstances; whether the party for whose benefit the relevant obligation was to be performed needed to participate in the performance, actively, in the sense of collaborating in what was needed to be done, or passively, in the sense of being in a position to receive performance, or not at all; whether it was necessary for third parties to collaborate with the performing party in order to enable it to perform; and what exactly was the cause, or were the causes of the delay to performance. The list is not intended to be exhaustive."[9]

However under the FIDIC conditions, the proper operation of the extension of time sub-cl.8.4, should ensure that time does not become at large and that the Contractor's obligation remains to complete within a specified time.

[7] [1981] 24 BLR 100.
[8] *Peregrine Systems Limited v Steria Ltd* [2005] EWCA Civ 239.
[9] *Astea (UK) Ltd v Time Group LTD* [2003] EWHC 725.

MDB HARMONISED EDITION

8-010   There has been a significant change. The first paragraph has been deleted. It has been replaced with the following:

> *Except otherwise specified in the Particular Conditions of Contract, the Commencement Date shall be the date at which the following precedent conditions have all been fulfilled and the Engineer's instruction recording the agreement of both Parties on such fulfilment and instructing to commence the Work is received by the Contractor:*
>
> *(a)   signature of the Contract Agreement by both Parties, and if required, the approval of the Contract by relevant authorities of the Country;*
>
> *(b)   delivery to the Contractor of reasonable evidence of the Employer's financial arrangements (under sub-clause 2.4[Employer's Financial Arrangement]);*
>
> *(c)   Except if otherwise specified in the Contract Data, possession of the Site given to the Contractor together with such permission(s) under (a) of Sub-clause 1.13[Compliance with Laws] as required for the commencement of the Works; (d) receipts by the Contractor of the Advance Payment receipts by the Contractor of the Advance Payment under sub-clause 14.2 [Advanced Payment] provided that the corresponding bank guarantee has been delivered by the Contractor. (e) If the Engineer's instruction is not received by the Contractor within 108 days from his receipt of the Letter of Acceptance, the Contractor shall be entitled to terminate the Contract under sub-clause 16.2 [Termination by Contractor].*

These new changes are, on the one-hand, of some benefit to the Contractor. The Project cannot commence unless the Contract Agreement has been signed by both parties, the Contractor has in its possession reasonable proof that the Employer can fund the works and the Contractor has received any advanced payments that it was entitled to.

All these conditions are stated to be conditions precedent. Significantly, the Contractor has the option of terminating the contract in accordance with sub-cl.16 if no instruction is received.

8-011   On the other hand, the actual time of commencement date is probably less clear. Under the old wording, there was a 42 day window. Now there appears to potentially be a 180 day window as there is no Commencement Date until the Engineer's Instruction has been received by the Contractor.

8.2 TIME FOR COMPLETION

The Contractor shall complete the whole of the Works, and each Section (if   **8–012**
any), within the Time for Completion for the Works or Section (as the case
may be) including:

(a)   achieving the passing of the Tests on Completion, and
(b)   completing all work which is stated in the Contract as being required for
the works or Section to be considered to be completed for the purposes of
taking-over under Sub-Clause 10.1 [Taking Over of the works and
Sections].

OVERVIEW OF KEY FEATURES

- The Contractor shall complete the whole of the Works within the time for   **8–013**
completion for the Works or Section.
- The obligation to complete includes the passing of any Tests on
Completion and the completing of all works stated in the Contract as
being necessary for the purpose of taking over under sub-cl.10.1.

COMMENTARY

Sub-clause 8.2 sets out the time within which the Contractor must complete   **8–014**
the Works.

The Time For Completion of the Works will be inserted in the Appendix
to Tender. It is calculated from the Commencement Date as defined in sub-
clause 8.1. The Appendix to Tender provides for the Time for Completion to
be expressed in days (and according to sub-cl.1.8.3.9 that means calendar not
working days). It is suggested that for certainty, the parties should agree a
calendar date.

By sub-cl.1.1.3.3, the "Time for completion" means the time for
completing the Works or a Section of the Works as stated in the Appendix to
Tender taking into account any extension under sub-cl.8.4 calculated from
the Commencement Date.

The Works (or Section of the Works) will not be complete until all the   **8–015**
necessary Tests on Completion (as defined within cl.9) have been successfully
carried out and all the Work required for the issuing of a Taking Over certificate as provided for by cl.10 has been completed.

The reference to the completion of the Works means the completion of all
the Permanent Works and the Temporary Works as set out in the Contract.

Provided these are adequately described, all the parties to the Contract will understand what work needs to be carried out.

However care is required when talking about sectional completion. Section is defined by sub-cl.1.1.5.6 as being a part of the works specified in the Appendix to Tender as a Section (if any). Therefore it is important that an appropriate description of any Section is set out in the Appendix. This description should include a separate Time for Completion of that Section together with details of the Delay Damages which may be allowable for a failure to meet the stated Time for Completion. This topic is described in more detail within sub-cl.8.7.

MDB HARMONISED EDITION

8–016   There is no change.

8.3 PROGRAMME

8–017   *The Contractor shall submit a detailed time programme to the Engineer within 28 days after receiving the notice under Sub-clause 8.1 [Commencement of Works]. The Contractor shall also submit a revised programme whenever the previous programme is inconsistent with actual progress or with the Contractor's obligations. Each programme shall include:*

(a)   the order in which the Contractor intends to carry out the works, including the anticipated timing of each stage of design (if any), contractor's Documents, procurement, manufacture of Plant, delivery to Site, construction, erection and testing,

(b)   each of these stages for work by each nominated Subcontractor (as defined in Clause 5 [Nominated Subcontractors]),

(c)   the sequence and timing of inspections and tests specified in the Contract, and

(d)   a supporting report which includes:

   (i)   a general description of the methods which the Contractor intends to   adopt, and of the major stages, in the execution of the works, and

   (ii)   details showing the Contractor's reasonable estimate of the number of each class of Contractor's Personnel and of each type of Contractor's Equipment, required on the Site for each major stage.

Unless the Engineer, within 21 days after receiving a programme, gives notice to the Contractor stating the extent to which it does not comply with the Contract, and Contractor shall proceed in accordance with the programme, subject to his other obligations under the Contract. The Employer's Personnel shall be entitled to rely upon the programme when planning their activities.

The Contractor shall promptly give notice to the Engineer of specific prob- **8–018**
able future events or circumstances which may adversely affect the work, increase the contract Price or delay the execution of the works. The Engineer may require the Contractor to submit an estimate of the anticipated effect of the future event or circumstances, and/or a proposal under Sub-Clause 13.3 [Variation Procedure].

If, at any time, the engineer gives notice to the Contractor that a programme fails (to the extent stated) to comply with the Contract or to be consistent with actual progress and the Contractor's stated intentions, the Contractor shall submit a revised programme to the Engineer in accordance with this Sub-Clause.

OVERVIEW OF KEY FEATURES

- The Contractor shall submit a detailed time programme to the Engineer **8–019** no later than 28 days after receiving the notice of the commencement of Works under sub-cl.8.1.
- The Contractor must submit revised and updated programmes when the original programme becomes out-of-date and inconsistent with actual progress.
- Whilst the format of the programme is left to the Contractor every programme shall include:

  (a)   the sequence in which the Contractor intends to carry out the work;
  (b)   details of any stage of work to be carried out by a nominated sub-contractor;
  (c)   details of the sequence and timing of any inspections and tests;
  (d)   supporting details of the methods the Contractor intends to use; and
  (e)   an estimate of the number of personnel and equipment required.

- If the Engineer believes that the programme does not comply with the Contract he should give notice to the Contractor within 21 days.
- Unless the Engineer gives such a notice, the Contractor shall proceed in accordance with the programme.
- The Employer's Personnel are entitled to rely upon the programme when planning their activities.

191

- The Contractor shall promptly give notice to the Engineer of any future events or circumstances which may increase the Contract Price or delay the completion of the Works.
- The Engineer may, if the Contractor gives such a notice, require the Contractor to submit an estimate of the anticipated effect of the future event and a proposal to rectify the potential effect under sub-cl.13.3.
- If the Engineer considers that a programme fails to comply with the Contract or is inconsistent with actual progress, he may give due notice to the Contractor.
- If the Engineer gives such a notice, the Contractor must submit a revised programme to the Engineer.

COMMENTARY

8–020 Sub-clause 8.3 provides that the Contractor must submit a programme containing the details set out above, to the Engineer within 28 days after receiving the Notice of the Commencement of the Works.

The basic purpose of the programme is to set out how the Contractor proposes to carry out the Works. The programme must be supported by a report setting out the methods the Contractor intends to adopt together with an estimate of the personnel and equipment required on site for carrying out the major stages of the Works.

Under cl.14 of the Old Red Book FIDIC 4th edn, programmes were submitted for approval to the Engineer. This is no longer the case. However the Engineer is able to give notice to the Contractor if he considers that the programme does not comply with the Contract. Therefore the Engineer should be careful to remember that it will accordingly be open to the Contractor to argue that, by not rejecting a programme, the Engineer has impliedly given his approval to it.

8–021 The Contractor is required to proceed as set out by his programme, unless he receives notice from the Engineer stating that the programme does not either reflect actual progress on site or the requirements of the Contract. Obviously this programme will not only be used to demonstrate progress, but also to demonstrate whether any delay may cause a delay to completion. By sub-cl.4.21, the Contractor, must submit monthly progress reports. These must include a comparison of planned and actual progress.

The Contractor is required to give advance notice or early warning to the Engineer of potential events which might adversely affect or delay the Works. There is no similar obligation on the Engineer or Employer. It is submitted that given the wording of this sub-clause, this requirement has a far wider application than just in relation to the programme. The presumed purpose behind the sub-clause is to enable the Contractor and Engineer to work together to minimise the effects of the potential delay event.

The notice must be given "promptly". Therefore consideration must be given when submitting such a notice to the potential impact if any on the claims procedure required by sub-cl.20. In other words, the giving of a notice under sub-cl.8.3 might not suffice as a notice as required by cl.20.

The Contractor should remember that he must revise the programme **8–022** whenever it is inconsistent with actual progress. This will include when the Contractor is ahead or behind schedule.

Whilst the programme must include the detail set out within sub-cl.8.3, there is no specific requirement as to the form the programme should take. Given the importance of the contract programme as a management tool to all parties, many contracts will no doubt specify exactly what form the programme is to take.

If the Engineer gives notice to the Contractor that a programme does not comply with the Contract or is not consistent with actual progress, then the Contractor is required to submit a revised programme to the Engineer. In addition, by sub-cl.8.6, the Engineer can require the Contractor to submit a revised programme where there is delay caused by reasons other than those set out in sub-cl.8.4.

For certainty, it is suggested that in addition, the Employer might want to **8–023** consider requiring the Contractor to provide a revised programme whenever an extension of time is granted.

The Contractor should remember that others, for example the Employer's Personnel, are entitled to rely upon the programme. For example, more staff might be required when work or testing was meant to be at a peak.

MDB HARMONISED EDITION

There is no change. **8–024**

8.4 EXTENSION OF TIME FOR COMPLETION

The Contractor shall be entitled subject to sub-clause 20.1 [Contractor's **8–025** *Claims] to an extension of time for Completion if and to the extent that completion for the purposes of sub-clause 10.1 [Taking Over of the Works and Sections] is or will be delayed by any of the following causes:*

(a)   a Variation (unless an adjustment to the Time for Completion has been agreed under sub-clause 13.3 [Variation Procedure]) or other substantial change in the quantity of an item of work included in the Contract,

(b)   a cause of delay giving an entitlement to extension of time under a sub-clause of these Conditions,

(c) *exceptionally adverse climatic conditions,*

(d) *unforeseeable shortages in the availability of personnel or Goods caused by epidemic or governmental actions, or*

(e) *any delay, impediment or prevention caused by or attributable to the Employer, the Employer's Personnel, or the Employer's other contractors on the Site.*

If the Contractor considers himself to be entitled to an extension of Time for Completion, the Contractor shall give notice to the Engineer in accordance with sub-clause 20.1 [Contractor's Claims]. When determining each extension of time under sub-clause 20.1, the Engineer shall review previous determinations and may increase, but shall not decrease, the total extension of time.

OVERVIEW OF KEY FEATURES

Conditions precedent to a claim for an extension of time

8–026   If (and only if) completion for the purposes of sub-cl.10.1 is or will be delayed by any one of the causes set out in sub-cl.8.4(a) to sub-cl.8.4(e), then the Contractor must give notice to the Engineer in accordance with sub-cl.20.1.

Assessment of an extension of time

8–027   In accordance with sub-cl.20.1, the Engineer shall review both the current claim and any previous determinations and may increase, but shall not decrease, the total extension of time awarded to the Contractor.

Events giving rise to an entitlement to an extension of time

8–028   • A Variation[10] or other substantial change in the quantity of an item of work included in the Contract under cl.12.
         • A cause of delay giving an entitlement to extension of time under a sub-clause of these Conditions. These are sub-cll 1.9 [Delayed Drawings or Instructions], 2.1[Right of Access to the Site], 4.7 [Setting Out], 4.12 [Unforeseeable Physical Conditions], 4.24 [Fossils], 7.4 [Testing], 10.3 [Interference with Tests on Completion], 13.7 [Adjustments for Changes in Legislation], 16.1 [Contractor's Entitlement to Suspend Work], 17.4 [Consequences of Employer's Risks], and 19.4 [Consequences of Force Majeure].

---

[10] Unless an adjustment to the Time for Completion has been agreed under sub-cl.13.3.

- Exceptionally adverse climatic conditions.
- Unforeseeable shortages in the availability of personnel or Goods caused by epidemic or governmental actions.
- Any act of delay, impediment or prevention caused by or attributable to the Employer, the Employer's Personnel, or the Employer's other contractors on the Site.

COMMENTARY

Overview

This is obviously a key sub-clause of the Contract. If the Contractor fails to complete his Works within the agreed time for completion then he will be in breach of contract. Sub-clause 8.4 provides the mechanism by which the time for completion can be extended, but only in certain clearly defined circumstances and only if the Contractor takes certain steps to give notice of his considered entitlement   **8–029**

A number of changes have been made to the general wording of this sub-clause from those contained in the extension of time provisions to be found at cl.44 of the Old Red Book, FIDIC 4th edn.

The material changes include modifications of the original paras (a), (b) and (e). Sub-paragraph (d) is a new event entitling the Contractor to an extension of time. Otherwise, the intent of this sub-clause remains largely the same as that contained in the extension of time provisions at sub-cl.44.1 of the Old Red Book, FIDIC 4th edn.

Sub-clauses 44.2 and 44.3 of the Old Red Book FIDIC 4th edn, set out how the Contractor was to notify and put forward any claim and detailed how the Engineer should determine that claim. These have now been modified in sub-cl.20.1.   **8–030**

The purpose of extensions of time

It has often been thought that extension of time provisions are solely for the benefit of the contractor. Under English law, this is not only wrong, but in reality the opposite of the true intent. The primary purpose of an extension of time provision is to preserve the contractor's obligation to complete within a specified time. Extending the completion date therefore preserves the Employer's right to liquidated damages known as Delay Damages under the FIDIC contract, even when by prevention, the employer has delayed the contractor and is responsible in part for late completion.[11] If there is no   **8–031**

---

[11] See *Holme v Guppy* (1838) 3 M & W 387 and *Peak Construction (Liverpool) Ltd v McKinney Foundations Ltd* (1970) 1 BLR 114.

completion date, there is no date from which "Delay Damages" can run. Sub-clause 8.4 only entitles the Contractor to an extension of time, not to additional payment.[12] Any extension of time granted will relieve the Contractor of the obligation to pay Delay Damages to the Employer in respect of the extension period.

Conditions precedent to a claim for an extension of time

8–032 There are two conditions precedent that the Contractor must comply with prior to making a submission to the Engineer for an extension of time. In other words, if these conditions are not complied with, then any entitlement to an extension of time is lost.

First, it is mandatory that one of the events listed in sub-cl.8.4(a) to 8.4(e) must have resulted in an actual delay in completion within the meaning of sub-cl.10.1. It is not good enough for the Contractor to say that there is a delay or disruption to the Contractor's programme. Second, the Contractor must give notice to the Engineer in accordance with the provisions and time limits required by sub-cl.20.1.

Quite often, the terms "delay" and "disruption" are used as if they are the same thing and it is not uncommon that disruption is understood in the same way as "delay and disruption". This is not correct. For example, the Society of Construction Law's "Delay and Disruption Protocol"[13] defines "disruption" as:

> "Disturbance, hindrance or interruption of a Contractor's normal work progress, resulting in lower efficiency or lower productivity than would otherwise be achieved. Disruption does not necessarily result in a Delay to Progress or a Delay to Completion."

8–033 The SCL Protocol then addresses the concept of delay by using two separate terms to distinguish them – "Delay to Completion" and "Delay to Progress". The former is a term used to describe delay to the planned date that the Contractor was to complete his works or a delay to the contract completion date. The latter term refers to a delay to the Contractor's progress, but does not affect the contract completion date.

The clear intent of FIDIC conditions is that only a "Delay to Completion" will afford the Contractor a right to an extension of time (subject always to compliance with the provisions of sub-cl.20.1).

---

[12] The grant of an extension of time does lead to an automatic entitlement of loss and/or expense. See for example – *H. Fairweather & Co. v London Borough of Wandsworth* (1987) 39 BLR 106.

[13] The Protocol launched in October 2002 is not intended to be a contract document. Further it does not purport to take precedence over the express terms of a contract or to be a statement of the law. Although expressed as a guidance note, it aspires to become the standard for best practice in dealing with delay and disruption issues.

The steps that the Contractor is required to take when giving notice under sub-cl.20.1 are discussed in detail below. In short there is in place a detailed and regimented procedure to ensure that the Contractor both gives early notice, and provides formal and detailed particulars of all necessary and relevant details to substantiate an extension of time claim. If the Contractor fails to takes these steps, he may lose the right to an extension, notwithstanding that he may have suffered a delaying event under this sub-clause. Once the Contractor has carried out these steps, the Engineer will then follow the procedures contained in sub-cl.3.5 in order to make a "fair determination" of any entitlement to an extension of time.

There is no specific requirement to ensure that the Contractor has made **8–034** reasonable and proper efforts to mitigate any delay and the prudent Employer might want to specifically insert such an obligation rather than try and suggest at a later date that such a term can be implied.

In certain jurisdictions the Employer would not be able to rely on a breach of these condition precedents as some local laws disallow any contractual right to restrict the right to bring a claim before a judge or arbitral tribunal.[14]

Where that is the case, the Employer would not be able to rely on the fact that the Contractor may have failed to comply with the notice requirements of sub-cl.8.3 or any other condition precedents that may be found within the Contract.

In addition, the Employer should take care to guard against waiving the **8–035** right to rely upon any breach of a condition precedent through its prior conduct during an earlier stage of the contract. In the Scottish case of *E&J Glasgow v UCG Estates Limited*[15], the contractor failed to follow the required contractual procedure for claiming an extension of time for a variation. However the Court of Session found that the contractor could still claim additional time as the employer had not insisted that the correct contractual procedures be complied with at the relevant time and so had waived his right to insist upon compliance when the matter came before the courts.

Events giving rise to an entitlement to an extension of time

The events listed in sub-cll 8.4(a) to 8.4(e) require careful consideration. **8–036**

The Contractor should note that the grant of an extension of time does not always lead to an award of Cost[16] and/or reasonable profit and as set out above, there is no automatic right to such an award. This is dependant on the event that gives rise to a claim for an extension of time and the relevant sub-clause concerned. If the Contractor also seeks a claim for additional payment, a claim has to be made under that particular sub-clause for that

---

[14] For example, in Taiwan, pursuant to art.58 of the Civil Procedural Code.
[15] [2005] CSOH 63.
[16] See definition of "Cost" at sub-cl.1.1.4.3 – it does not include any profit element.

additional payment in addition to the claim being made for an extension of time.

Sub-clause 8.4(a) refers to cl.13, which provides that the Engineer is empowered to issue either an instruction or request for the Contractor to submit a proposal at any time prior to issuing the Taking-Over certificate for the Works. However, the Contractor should not make any alteration and/or modification of the Permanent Works unless and until the Engineer instructs or approves a Variation.

8–037   Assuming that there is a substantial change in the quantity of an item of work, then under cl.12, such extra quantities have to be measured pursuant to the measurement procedure at cl.12.

By virtue of sub-cl.4.8(b), the following sub-clauses[17] give rise to a claim for an extension of time:

1.9	Delayed Drawings or Instructions
2.1	Right of Access to the Site
4.7	Setting Out
4.12	Unforeseeable Physical Conditions
4.24	Fossils
7.4	Testing
8.5	Delays caused by Authorities
8.9	Suspension initiated by the Employee
10.3	Interference with Tests on Completion
13.7	Adjustments for Changes in Legislation
16.1	Contractor's Entitlement to Suspend Work
17.4	Consequences of Employer's Risks
19.4	Consequences of Force Majeure

As noted above, the granting of an extension of time does not necessarily entitle the Contractor to cost or cost plus reasonable profit.

8–038   The following sub-clause carries an entitlement to an extension of time only:

8.5   Delays caused by Authorities

The following sub-clauses carry an entitlement to an extension of time plus cost only:

4.12   Unforeseeable Physical Conditions
4.24   Fossils
8.9   Suspension initiated by the Employee
13.7   Adjustments for Changes in Legislation

---

[17] See the separate commentary to each of these sub-clauses.

17.4 Consequences of Employer's Risks
19.4 Consequences of Force Majeure

The following sub-clauses carry an entitlement to an extension of time plus cost and a reasonable profit:

1.9 Delayed Drawings or Instructions
2.1 Right of Access to the Site
4.7 Setting Out
7.4 Testing
10.3 Interference with Tests on Completion
16.1 Contractor's Entitlement to Suspend Work

The entitlement to an extension of time is assessed, reviewed and determined **8–039** in accordance with the sub-cl.8.4.

The Contractor must provide contemporaneous documentation[18] to substantiate a claim such as under sub-cl.4.8(c) for exceptionally adverse climatic conditions. The Contractor is normally required to have records for normal weather and will have made use of these at the tender stage. The Contractor might want to consider putting mechanisms in place at the commencement of works so that changes in conditions are recorded as they occur.

The FIDIC Guide suggests that to establish whether such exceptionally adverse climatic conditions occurred, it would be sensible to compare the frequency with which events of a similar adversity have taken place. This sounds sensible. The Guide then suggests that an exceptional degree of adversity might be one which has a probability of occurrence of four or five times the Time for Completion of the Works.[18a]

This suggestion is not contractually binding but it might stand as a useful **8–040** starting point for consideration. It is also something the Contractor might want to think about when assessing the risks in establishing a tender figure.

Sub-clause 8.4(d) is new. Under this sub-clause, any shortage of personnel or goods would have to be shortage of the nature unforeseen by the experienced contractor.[19] This could cause some difficulties for the Contractor. Would the experienced Contractor make provision for the possibility of epidemic or government interference causing delay? This will, of course, largely depend on the particular circumstances of each project.

No definition is provided of epidemic. It will be a question of fact. SARS, caused mayhem in Asia and had a serious impact on a number of countries.

---

[18] For further detail of what this means see discussion of the *Attorney General for the Falklands Island v Gordon Forbes Construction(Falklands) Limited* case under sub-cl.6.11 above.
[18a] i.e. once every eight to ten years in a two-year project.
[19] For further discussion on "foreseeable" and the "experienced contractor" see sub-cll 1.1.6.8, 4.7 and 4.12 above.

Was it an epidemic? Probably. Avian flu, about which as at the Autumn of 2006 there is serious concern, would quite probably qualify as an epidemic should the concerns of a pandemic prove to be correct. However, the question as to whether it could be deemed foreseeable given the widespread publicity it has received is much more difficult.

8–041   The question of unforeseeable delay caused by government action would again be a question of fact and depend on the specific circumstances of each project. However, the Contractor should bear in mind the provisions of sub-clause 8.5 which provide that the Contractor is entitled to an extension of time for delays caused by Authorities. The provisions of this sub-cl. might make a claim easier to make, although as set out above, a claim for an extension of time under sub-cl.8.5 does not carry an entitlement to Cost.

Sub-clause 8.4(e) provides an overall right to an extension of time for the Contractor because of delay caused by the Employer or the Employer's Personnel provided such act or omission occurs on the Site. However, some of these events are covered by other sub-clauses such as sub-cll 17.3 and 17.4. The reference to the Employer's other contractors must be read in conjunction with sub-cl.4.6, which makes reference to the reimbursement of Unforeseeable Cost, but not delays.

By sub-cl.1.1.2.6, the definition of Employer's Personnel includes the Engineer. Of course, by sub-cl.8.3 it is the Engineer who is responsible for determining the entitlement, if any, of the Contractor to an extension of time. There is therefore the potential for conflict, as the Engineer might be called upon to consider a claim for an extension based on his own default.

8–042   Sub-clause 8.4 does not, unlike cl.44 of the Old Red Book, FIDIC 4th edn provide a general clause referring to "other special circumstances". The Contractor's entitlement to an extension of time is that set out in cl.8.4 and that is that.

Assessment of an extension of time

8–043   The Engineer must make an assessment of the application for an extension of time in accordance with cll 3.5 and 20. This means that the Engineer must respond and must provide detailed comments. Before making a determination, the Engineer must consult with both parties to see if an agreement can be reached. In carrying out his assessment, the Engineer is required to review any previous assessments. However, he is not allowed to decrease the total extension period. This means that the Engineer must not consider applications for an extension in isolation.

Concurrency

8–044   Concurrent delay is a frequently encountered problem with regards to construction contracts. By concurrent delay we mean a situation whereby if the Contractor is in delay, but the Employer also appears to be delaying the

Contractor, then is the Contractor entitled to receive an extension of time in any event? To be a concurrent delay, the event for which the employer is responsible and an event for which the contractor is responsible should commence at precisely the same time. Therefore it is likely that concurrency will only rarely arise. For example, the Employer might be late in giving possession of the site, but at the same time the Contractor might not have been ready to start in any event due to a lack of labour or materials.

However, the term concurrency is also used to refer to situations where the two events occur at different times but continue such that some of the effects are felt during an overlapping period. In practical terms is the Contractor entitled to an extension of time and loss and/or expense or is the Employer entitled to Delay Damages? That question is not dealt with by sub-cl.8.4.

The law in relating to concurrency (under UK jurisdictions at least) is unclear.[20] The preferred approach is known as the "dominant cause". This is defined as follows:

"If there are two causes, one the contractual responsibility of the Defendants and the other the contractual responsibility of the Claimant, the Claimant succeeds if he establishes that the cause for which the Defendant is responsible is the effective, dominant cause . . ."[20a]

In respect of claims for an extension of time, there is precious little by way of assistance from the Courts. In *Balfour Beatty v Chestermount*[21] a case under the JCT form of contract, the Court had to decide whether the architect could grant an extension of time during a period when the contractor was in culpable delay. The Court held that the architect can consider and award an extension of time if in his opinion he considers it fair and reasonable to do so. Judge Coleman said that the architect when approaching this task should: **8–045**

". . . assess whether any of the relevant events has caused delay to the progress of the works and if so, how much . . .

Fundamental to this exercise is an assessment of whether the relevant event occurring during a period of culpable delay has caused delay to the completion of the works and, if so, how much delay."

In other words the architect is looking for the dominant cause. This theory can perhaps be better demonstrated by reference to claims for loss and/or expense. In the Scottish case of *John Doyle Construction v Laing Management (Scotland) Limited*,[22] the Scottish Inner House had to consider the question of concurrency. Lord Drummond Young said this:

[20] See for example *Keating on Construction Contracts* 8th edn, paras 8–14–22 for a full discussion.
[20a] Keating, para.8–018.
[21] 1993 62 BLR.
[22] (2004) CILL 2135.

"... it is frequently possible to say that an item of loss has been caused by a particular event notwithstanding that other events played a part in its occurrence. In such cases, if an event or events for which the employer is responsible can be described as the dominant cause of an item of loss, that will be sufficient to establish liability, notwithstanding the existence of other causes that are to some degree at least concurrent ... If an item of loss results from concurrent causes, and one of those causes can be identified as the proximate or dominant cause of the loss, it will be treated as the operative cause, and the person responsible for it will be responsible for the loss."

8–046 In other words if loss (or delay) resulted from concurrent causes then it might be possible to identify one of those causes as being dominant in respect of the loss (or delay). If that were the case then it would be treated as the operative cause and the person responsible for it would be responsible for the entirety of that loss (or delay).

It would be possible for the Employer to seek to amend the contract to make it clear that where there are two causes of delay during a single period (one of which is the liability of the Contractor and the other of the Employer) then the Contractor will not receive an extension of time. Such an amendment could be achieved as follows:

"Where more than one event causes concurrent delays and the cause of at least one of those events, but not all of them, is a cause of delay which would not entitle the Contractor to an extension of time under sub-clause ... then to the extent that the delays are concurrent, the Contractor will not be entitled to an extension of time."

Alternatively, contracts such as the JCT Major Projects[23] (cl.12.7.3) favour the Contractor, and provide that the Contractor is entitled to an extension of time, notwithstanding that completion of the project may have been delayed due to the concurrent effects of a cause that is not listed as a relevant event.

8–047 A further alternative or middle way would be to try and apportion delay. This is attempted by cl.34.4 of the the Australian Standard construction contract AS4000 which states:

"34.4 Assessment

When both non qualifying and qualifying causes of delay overlap, the Superintendent shall apportion the resulting delay to WUC according to the respective causes' contribution.

---

[23] A solution preferred by the SCL Protocol.

In assessing each EOT the Superintendent shall disregard questions of whether:

a)   WUC can nevertheless reach practical completion without an EOT; or

b)   the Contractor can accelerate,

but shall have regard to what prevention and mitigation of the delay has not been effected by the Contractor."

An example from the Courts

There is very little by way of assistance from the courts in the interpretation **8–048** of the FIDIC clauses. In one such case the parties to a contract found themselves bound by the FIDIC extension of time procedures and rules without them necessarily having signed up to them. The case of *Motherwell Bridge Construction v Micafil Vakuumtechnik*[24] demonstrates both the difficulty in working out what the parties have contracted, if no contract is agreed and signed and the extent to which the English Courts will often go in order to establish that a contract has been agreed. In August 1997 Micafil wrote to Motherwell Bridge saying that it had been awarded the subcontract *"conditional upon . . . both parties signing the formal contract consolidating all necessary commercial technical and operational requirements . . .".* Nevertheless, no formal contract was executed.

The court concluded, on the basis of the conduct of the parties, that the contract was formed by correspondence and discussions culminating in that August 1977 letter. As a consequence of their conduct, Judge Toulmin CMG Q.C. held that the parties had agreed to conduct their relations within the spirit of FIDIC terms but not to be bound by the strict terms. This was even though there was no FIDIC subcontract available at the time. What this meant was that whilst Motherwell Bridge were entitled to claim an extension of time on grounds available under the FIDIC conditions they did not have to comply with the FIDIC procedural time limits.

Motherwell Bridge had accelerated and resequenced his Work to overcome the effects of delay by Micafil. The court approached the problem in this way. Was the delay on the critical path and if so, was it caused by Motherwell Bridge. On the evidence, the delays were on the critical path, but were not the fault of Motherwell Bridge. Accordingly Motherwell Bridge was entitled to an extension of time to the date it would have been likely to have completed the Work had it not accelerated and attempted to mitigate the Employer delay.[25] It was also entitled to claim for prolongation costs.

---

[24] (2002) CILL 1913 Probably a unique and certainly a slightly surprising case, but one which demonstrates the dangers of not signing up to an agreed contract.

[25] It was not entitled to an extension to the date it actually did complete.

A Coda

8–049 Finally, all parties should ensure that wherever an extension of time has been granted, the new completion date (which supersedes the original or previous contract completion date) is clearly set out. There is no contractual requirement under sub-cl.8.3 for the Contractor to prepare a new programme whenever an extension of time is granted. He merely has to submit a revised programme whenever the previous programme is inconsistent with actual progress or his obligations under the Contract. Whilst you could argue that this does amount to a requirement to issue a new programme upon the award of an extension of time, it is suggested that this is not sufficiently clear to amount to a contractual requirement.

MDB HARMONISED EDITION

8–050 The words "on the Site" have been deleted from sub-cl.8.4(e) which therefore reads as follows:

> *(e) any delay, impediment or prevention caused by or attributable to the Employer, the Employer's Personnel, or the Employer's other contractors.*

The deletion of the words "on the site" therefore expands the grounds for which a Contractor might be entitled to an extension of time as a delay caused by the Employer need no longer be confined to an act or omission which takes place on the Site.

8.5 DELAYS CAUSED BY AUTHORITIES

8–051 *If the following conditions apply, namely:*

> *(a) the Contractor has diligently followed the procedures laid down by the relevant legally constituted public authorities in the Country,*
> *(b) these authorities delay or disrupt the Contractor's work, and*
> *(c) the delay or disruption was Unforeseeable,*

then this delay or disruption will be considered as a cause of delay under sub-paragraph (b) of Sub-Clause 8.4 [Extension of Time for Completion].

OVERVIEW OF KEY FEATURES

- The Contractor is required to follow any applicable procedures required **8–052** by the public authorities in the Country where the project is taking place.
- If he has been delayed by the local authorities, the Contractor may be entitled to an extension of time (not Cost) pursuant to sub-cl.8.4(b), provided:

  (a)   the Contractor has "diligently" followed the procedures of the local authorities; and

  (b)   the delay or disruption was unforeseeable.

COMMENTARY

Sub-clause 8.4(b) refers to causes of delay giving an entitlement to an exten-   **8–053** sion of time under a sub-clause of the contract. Sub-clause 8.5 is one such example. By this sub-clause, the Contractor may claim an extension of time if a delay has been caused by a public authority in the country in which the Site (or most of it) is located.

The Contractor should note subss.(a)–(c) carefully. Unless all three conditions apply, the Contractor will not be entitled to an extension of time.

Unlike the other sub-clauses of the Contract which carry the right to both an extension of time and cost (with or without reasonable profit), this sub-clause is silent as to whether there is any such entitlement to cost or profit if additional time is granted. Therefore it is submitted that there cannot be any entitlement here. The Contractor is entitled to additional time but no more.

The FIDIC Guide says that no mention of financial consequences of delay   **8–054** was made because they would depend on the particular circumstances. That is, it is suggested, something which is true of any event which causes delay. In any event, whatever the reason why, the fact that no mention is made of any entitlement to cost or profit does have the result that there is no such entitlement.

This sub-clause is another example of the importance to the Contractor of making proper use of the right conferred to sub-cl.2.2 to request copies of Laws of the Country which are relevant to the country.

The delay or disruption must be unforeseeable, which by sub-cl.1.1.6.8 means not "reasonably foreseeable by an experienced contractor" at the date from the submission of tender. For a discussion of the difficulties caused by this definition see sub-cll 4.7 and 4.12.

MDB HARMONISED EDITION

8–055   There is no change.

8.6 RATE OF PROGRESS

8–056   *If, at any time:*

(a)   actual progress is too slow to complete within the Time for Completion, and/or

(b)   progress has fallen (or will fall) behind the current programme under Sub-Clause 8.3 [Programme].

Other than as a result of a cause listed in Sub-clause 8.4 [Extension of time for Completion], then the Engineer may instruct the contractor to submit, under Sub-Clause 8.3 [Programme], a revised programme and supporting report describing the revised methods which the Contractor proposes to adopt in order to expedite progress and complete within the Time for Completion.

Unless the Engineer notifies otherwise, the Contractor shall adopt these revised methods, which may require increases in the working hours and/or in the numbers of Contractor's Personnel and/or Goods, at the risk and cost of the Contractor. If these revised methods cause the Employer to incur additional costs, the Contractor shall subject to Sub-Clause 2.5 [Employer's Claims] pay these costs to the Employer, in addition to delay damages (if any) under Sub-Clause 8.7 below.

OVERVIEW OF KEY FEATURES

8–057   • Where actual progress is too slow to enable the Contractor to complete the project within the Time for Completion or where progress has fallen behind the programme, the Engineer may instruct the Contractor to submit a revised programme.

• The Engineer can only do this, where the reason for the delay is a reason other than those listed in sub-cl.8.4.

• The revised programme must set out the way the Contractor intends to deal with the situation and complete within the Time for Completion.

• If the Engineer does not object, the Contractor shall adopt these proposals at his own risk and cost.

- If the Employer incurs increased costs as a result of the revised proposals, he may be entitled to claim these costs from the Contractor pursuant to sub-cl.2.5.
- Any costs the Employer receives as a consequence of this sub-clause shall be in addition to any entitlement for delay damages described in sub-cl.8.7.

COMMENTARY

The purpose of this sub-clause is to enable the Engineer to instruct the Contractor to accelerate his work to ensure that the project is completed on time.   **8–058**

Sub-clause 8.1 provides that the Contractor must proceed "with due expedition and without delay", whilst sub-cl.8.3 already provides that whenever the project falls into delay, the Contractor must, if required by the Engineer, submit a revised programme setting out proposals to rectify the delay. Sub-clause 8.6 specifically deals with what happens if the cause of delay is not a reason which pursuant to sub-cl.8.4 would entitle the Contractor to an extension of time.

The difference between sub-cll 8.3 and 8.6 is that as the Contractor is not entitled to an extension of time and cost, he is responsible for any increased costs and risk caused by the acceleration or other measures which it is necessary to take. In addition to this, the Contractor is responsible for any additional costs which the Employer might incur as a result of the accelerative measures. Costs here, refer to actual costs incurred and not any entitlement the Employer might have to Delay Damages. This is dealt with in sub-cl.8.7.

If incentives are proposed for early completion, the Particular Conditions propose the following additional paragraph: **8–059**

Sections are required to be completed by the dates given in the Appendix to Tender in order that these sections may be occupied and used by the Employer in advance of the completion of the whole of the Works. Details of the work requires to be executed to entitle the Contractor to bonus payments and the amount of the bonuses are stated in the Specification.

For the purpose of calculating bonus payments, the dates given in the Appendix to Tender for completion of Sections are fixed. No adjustments of the dates by reason of granting an extension of time for Completion will be allowed.

MDB HARMONISED EDITION

8–060   The words "*Notice under*" have been added to go before sub-cl.2.5 in the third paragraph. These words have been added for clarification.

In addition, the following additional paragraph has been added at the end of this sub-clause:

> *Additional costs of revised methods including acceleration measures, instructed by the Engineer to reduce delays resulting from causes listed under sub-clause 8.4 [Extension of Time for Completion] shall be paid by the Employer, without generating, however, any other additional payment benefit to the Contractor.*

In other words, the Contractor is entitled to payment for the costs of acceleration measures provided they are instructed by the Engineer. However, the words "without any other additional payment benefit" strongly suggest that the Contractor is not entitled to the payment of any profit on top of these costs.

8.7 DELAY DAMAGES

8–061   *If the Contractor fails to comply with Sub-Clause 8.2 [Time for Completion], the Contractor shall subject to Sub-Clause 2.5 [Employer's Claims] pay delay damages to the Employer for this default. These delay damages shall be the sum stated in the Appendix to Tender, which shall be paid for every day which shall elapse between the relevant Time for Completion and the date stated in the Taking-Over Certificate. However, the total amount due under this Sub-Clause shall not exceed the maximum amount of delay damages (if any) stated in the Appendix to Tender.*

These delay damages shall be the only damages due from the Contractor for such default, other than in the event of termination under Sub-Clause 15.2 [Termination by Employer] prior to completion of the works. These damages shall not relieve the contractor from his obligation to complete the works, or from any other duties, obligations or responsibilities which he may have under the Contract.

OVERVIEW OF KEY FEATURES

8–062   • If the Contractor fails to complete his Works within the Time for Completion, the Employer is entitled to levy Delay Damages.

- The Delay Damages shall be deducted at the rate provided for in the Appendix to Tender.
- The Appendix to Tender provides for the maximum amount of Delay Damages to be capped.
- Unless the Contract is terminated, Delay Damages are the only remedy available to the Employer prior to Completion.
- The Delay Damages do not relieve the Contractor of his other contractual obligations.
- To recover the Delay Damages, the Employer must make an application in accordance with sub-cl.2.5.

COMMENTARY

This sub-clause entitles the Employer to levy Delay Damages (often known as Liquidated & Ascertained damages) if the Contractor fails to complete the Works by the Time for Completion (as extended in accordance with sub-cl.8.4).   **8–063**

To be able to levy such damages, the Employer must make an application in accordance with sub-cl.2.5. The appropriate rates must be set out in the Appendix to Tender. The Appendix to Tender provides for the levying of Delay Damages on a daily basis until the date set out in the Taking-over Certificate. The Delay Damages are expressed in the Appendix to Tender as a percentage of the final Contract Price which is calculated according to sub-cl.14.15(b). These damages can also be capped at a maximum percentage of the final Contract Price.

As a consequence, it is submitted that no delay damages can actually be deducted until the Final Contract price has been ascertained. Until that has been done, the actual amount of the Delay Damages cannot be known.

The law under which the Contract operates will have particular significance here. Under English law, a liquidated damage clause will not be enforceable where it constitutes a "penalty".[26] However under many other jurisdictions penalty clauses are not only valid but common.   **8–064**

In England and other common law jurisdictions, the approach the Courts will take was set out by Lord Woolf who was sitting as part of the Privy Council in *Philips Hong Kong Limited v The Attorney General of Hong Kong* (1993)[27]:

"Except possibly in the case of situations where one of the parties to the contract is able to dominate the other as to the choice of the terms of a contract, it will normally be insufficient to establish that a provision is

---

[26] *Dunlop Ltd v New Garage Co Ltd* (1915) A.C. 79.
[27] 61 BLR 41.

objectionably penal to identify situations where the application of the provision could result in a larger sum being recovered by the injured party than his actual loss. Even in such situations so long as the sum payable in the event of non-compliance with the contract is not extravagant, having regard to the range of losses that it could reasonably be anticipated it would have to cover at the time the contract was made, it can still be a genuine pre-estimate of the loss that would be suffered and so a perfectly valid liquidated damage provision. The use in argument of unlikely illustrations should therefore not assist a party to defeat a provision as to liquidated damages. As the Law Commission stated in Working Paper No.61 (page 30):

The fact that in certain circumstances a party to a contract might derive a benefit in excess of his loss does not . . . outweigh the very definite practical advantages of the present rule upholding a genuine estimate, formed at the time the contract was made of the probable loss."

8–065 More recently, Mr Justice Jackson reviewed the position in *Alfred McAlpine Capital Projects Ltd v Tilebox*[28]. He made four general observations:

1.   There seem to be two strands in the authorities. In some cases judges consider whether there is an unconscionable or extravagant disproportion between the damages stipulated in the contract and the true amount of damages likely to be suffered. In other cases the courts consider whether the level of damages stipulated was reasonable. Mr Darling submits, and I accept, that these two strands can be reconciled. In my view, a pre-estimate of damages does not have to be right in order to be reasonable. There must be a substantial discrepancy between the level of damages stipulated in the contract and the level of damages which is likely to be suffered before it can be said that the agreed pre-estimate is unreasonable.

2.   Although many authorities use or echo the phrase "genuine pre-estimate", the test does not turn upon the genuineness or honesty of the party or parties who made the pre-estimate. The test is primarily an objective one, even though the court has some regard to the thought processes of the parties at the time of contracting.

8–066 3.   Because the rule about penalties is an anomaly within the law of contract, the courts are predisposed, where possible, to uphold contractual terms which fix the level of damages for breach. This predisposition is even stronger in the case of commercial contracts freely entered into between parties of comparable bargaining power.

4.   Looking at the bundle of authorities provided in this case, I note only four cases where the relevant clause has been struck down as a penalty.

---

[28] [2005] E.W.H.C. 281 (TCC).

These are *Commissioner of Public Works v Hills* [1906] A.C. 368, *Bridge v Campbell Discount Co Ltd* [1962] A.C. 600, *Workers Trust and Merchant Bank Limited v Dojap Investments Limited* [1993] A.C. 573, and *Ariston SRL v Charly Records* (Court of Appeal, March 13, 1990). In each of these four cases there was, in fact, a very wide gulf between (a) the level of damages likely to be suffered, and (b) the level of damages stipulated in the contract.

Mr Justice Jackson's judgment provides a reminder that in addressing this question, the test is the reasonableness of the estimate when considered objectively at the time of entering into the contract.

8–067   Therefore under English jurisdictions, the Employer would be advised to take steps to ensure that he can demonstrate that the percentage of the Contract Price (or any alternative formulation which may be agreed between the parties) can be demonstrated to be a genuine pre-estimate of the likely losses where completion is delayed. To be a genuine pre-estimate, the sum must bear a reasonable relationship to the likely level of damages that would result from a breach of contract and be a pre-estimate of damages that would otherwise be recoverable as damages for breach of contract. The fact that the delay damages here are expressed as a proportion of the Contract Price suggest a lump sum rather than a carefully calculated approach. This may well mean that the Delay Damages will be considered by the English courts to be a penalty.[29]

If the Delay Damages clause was found to be a penalty, then the Employer may still be entitled to damages to compensate for any actual loss which may have been suffered.

There is a further potential difficulty here caused by the fact that the Contract Price will not be known with any certainty until the Works are, at least, nearly complete. How can the Employer make a reasonable pre-estimate based in percentage terms of a figure which is unclear? There are two answers to this. First, amend the Appendix to Tender and use a different basis for the calculation of the Delay Damages. Alternatively, ensure that the percentage figure entered into the Appendix to Tender is based on a genuine pre-estimate of the Contract Price as anticipated at the time the Contract is entered into.

8–068   Care must be taken when filling in the Appendix to Tender. In *Temloc v Ermill* Properties (1988) 39 BLR 30, the liquidated damages were expressed to be "£nil". This, it turned out, did not mean that there was no liquidated

---

[29] Note however the case of *Decoma UK Limited v Haden Drysys International Limited* [2005] E.W.H.C. 2429 (TCC) where the parties agreed by Article 12 of the contract that if Haden failed to achieve the final completion date, then liquidated and ascertain damages could be levied up to a maximum amount of 5 per cent of the contract price. According to the judgment, both parties agreed that these rates were a genuine pre-estimate of loss. See also sub-cl.17.6, below, for the discussion on its liability cap.

damages clause so that damages became "at large". Instead it meant that the Employer lost all rights to damages for delay.

Sub-clauses 1.1.3.3 and 8.2 refer to the possibly of sectional completion. Where sectional completion is required by the Contract, the Delay Damages provision should make clear how the damages payable in the event of late completion are apportioned between the different stages of work. Where there is provision in a contract for partial possession or sectional completion prior to takeover, but no equivalent provision for the reduction of Delay Damages in such circumstances, the right to deduct Delay Damages will be lost.

The most straightforward way to do this is to reduce the sum by reference to the section of the Works which is complete. The Contract must identify the value to be ascribed to each section. A failure to do so can be fatal to the Employer's claim for liquidated damages. The Employer should, in particular, be aware that the English Courts will almost certainly construe such clauses by operating the *contra proferentum rule,* in other words, against the Employer who will be the party seeking to rely on the particular clause. [30]

8–069   If the appropriate part of the Appendix to Tender is not filled out, then the Employer will not be entitled to levy Delay Damages. However, under English law, a term will normally be implied into any building contract which provides that the contractor shall complete the works within a reasonable time. If, without sufficient excuse, the contractor fails to achieve this then he is liable to pay damages at common law assessed under the common law principles that derive from *Hadley v Baxendale* and *Victoria Laundry (Windsor) Ltd v Newman Industries Ltd.*[31]

Sub-clause 8.7 does not deal with what happens if delay damages are paid out but the Contractor is subsequently awarded an extension of time.[32] The proper course is that the Contractor is entitled to a refund based on the extension awarded. In the interests of clarity the parties should consider inserting the following sub-clause:

"If the Employer extends the Time for Completion under Sub-Clause 8.4 after delay damages have been paid or otherwise accounted for under this Sub-Clause, the extension of the Time for Completion shall not invalidate the Employer's claim for the delay damages. In such circumstances the Contractor's liability to pay delay damages shall be limited to the sum due on the basis of the new Time for Completion, and the Employer shall repay or otherwise account for the difference between that sum and the sum which has been paid or otherwise accounted for on the basis of the

---

[30] *Bramall & Ogden Ltd v Sheffield City Council* (1983) 29 BLR 73, *Bruno Zornow v Beechcroft Development* (1989) 51 BLR 16; *Stanor Electric v Mansell* (1988) CILL 399.
[31] (1854) 9 Ex. 341.
[32] This problem was highlighted by the case of *DOE v Farrans (Construction) Limited (1982)* 19 BLR 1.

Employer's claim before the extension of the Time for Completion. Interest shall [or shall not] be payable by the Employer on any amounts payable or repayable under this Sub-Clause."

Sub-clause 8.7 deals with the deduction of damages if completion is late. The contract conditions do not make provision for a bonus for early completion. Employers sometimes are prepared to incentivise the Contractor to complete by a certain time through the payment of a further bonus. If this is something which the Parties agree upon, then it is important that the appropriate clause is drafted with sufficient clarity.

MDB HARMONISED EDITION

The words *"notice under"* have been inserted before *"Sub-Clause 2.5"* in the **8–070** first paragraph. In addition, the words *"Appendix to Tender"* have been replaced by *"Contract Data."*

8.8 SUSPENSION OF WORK

The Engineer may at any time instruct the Contractor to suspend progress of **8–071** *part or all of the Works. During such suspension, the Contractor shall protect, store and secure such part of the works against any deterioration, loss or damage.*
The Engineer may also notify the cause for the suspension. If and to the extent that the cause is notified and is the responsibility of the Contractor, the following Sub-Clauses 8.9, 8.10 and 8.11 shall not apply.

OVERVIEW OF KEY FEATURES

- The Engineer may at any time instruct the Contractor to suspend progress **8–072** of his Works or any part of them.
- The Engineer may, but does not have to, notify the Contractor for the reason for the suspension.
- During the period of the suspension it is the responsibility of the Contractor to protect the Works.
- If the cause of the suspension is something for which the Contractor is responsible, sub-cll 8.9–8.11 shall not apply.

COMMENTARY

8–073 This sub-clause details what is to happen should the Engineer instruct the Contractor to suspend progress of his Works.

The Engineer can issue such an instruction at any time. There is no obligation on the Engineer to inform the Contractor of the reason for the suspension, but it is suggested that the Engineer would be wise to do so.

There is also no restriction on when the Engineer can issue such a notice. The words "at any time"[33] mean exactly what they say. There is also no obligation on the Engineer to give any advance notice of the suspension. Presumably therefore, a notice can be served with immediate effect.

8–074 During the period of the suspension, it is the responsibility of the Contractor to "protect, store and secure" the Works (or part of the Works if the suspension only relates to a part) against any "deterioration, loss or damage." The clause does not say who is responsible for the costs of undertaking this protection work. However, from the following sub-cll 8.9–8.11, it seems clear that if the reason for the suspension is related to a reason for which the Contractor is entitled to an extension of time under sub-cl.8.4, then the Contractor will be entitled to make a claim for these costs. However, if the reason for the suspension is the responsibility of the Contractor, he will equally be responsible for these protection costs.

If the Contractor fails to provide adequate protection which leads to a delay on resumption of the works, then in accordance with sub-cl.8.9 he will not be entitled to an extension of time.

In addition to this, once work resumes, the Contractor might, again if the reason for the suspension is related to a reason for which the Contractor is entitled to an extension of time under sub-cl.8.4, be entitled to make a claim for the costs of resumption. These can often be quite considerable.

8–075 One thing which sub-cll 8.8–8.12 do not do, is to require the Contractor to maintain its staff and labour resources ready to commence work when the period of suspension his lifted. The Contractor has to take steps to protect the Works, but an Employer might want to consider how he can ensure that key personnel in addition to the Contractor's Representative (who will presumably be required by sub-cl.4.3 to remain available) are not moved from the project. Obviously, at the very least the Contractor would look to recover its costs in these circumstances.

---

[33] Something that was confirmed by Mr Justice Dyson in *Herschel v Breen* (2002) BLR 272 in a case involving the enforcement of an adjudicator's decision under the Housing Grants Construction & Regeneration Act 1996.

MDB HARMONISED EDITION

There is no change.   **8–076**

8.9 CONSEQUENCE OF SUSPENSION

If the Contractor suffers delay and/or incurs Cost from complying with the   **8–077**
Engineer's instructions under Sub-Clause 8.8 [Suspension of Work] and/or
from resuming the work, the Contractor shall give notice to the Engineer and
shall be entitled subject to Sub-Clause 20.1 [Contractor's Claims] to:

(a)   an extension of time for any such delay, if completion is or will be
delayed, under Sub-Clause 8.4 [Extension of Time for Completion], and
(b)   payment of any such Cost, which shall be included in the Contract Price.

After receiving this notice, the Engineer shall proceed in accordance with
Sub-Clause 3.5 [Determinations] to agree or determine these matters.
*   The Contractor shall not be entitled to an extension of time for, or to*
payment of the Cost incurred in, making good the consequences of the
Contractor's faulty design, workmanship or materials, or of the Contractor's
failure to protect, store or secure in accordance with Sub-Clause 8.8
[Suspension of Work].

OVERVIEW OF KEY FEATURES

- If the Contractor suffers delay or increased costs as a result of a suspen-   **8–078**
sion notified by the Engineer, he is entitled to make a claim and must give
an appropriate notice to the Engineer as required by cl.20.
- The Contractor will not be entitled to an extension of time and/or additional cost, if the reason for the suspension is not covered by sub-cl.8.4
and/or was a result of the Contractor's poor design, workmanship or
materials.
- The Contractor will not be entitled to an extension of time and/or
payment for cost incurred as a consequence of any failure on the part of
the Contractor to protect the Works during the period of suspension in
accordance with sub-cl.8.8.

COMMENTARY

8–079   This sub-clause confirms that the Contractor is entitled to submit a claim for an extension of time and additional cost as a consequence of any suspension notice issued by the Engineer in accordance with sub-cl.8.8.

However, a Contractor will not be entitled to an extension of time and/or additional cost if the reason for the suspension was the responsibility of the Contractor and not an item covered by the sub-cl.8.4.

In addition, the Contractor will not be entitled to an extension of time and/or additional cost, if he fails to take adequate steps to protect the Works during the period of the suspension.

MDB HARMONISED EDITION

8–080   There is no change.

8.10 PAYMENT FOR PLANT AND MATERIALS IN EVENT OF SUSPENSION

8–081   *The Contractor shall be entitled to payment of the value (as at the date of suspension) of Plant and/or Materials which have not been delivered to Site, if:*

(a)   *the work on Plant or delivery of Plant and/or materials has been suspended for more than 28 days, and*
(b)   *the Contractor has marked the Plant and/or Materials as the Employer's property in accordance with the Engineer's instructions.*

OVERVIEW OF KEY FEATURES

8–082   • If Work or the delivery of Plant and Materials has been suspended for more than 28 days and if the Contractor has marked the Plant and Materials as the Employer's property then the Contractor shall be entitled to payment for Plant and Materials which have not been delivered to site.
• The value of the payment will be the value of the Plant and Material as at the date of the suspension.

COMMENTARY

This sub-clause recognises that some items required on site can have long lead-in times. The sub-clause is intended to ensure that the Contractor has an entitlement to payment when suspension impacts on any such times.   **8–083**

MDB HARMONISED EDITION

There is no change.   **8–084**

8.11 PROLONGED SUSPENSION

If the suspension under Sub-Clause 8.8 [Suspension of Work] has continued   **8–085**
for more than 84 days, the Contractor may request the Engineer's permission to
proceed. If the Engineer does not give permission within 28 days after being
requested to do so, the Contractor may, by giving notice to the Engineer, treat
the suspension as an omission under Clause 13 [Variations and Adjustments]
of the affected part of the Works. If the suspension affects the whole of the
Works, the contractor may give notice of termination under Sub-Clause 16.2
[Termination by Contractor].

OVERVIEW OF KEY FEATURES

* If the period of suspension runs for more than 84 days, the Contractor   **8–086**
  may seek permission to proceed with the Works from the Engineer.
* If the Engineer does not give permission within 28 days of being requested
  to do so, in respect of a suspension of the entire Works, the Contractor will
  be entitled to give notice of termination under sub-cl.16.2.
* If the Engineer fails to respond within 28 days of any such request, in
  respect of a suspension of a part of the Works, the Contractor may treat
  the suspension as an omission under sub-cl.13.

COMMENTARY

Obviously it cannot be in the Contractor's interest for the periods of suspen-   **8–087**
sion to continue indefinitely. During the period of suspension the Contractor

217

will incur potentially significant costs in protecting the site and also disruption to the resourcing of other projects. The Contractor will have also had to maintain his insurance during the period of insurance to the extent required by cl.18.

Sub-clause 8.11 provides the Contractor with the mechanism to bring the period of suspension to an end through the provision of a notice the Engineer. The Contractor should note that the notice should specifically request permission from the Engineer to proceed with the suspended Works.

However the notice period is a long one, the Contractor must wait for 84 days from the date of the suspension notice before being able to take action. The Contractor must then wait for a further period of up to 28 days (i.e. a total of 112 days) before being able to take any action to bring the suspension period to an end.

8–088   If the area affected by the suspension is a part of the Works, then the Contractor can treat the suspension as an omission under sub-cl.13.1(d).

If the whole of the Works were suspended, then the Contractor is entitled to give immediate notice of termination in accordance with sub-cl.16.2(f).

MDB HARMONISED EDITION

8–089   There is no change.

8.12 RESUMPTION OF WORK

8–090   *After the permission or instruction to proceed is given, the Contractor and the Engineer shall jointly examine the Works and the Plant and Materials affected by the Suspension. The Contractor shall make good any deterioration or defect in or loss of the works or Plant or Materials, which has occurred during the suspension.*

OVERVIEW OF KEY FEATURES

8–091
- Once permission to proceed has been given, the Contractor and the Engineer shall carry out a joint inspection of the Works affected by the suspension.
- It is the responsibility of the Contractor to make good any damage in the Works which has occurred during the period of suspension.

COMMENTARY

This sub-clause sets out the procedure for the resumption of work after
suspension. Following a joint inspection, the Contractor must rectify any
defects that may have occurred to the Works during the period of the
suspension.

8–092

In accordance with sub-cll 8.8–8.10, the Contractor will not be able to
recover the costs of so doing if he was responsible for the suspension in the
first place or if the defects occurred as a result of any failure by the
Contractor during the period of suspension to protect the Works properly.

MDB HARMONISED EDITION

The words *"after receiving from the Engineer an instruction from the Engineer
an instruction to this effect under Clause 13 [Variations and Adjustments]"*
have been added to the end of the final paragraph. In other words, the
Contractor should not carry out any such work, unless he has received an
appropriate instruction from the Engineer.

8–093

CLAUSE 9 – TESTS ON COMPETITION

9.1 CONTRACTOR'S OBLIGATIONS

The Contractor shall carry out the Tests on Completion in accordance with this **9–001**
Clause and Sub-Clause 7.4 [Testing], after providing the documents in accordance with sub-paragraph (d) of Sub-Clause 4.1 [Contractor's General Obligations].

The Contractor shall give to the Engineer not less than 21 days' notice of the date after which the Contractor will be ready to carry out each of the Tests on Completion. Unless otherwise agreed, Tests on Completion shall be carried out within 14 days after this date, on such day or days as the Engineer shall instruct.

In considering the results of the Tests on Completion, the Engineer shall make allowances for the effect of any use of the Works by the Employer on the performance or other characteristics of the works. As soon as the works, or a Section, have passed any Tests on Completion, the Contractor shall submit a certified report of the results of these Tests to the Engineer.

OVERVIEW ON KEY FEATURES

- It is the responsibility of the Contractor to carry out the Tests on **9–002**
  Completion.
- Before the Tests are carried out, the Contractor must provide the as-built drawings and operations and maintenance manuals.
- The Contractor must give at least 21 days' notice to the Engineer of the date on which they will be ready to carry out the Tests.
- Unless otherwise agreed, the Tests will be carried out within 14 days after this date.
- The Engineer must consider the Tests and decide whether the works tested have passed.

COMMENTARY

The requirement for "Tests on Completion" is not new. The Old Red Book **9–003**
FIDIC 4th edn did provide for "Tests on Completion" noting that they were the tests specified in the Contract or otherwise agreed by the Engineer and the Contractor which were to be made by the Contractor before the

Works or any Section or part thereof could be taken over by the Employer. Clause 9 here sets out the specific way in which the tests are now required to be carried out.

By sub-cl.1.1.3.4 "Tests on Completion" are defined quite simply as the tests which are specified in the Contract or agreed by both Parties or instructed as a Variation, and which are carried out under cl.9 to determine whether the works or a Section (as the case may be) are ready to be taken over by the Employer.

The FIDIC Guide sensibly suggests that the Specification should describe the tests which the Contractor has to carry out. It also notes that typically such tests are described in some considerable detail.

9–004   The Contractor must first provide the documents set out in sub-cl.4.1(d) before carrying out the Tests on Completion. The documents so required by sub-cl.4.1(d) are the "as-built" documents and the operation and mainte-nance manuals. These documents must be "in accordance with the Specification and in sufficient detail for the Employer to operate, maintain, dismantle, reassemble, adjust and repair this part of the Works."

Sub-clause 4.1(a) also provides that until the documents have been provided the Works cannot be taken over in accordance with clause 10. There is no specific obligation on the Engineer to check these documents. However, given their importance, it is assumed that the Employer will want to ensure that they fulfil their purpose.

The Tests must be carried out in accordance with both this sub-clause and the detailed provisions of sub-cl.7.4.[1] For example the Contractor must supply all the apparatus and staff that may be necessary to carry the tests out efficiently.

9–005   Although this sub-clause requires the Contractor to submit the results of the Tests on Completion to the Engineer, it does not specifically require the Engineer to approve the Tests. However by sub-cl.7.4, when tests are passed the Engineer must endorse the test certificate accordingly. Equally by sub-cl.7.4, the Engineer can reject any item which fails a test and order a retest in accordance with sub-cl.9.3. The FIDIC Guide notes that if the Works are being tested and taken over in stages, then allowance may need to be made to take account of the fact that part of the Works will be incomplete at the time of testing.

It is important that the Contractor factors in the time that might be taken to organise the Tests For Completion into his programme. By sub-cl.8.2, the Contractor must complete the Works in accordance with the Time for Completion. This includes passing the Tests on Completion. This might be important as the Contractor must give 21 days' notice of the test to enable the Engineer to make his preparations for the test. The Engineer then has 14 days to set a date.

---

[1] See above.

Therefore, whilst by virtue of the Contractor's programme prepared in accordance with sub-cl.8.3 and the Progress Reports required by sub-clause 4.21, the Engineer will already have been provided with the Contractor's estimate of when the particular tests were going to be carried out, it might take 35 days before the tests are carried out.

MDB HARMONISED EDITION 9–006

There is no change.

9.2 DELAYED TESTS

If the Tests on Completion are being unduly delayed by the Employer, Sub- 9–007
Clause 7.4 [Testing] (fifth paragraph) and/or Sub-Clause 10.3 [Interference with Tests on Completion] shall be applicable.

If the Tests on Completion are being unduly delayed by the Contractor, the Engineer may by notice require the Contractor to carry out the Tests within 21 days after receiving the notice. The Contractor shall carry out the Tests on such day or days within that period as the Contractor may fix and of which he shall give notice to the Engineer.

If the Contractor fails to carry out the Tests on Completion within the period of 21 days, the Employer's Personnel may proceed with the Tests at the risk and cost of the Contractor. The Tests on Completion shall then be deemed to have been carried out in the presence of the Contractor and the results of the Tests shall be accepted as accurate.

OVERVIEW OF KEY FEATURES

- If the Employer unduly delays the Tests on Completion then sub-cll 7.4 9–008 or 10.3 apply and the Contractor may be entitled to additional time and costs.
- If the Tests on Completion are unduly delayed by the Contractor, the Engineer may give notice requiring the Contractor to carry out the Tests within 21 days.
- If the Contractor fails to carry out the tests within the 21 days specified, the Employer's Personnel can complete the Tests.
- If the Employer's Personnel carry out the tests in these circumstances, they do so at the Contractor's risk and cost.

COMMENTARY

9–009   If Tests on Completion are unduly delayed by the Employer, the Contractor may give notice under sub-cl.7.4 and make a claim in accordance with the provisions of cl.20. No definition is provided of "undue delay". Sub-clause 7.4 only refers to "delay". The most likely definition will be one based on the factual circumstances of the delay and in particular in relation to the impact on the Contractor's programme.

Alternatively the Contractor may be able to make use of sub-cl.10.3 which provides that if the Contractor is prevented for more than 14 days from carrying out the Tests on Completion by the Employer, the Employer is deemed to have taken over the Works on the date when the Tests on Completion should have been completed. Thereafter the Contractor will be able to make a claim for an extension of time together with associated costs (including carrying out the tests) and reasonable profit.

The advantage to the Contractor of proceeding in accordance with sub-cl.10.3 is that it provides a certain time frame. Sub-clause 10.3 specifically comes into effect if the Contractor is prevented from carrying out the tests for more than 14 days. Sub-clause 7.4 refers to "delay" whilst sub-cl.9.2 refers to more pressing "undue delay". A prudent Contractor would be able to use his Programme to demonstrate that the Tests were being delayed. However, it is suggested that the maximum period of delay before the Contractor could take advantage of sub-cll 7.4 and 9.2 should correlate with the 14 day time frame set out in sub-cl.10.3.

9–010   This must mean that the Employer too would have to wait for 14 days before being able to submit a notice requiring the Contractor to carry out the Tests. The time reference is the same, the Tests being "unduly delayed".

If the Employer submits a notice, the Contractor must still give notice to the Engineer. This potentially gives the Contractor some difficulty, as the Contractor is supposed by sub-cl.9.1 to give the Engineer 21 days' notice. It is suggested that the time provisions set out within sub-cl.9.1 cannot apply when the Tests are in delay.

It is important for the Contractor to respond to any notice from the Employer requiring him to carry out the Tests. If he does not, the Employer is entitled to allow his personnel to carry out the tests, at the Contractor's risk and cost. In addition, the Tests will be deemed both to be accurate and to have taken place in the presence of the Contractor.

MDB HARMONISED EDITION

9–011   There is no change.

9.3 RETESTING

If the Works, or a Section, fail to pass the Tests on Completion, Sub-Clause 7.5 **9–012**
[Rejection] shall apply, and the Engineer or the Contractor may require the
failed Tests, and Tests on Completion on any related work, to be repeated under
the same terms and conditions.

OVERVIEW OF KEY FEATURES

If the Works do not pass the Tests on Completion, the Engineer can reject the   **9–013**
Works and require the Tests to be repeated.

COMMENTARY

This sub-clause here simply provides that if the Works fail the Tests on   **9–014**
Completion, the Engineer or the Contractor can require the tests to be re-
taken. The new tests must be carried out under the same conditions as when
the tests were originally carried out.

   This sub-clause should be read in conjunction with sub-cl.7.5 which
provides that the Contractor shall make good any defect and ensure that the
rejected item complies with the Contract. The tests can then be repeated.

   There is no limit set on how many times the works can be carried out.
However sub-cl.15.2(c) provides that the Employer is entitled to terminate the Contract, if the Contractor fails to comply with a notice issued
under sub-cl.7.5. This is the same ultimate sanction as set out in
sub-cl.9.4.

MDB HARMONISED EDITION

There is no change.   **9–015**

9.4 FAILURE TO PASS TESTS ON COMPLETION

If the Works, or a Section, fail to pass the Tests on Completion repeated under   **9–016**
Sub-Clause 9.3 [Retesting], the Engineer shall be entitled to:

(a)   order further repetition of Tests on Completion under Sub-Clause 9.3;

(b)   if the failure deprives the Employer of substantially the whole benefit of the Works or Section, reject the Works or Section (as the case may be), in which event the Employer shall have the same remedies as are provided in sub-paragraph (c) of Sub-Clause 11.4 [Failure to Remedy Defects]; or

(c)   issue a Taking-Over Certificate, if the Employer so requests.

In the event of sub-paragraph (c), the Contractor shall proceed in accordance with all other obligations under the Contract, and the Contract Price shall be reduced by such amount as shall be appropriate to cover the reduced value to the Employer as a result of this failure. Unless the relevant reduction for this failure is stated (or its method of calculation is defined) in the Contract, the Employer may require the reduction to be (i) agreed by both Parties (in full satisfaction of this failure only) and paid before this Taking-Over Certificate is issued, or (ii) determined and paid under Sub-Clause 2.5 [Employer's Claims] and Sub-Clause 3.5 [Determinations].

OVERVIEW OF KEY FEATURES

9–017   • If the Works do not pass the Tests on Completion, the Engineer can either:

(i)   order the Tests on Completion to be repeated;
(ii)   reject the Works; or
(iii)   issue a Taking-Over Certificate.

• There is no limit on the number of repetitions.
• If the works are rejected, the Contractor's contract could be terminated.
• If a Taking-Over Certificate is issued in these circumstances, the Contract Price shall be reduced to take account of the consequential reduced value to the Employer.

COMMENTARY

9–018   If the Works being tested fail the test then the Engineer can either order a re-test or reject the Works. There is no limit to the number of times a retest can be ordered.

The Engineer can only chose to reject the Works if "the failure" deprives the Employer of substantially the "whole benefit" of the Works being tested. If the Engineer takes this option, then the Employer shall have the same

226

remedies as are available for a Failure to Remedy Defects as provided for at sub-cl.11.4. In other words, the Employer may:

(i)   make other arrangements to carry out the Works and require the Contractor to pay for these Works;
(ii)   accept the defective work and seek a reduction in price; or
(iii)   terminate the contract and seek to recover all sums paid to date plus financing costs and the costs of clearing the site.

The third option is quite obviously a very serious one for the Contractor. An **9–019** Employer should consider quite carefully the risks before taking the final option. There might be a considerable delay and increase in costs in appointing a new Contractor at a late stage in the Contract.

The Engineer cannot issue a Taking Over Certificate. This can only be done at the request of the Employer. If the Employer chooses to take over the Works even though they have failed the Tests, the Employer is entitled to seek a reduction in the Contract Price in order to cover the reduced value. If the amount of the reduction cannot be agreed and paid before the Taking Over Certificate is issued, then the amount to be deducted will be worked out accordance with sub-cl.2.5.

MDB HARMONISED EDITION

There is no change.   **9–020**

CLAUSE 10 – EMPLOYER'S TAKING OVER

10.1 TAKING OVER OF THE WORKS AND SECTIONS

Except as stated in Sub-Clause 9.4 [Failure to Pass Tests on Completion], the **10–001**
Works shall be taken over by the Employer when (i) the Works have been
completed in accordance with the Contract, including the matters described in
Sub-Clause 8.2 [Time for Completion] and except as allowed in sub-paragraph
(a) below, and (ii) a Taking-Over Certificate for the Works has been issued, or
is deemed to have been issued in accordance with this Sub-Clause.

    The Contractor may apply by notice to the Engineer for a Taking-Over
Certificate not earlier than 14 days before the Works will, in the Contractor's
opinion, be complete and ready for taking over. If the Works are divided into
Sections, the Contractor may similarly apply for a Taking-Over Certificate for
each Section.

    The Engineer shall, within 28 days after receiving the Contractor's application:

(a)   issue the Taking-Over Certificate to the Contractor, stating the date on
which the Works or Section were completed in accordance with the
Contract, except for any minor outstanding work and defects which will
not substantially affect the use of the Works or Section for their intended
purpose (either until or whilst this work is completed and these defects are
remedied); or

(b)   reject the application, giving reasons and specifying the work required to
be done by the Contractor to enable the Taking-Over Certificate to be
issued. The Contractor shall then complete this work before issuing a
further notice under this Sub-Clause.

If the Engineer fails either to issue the Taking-Over Certificate or to reject the **10–002**
Contractor's application within the period of 28 days, and if the Works or
Section (as the case may be) are substantially in accordance with the Contract,
the Taking-Over Certificate shall be deemed to have been issued on the last day
of that period.

OVERVIEW OF KEY FEATURES

- The Works will be taken over by the Employer when: **10–003**

229

(i)　the Works have been completed; and
(ii)　a Taking-Over Certificate has been issued.

- The Contractor may apply to the Engineer for a Taking-Over Certificate not earlier than 14 days before he anticipates completing the Works.
- Within 28 days of receiving the Contractor's application, the Engineer may either:

(i)　issue the Taking-Over Certificate; or
(ii)　reject (with reasons) the application.

- If the Engineer fails to accept or reject the application within 28 days and the works are substantially in accordance with the Contract, then the Taking-Over Certificate will be deemed to have been issued on the last day of the 28-day period.

COMMENTARY

10–004　Sub-clause 10.1 sets out the procedure whereby the Works can be taken over by the Employer. This sub-clause refers to either the Works in their entirety or Sections of the Works (if any are defined in the Appendix to Tender[1]). This sub-clause does not refer to parts of the Works. These are dealt with in sub-cl.10.2.

Prior to commencement of the Tests on Completion, as required by sub-cll 4.1 and 9.1, the Contractor must submit to the Engineer, the operations and maintenance manuals in sufficient detail to enable the Engineer to operate the relevant part of the works. The Works or part thereof will not be considered complete for the purposes of Taking Over of the Works, until these manuals have been submitted in an acceptable form.

For the Works to be taken over, the Works must be complete and a Taking Over certificate must have been issued. The Contractor commences the taking over process by applying for a Taking-Over Certificate.

10–005　If the Works have failed the Tests on Completion, then in accordance with sub-cl.9.4, they will not be taken over unless the Employer specifically requests a Taking Over Certificate.

Sub-clause 10.1 cross-refers to sub-cl. 8.2 and mirrors its requirements. Completion of the Works (or a Section of the Works) requires that:

(i)　the Works have passed the Tests of Completion; and
(ii)　all the work stated under the Contract as being required before taking over have been completed.

---

[1] And for clarity, if there are any, they should be so defined.

The Taking Over of the Works follows a three-stage process:

(i)   the Contractor gives up to 14-days' notice[2] that the Works are suitably complete;

(ii)   the Engineer has 28 days to accept or reject the Works;

(iii) if the Engineer neither rejects nor accepts, then the Works will be deemed to have been taken over at the end of that period, provided they are substantially in accordance with the Contract.

By sub-cl.10.1(a), the Engineer cannot refuse to issue the Taking-Over **10–006** Certificate because of *"any minor outstanding work"* provided the minor defect does not *"substantially affect the use of the Works"*. However, there is no guidance given as to the meaning of "substantially" or "substantially affect" and it will be left to the judgment of the Engineer. This could potentially lead to disputes.

If there is any minor outstanding Work, this must be completed by the Contractor during the Defects Notification Period as set out in sub-cl.11.1.

If the Engineer decides to reject the Works, he must give reasons. The Engineer must also specify the work required to be done by the Contractor to enable the Taking-Over Certificate to be issued. It is the responsibility of the Contractor to complete this work before applying for a further certificate.

If the Engineer fails to approve or reject the Works within the 28-day **10–007** period, they will be deemed to be complete provided they are "substantially in accordance with the Contract". As noted above, there is no definition within the Contract of "substantially".

Under the Society Of Construction Law Delay And Disruption Protocol, "Substantial Completion" is bracketed with Practical Completion and both are defined as the completion of all the construction work that has to be done, subject only to very minor items of work left incomplete. This definition is close to the words of Salmon L.J. who in *J. Jarvis & Son Ltd v Westminster Corporation*[3] said of Practical Completion that it was completion:

"for all practical purposes, that is to say for purpose of allowing the employer to take possession of the works and use them as intended, but not "completion" down to the last detail, however trivial and unimportant".

It is accordingly suggested that the works will be substantially in accordance with the Contract if they are free from known defects which would prevent the Employer from taking over and making use of the Project.

---

[2] The Contract says "may", the FIDC Guide says "should".
[3] [1981] CA 1 W.L.R. 448.

MDB HARMONISED EDITION

10–008 There is no change.

10.2 TAKING OVER OF PARTS OF THE WORKS

10–009 *The Engineer may, at the sole discretion of the Employer, issue a Taking-Over Certificate for any part of the Permanent Works.*

The Employer shall not use any part of the Works (other than as a temporary measure which is either specified in the Contract or agreed by both Parties) unless and until the Engineer has issued a Taking-Over Certificate for this part. However, if the Employer does use any part of the Works before the Taking-Over Certificate is issued:

(a)   the part which is used shall be deemed to have been taken over as from the date on which it is used,
(b)   the Contractor shall cease to be liable for the care of such part as from this date, when responsibility shall pass to the Employer, and
(c)   if requested by the Contractor, the Engineer shall issue a Taking-Over Certificate for this part.

After the Engineer has issued a Taking-Over Certificate for a part of the Works, the Contractor shall be given the earliest opportunity to take such steps as may be necessary to carry out any outstanding Tests on Completion. The Contractor shall carry out these Tests on Completion as soon as practicable before the expiry date of the relevant Defects Notification Period.

10–010 *If the Contractor incurs Cost as a result of the Employer taking over and/or using a part of the Works, other than such use as is specified in the Contract or agreed by the Contractor, the Contractor shall (i) give notice to the Engineer and (ii) be entitled subject to Sub-Clause 20.1 [Contractor's Claims] to payment of any such Cost plus reasonable profit, which shall be included in the Contract Price. After receiving this notice, the Engineer shall proceed in accordance with Sub-Clause 3.5 [Determinations] to agree or determine this Cost and profit.*

If a Taking-Over Certificate has been issued for a part of the Works (other than a Section), the delay damages thereafter for completion of the remainder of the Works shall be reduced. Similarly, the delay damages for the remainder of the Section (if any) in which this part is included shall also be reduced. For any period of delay after the date stated in this Taking-Over Certificate, the proportional reduction in these delay damages shall be calculated as the proportion which the value of the part so certified bears to the value of the Works or Section (as the case may be) as a whole. The Engineer shall proceed in

232

accordance with Sub-Clause 3.5 [Determinations] to agree or determine these proportions. The provisions of this paragraph shall only apply to the daily rate of delay damages under Sub-Clause 8.7 [Delay Damages], and shall not affect the maximum amount of these damages.

OVERVIEW OF KEY FEATURES

- The Engineer may, at the request of the Employer, issue a Taking-Over   **10–011**
Certificate for any part of the Permanent Works.
- The Employer may not use any part of the Works until the Engineer has issued a Taking-Over Certificate.
- If the Employer does use any part of the Works prior to the issue of a Taking Over Certificate then:

  (i)   that part shall be deemed to have been taken over;
  (ii)   the Employer and not the Contractor will be responsible for that part of the Works which has been taken over; and
  (iii) if the Contractor incurs any costs as a result, the Contractor shall give notice to the Engineer and may be entitled to payment of these costs to include reasonable profit.

- If a Taking-Over Certificate has been issued for part of the Works, the rate of Delay Damages for the remainder of the Works will be reduced.

COMMENTARY

Sub-clause 10.2 refers to taking over of parts of the Works rather than the   **10–012**
whole of the Works (or Sections of the Works, if any, as defined in the Appendix to Tender).

The Contractor does not have a right to request that the Employer take over part of the Works under this Sub-Clause. However, the Employer has the discretionary right to request that the Engineer issue a Taking-Over Certificate for any part of the Works. The Employer is not entitled to use any part of the Works (unless agreed as a temporary measure), unless the Engineer has issued a Taking-Over Certificate.

The consequences of the Employer using any part of the Works before a Taking-Over certificate is issued as set out in sub-cll 10.2(a) to (c) are:

(i)   that part shall be deemed to have been taken over;
(ii)   the Employer will be responsible for that part of the Works which have been taken over;

(iii) the Contractor may request that the Engineer issue a Taking-Over Certificate; and

(iv) if the Contractor incurs any costs as a result, the Contractor shall give notice to the Engineer and may be entitled to payment of these costs to include reasonable profit.

10–013 One potential difficulty of the Employer taking over a part of the Works early might be that an area will have been taken over before all the necessary Tests on Completion have been carried out. Sub-clause 10.2 makes no provision for this.

It may be that it is no longer necessary to carry out those tests in these circumstances. The reason for this is that the Contractor has the option under this sub-clause of requesting that the Engineer issue a Taking-Over Certificate. The obligation on the Engineer is a positive one and he must issue the certificate. In accordance with sub-cll 8.2 and 10.1, it can be argued that the issuing of the certificate represents an acknowledgment that the Tests on Completion have been carried out.

If it is still necessary to carry out the tests, the parties (and particularly the Employer) should recognise that this is likely to cause some difficulties for the Contractor as it is likely the Contractor will have prepared his programme for the Works on the understanding that parts of the Works would not be taken over early by the Employer. Therefore the Contractor may need to re-organise his programme and resources which could lead to delay and increased costs.

10–014 This will be one of the reasons why sub-cl.10.2 provides that the Contractor, if he incurs additional cost in these circumstances, may make a claim in accordance with clause 20 for payment of such cost to include reasonable profit.

The other potential problem caused by the Employer taking over parts of the Works at an early (and unanticipated stage) relates to Delay Damages. We discussed the difficulties caused by sectional completion in sub-cl.8.7. Sub-clause 10.2 recognises that where a Taking-Over Certificate is issued in respect of a part of the Works, then the Employer's entitlement to Delay Damages is reduced proportionally.

MDB HARMONISED EDITION

10–015 The word "*reasonable*" has been deleted in the fourth paragraph, so that the Contractor shall be entitled to its "*profit*" not "*reasonable profit*".

10.3 INTERFERENCE WITH TESTS ON COMPLETION

If the Contractor is prevented, for more than 14 days, from carrying out   **10–016**
*the Tests on Completion by a cause for which the Employer is responsible, the
Employer shall be deemed to have taken over the Works or Section (as the case
may be) on the date when the Tests on Completion would otherwise have been
completed.*

*The Engineer shall then issue a Taking-Over Certificate accordingly, and the
Contractor shall carry out the Tests on Completion as soon as practicable,
before the expiry date of the Defects Notification Period. The Engineer shall
require the Tests on Completion to be carried out by giving 14 days' notice and
in accordance with the relevant provisions of the Contract.*

*If the Contractor suffers delay and/or incurs Cost as a result of this delay in
carrying out the Tests on Completion, the Contractor shall give notice to the
Engineer and shall be entitled subject to Sub-Clause 20.1 [Contractor's
Claims] to:*

(a)   *an extension of time for any such delay, if completion is or will be
delayed, under Sub-Clause 8.4 [Extension of Time for Completion], and*
(b)   *payment of any such Cost plus reasonable profit, which shall be included
in the Contract Price.*

*After receiving this notice, the Engineer shall proceed in accordance with
Sub-Clause 3.5 [Determination] to agree or determine these matters.*

OVERVIEW OF KEY FEATURES

- If the Employer prevents the Contractor from carrying out Tests on   **10–017**
Completion for more than 14 days, the Employer is deemed to have taken
over the Works.
- In these circumstances, the Engineer shall issue a Taking-Over Certificate.
- However, the Engineer shall still require the Contractor to carry out the
Tests of Completion.
- If the Contractor suffers delay or increased costs he is entitled to submit
a claim in accordance with the provisions of clause 20.

COMMENTARY

This sub-clause deals with the situation where the Employer prevents the   **10–018**
Contractor from carrying out Tests on Completion. Provided the cause of

the delay is a cause for which the Employer is responsible, then if the Contractor is prevented from carrying out the tests for a period of more than 14 days, the Works will be deemed to have been taken over on the date on which the Tests would have been completed but for the interference of the Employer.

However, even if the Contractor is prevented from carrying out the Tests by the Employer, this does not release the Contractor from the requirement of carrying out the Taking-Over Tests and ensuring that the Works pass those tests. Arrangements for the Tests must be made with the Engineer. The Tests must be carried out as soon as practicable and in any event no later than the expiry of the Defects Notification Period.

If the Contractor suffers delay and/or increased costs he may be entitled to submit a claim in accordance with sub-cl.20. Equally, it is possible that a dispute might arise over whether the Employer is responsible for the interference. That dispute will presumably be determined under the provisions of cl.20.

MDB HARMONISED EDITION

10–019   The word "*reasonable*" has been deleted in the third paragraph, so that the Contractor shall be entitled to its "*profit*" not "*reasonable profit*".

10.4 SURFACES REQUIRING REINSTATEMENT

10–020   *Except as otherwise stated in a Taking-Over Certificate, a certificate for a Section or part of the Works shall not be deemed to certify completion of any ground or other surfaces requiring reinstatement.*

OVERVIEW OF KEY FEATURES

10–021   A Taking-Over Certificate must expressly state if it certifies the completion of any ground or other surfaces requiring reinstatement.

COMMENTARY

10–022   This sub-clause acknowledges that ground and other surfaces typically cannot be reinstated until right at the end of a project, often not until the end of the Defects Notification Period, which, of course, takes place after the Taking-Over Certificates have been issued.

MDB HARMONISED EDITION

There is no change.   **10–023**

CLAUSE 11 – DEFECTS LIABILITY

11.1 COMPLETION OF OUTSTANDING WORK AND REMEDYING DEFECTS

In order that the Works and Contractor's Documents, and each Section, shall **11–001**
*be in the condition required by the Contract (fair wear and tear excepted) by
the expiry date of the relevant Defects Notification Period or as soon as
practicable thereafter, the Contractor shall:*

*(a) complete any work which is outstanding on the date stated in a Taking-
Over Certificate, within such reasonable time as instructed by the
Engineer, and*

*(b) execute all work required to remedy defects or damage, as may be noti-
fied by (or on behalf of) the Employer on or before the expiry date of
the Defects Notification Period for the Works or Section (as the case
may be).*

*If a defect appears or damage occurs, the Contractor shall be notified
accordingly, by (or on behalf of) the Employer.*

OVERVIEW OF KEY FEATURES

- The aim is that all outstanding work should have been completed by the **11–002**
  end of the Defects Notification Period (or as soon as is practicable
  thereafter).
- The Contractor is obliged to complete the work identified in the Taking-
  Over Certificate within the reasonable time set by the Engineer.
- The Contractor is also obliged to carry out all work required to remedy
  defects or damage identified by (or on behalf of) the Employer, on or
  before the end of the Defects Notification Period.

COMMENTARY

11–003   It has already been noted under sub-cl.1.1.3.7 that "Defects Notification Period" is a more helpful concept than "Defect Liability Period", the term used in the Old Red Book FIDIC 4th edn.[1] Defects must be *notified* to the Contractor within this period, although the Contractor may require longer than the period itself for their execution and completion.

The provisions of cl.11 impose obligations relating to the making good of defects as opposed to a wider duty to maintain the Works – the difference being that "wear and tear" is excepted.[2]

It is believed that the reference to the expression "fair wear and tear" is taken directly from the English law of landlord and tenant.[3] This term may lead to different interpretations depending on the law governing the contract and/or the location of the site of the project.

11–004   Importantly, the obligations upon the Contractor under Clause 11 depend upon *notice* being given either by the Employer or alternatively by a request from the Engineer under sub-cl.11.1(a). The Employer cannot simply rectify any defect itself and then seek to recover the costs from the Contractor. If the Employer were to proceed to rectify the defects himself without giving notice to the Contractor, then it seems that the Employer will not be entitled to rely on the defects clause against the Contractor.

In the case of *Pearce and High v Baxter*,[4] the Employer's failure to comply with the notice requirements of defects liability provisions, whether by refusing to allow the contractor to carry out the repairs or by failing to give notice of the defects, was held to limit the amount of damages which it was entitled to recover.

It is to be noted that the decided cases in this area proceed on the basis that the Employer, but not the Contractor, has knowledge of the defects.[5] The liability of a Contractor with actual knowledge depends upon the words of the contract, but it is doubted whether the words in sub-cl.11.1 are strong enough to create notice as a condition precedent to liability of the Contractor for defects. See however sub-cl.11.4 for further discussion on this point.

MDB HARMONISED EDITION

11–005   There is no change.

---

[1] The 3rd edn, even less helpfully referred to the "Period of Maintenance."
[2] See *Sevenoaks Railways v London and Chatham Railway* (1879) 11 Ch. D. 625.
[3] For guidance on the legal meaning of "fair wear and tear", see cases such as *Haskell v Marlow* [1928] 2 K.B. 45: 97 LJKB 311 or *Warren v Keen* [1954] 1 QB 15, 19, 20.
[4] (1999) BLR 101 (CA).
[5] *London and South Western Railways v Flower* (1875) 1 CPD 77 at 85.

11.2 COST OF REMEDYING DEFECTS

All work referred to in sub-paragraph (b) of Sub-Clause 11.1 [Completion of **11–006**
*Outstanding Work and Remedying Defects] shall be executed at the risk and
cost of the Contractor, if and to the extent that the work is attributable to:*

(a) any design for which the Contractor is responsible,
*(b) Plant, Materials or workmanship not being in accordance with the
        Contract, or*
(c) failure by the Contractor to comply with any other obligation.

*If and to the extent that such work is attributable to any other cause, the
Contractor shall be notified promptly by (or on behalf of) the Employer, and
Sub-Clause 13.3 [Variation Procedure] shall apply.*

OVERVIEW OF KEY FEATURES

- The risk and cost of remedying defects is placed upon the Contractor in   **11–007**
  the case of:

    (i)   design work for which the Contractor is responsible;
    (ii)   plant, materials and workmanship not in accordance with the
           Contract;
    (iii) any other failure of the Contractor to comply with any obligation.

- Remedial work attributable to any other cause is dealt with under the
  Variation Procedure (sub-cl.13.3).

COMMENTARY

Sub-clause 11.1 makes the Contractor responsible for remedying defects.   **11–008**
   The scheme used here achieves the clear and sensible result that, save in the
case of design,[6] workmanship or other default by the Contractor, the work is
treated, and valued, as a Variation.
   The Contractor's liability in damages is not removed by the existence of a
defects clause unless in the clearest words,[7] and the words of cl.11 do not have
such an effect. Again, it is to be noted that the Contractor's liability for not

---

[6] See sub-cl.4.1 for comment on the extent of the Contractor's responsibility.
[7] *Hancock v Brazier* [1966] 1 W.L.R. 1317 (CA).

241

completing the Works in accordance with the provisions of the contract continues, under English law, until it becomes barred by the Limitation Act 1980. A cause of action at common law for failure to comply with the defects liability obligations will arise at the time when the contract terms provides that the defects liability obligations are to be carried out.

MDB HARMONISED EDITION

11–009   There is no change.

11.3 EXTENSION OF DEFECTS NOTIFICATION PERIOD

11–010   *The Employer shall be entitled subject to Sub-Clause 2.5 [Employer's Claims] to an extension of the Defects Notification Period for the Works or a Section if and to the extent that the Works, Section or a major item of Plant (as the case may be, and after taking over) cannot be used for the purposes for which they are intended by reason of a defect or damage. However, a Defects Notification Period shall not be extended by more than two years.*

If delivery and/or erection of Plant and/or Materials was suspended under Sub-Clause 8.8 [Suspension of Work] or Sub-Clause 16.1 [Contractor's Entitlement to Suspend Work], the Contractor's obligations under this Clause shall not apply to any defects or damage occurring more than two years after the Defects Notification Period for the Plant and/or Materials would otherwise have expired.

OVERVIEW OF KEY FEATURES

11–011
- The Defects Notification Period for the Works can be extended if the Works, a Section of the Works or a major item of Plant cannot be used for the reason for which they are intended as a result of defect or damage.
- The Defects Notification Period may not be extended for more than two years.

COMMENTARY

In the case of a complex project, the ability to extend the Defects Notification **11–012**
Period is of considerable importance. This provision is new. It is to be noted
that the test to be applied here is that there must be a "major" item of Plant,
and this major item of Plant must be such that it "cannot be used" for the
reason or purpose intended. It is suggested that when considering the
meaning to be given to the word "major" it will be necessary to look at the role
played by the item of Plant in the project or process as a whole, rather than
focussing upon the piece of Plant in isolation.

The ability to extend the period is limited to two years. This should be
noted as providing a long-stop which cannot be further extended.

Sub-clause 11.3 allows the Employer to extend the Defects Notification
Period in the event of a "default or defect". The sub-clause does not say
whether it is to be implied that it must be a "default or defect" on the part of
the Contractor for the Employer to be able to rely on this clause. For
example, it is not clear whether the Contractor should only be allowed to take
account of suspensions if those suspensions were not of the Contractor's
own making. If this is not the case, then there may be difficulties with
applying this sub-clause in civil law countries where the Contractor would be
able to rely on the principle of *nemo auditur propriam turpitudinem allegans*
whereby "no one is heard when alleging one's own wrong." In other words if
the Contractor is not responsible for the defects, then the Defects
Notification Period cannot be extended.

MDB HARMONISED EDITION

The words "*or by reason of damage attributable to the Contractor*" have been **11–013**
added to the end of the long sentence which makes up the majority of the
first paragraph. This serves to deal with the point raised above as to whether
or not the Employer can only seek an extension when the defect or damage
is attributable to the Contractor. It would seem that under the MDB version,
the answer is yes.

11.4 FAILURE TO REMEDY DEFECTS

If the Contractor fails to remedy any defect or damage within a reasonable time, **11–014**
a date may be fixed by (or on behalf of) the Employer, on or by which the defect
or damage is to be remedied. The Contractor shall be given reasonable notice of
this date.

If the Contractor fails to remedy the defect or damage by this notified date and this remedial work was to be executed at the cost of the Contractor under Sub-Clause 11.2 [Cost of Remedying Defects], the Employer may (at his option):

(a) *carry out the work himself or by others, in a reasonable manner and at the Contractor's cost, but the Contractor shall have no responsibility for this work; and the Contractor shall subject to Sub-Clause 2.5 [Employer's Claims] pay to the Employer the costs reasonably incurred by the Employer in remedying the defect or damage;*

11–015 (b) *require the Engineer to agree or determine a reasonable reduction in the Contract Price in accordance with Sub-Clause 3.5 [Determinations]; or*

(c) *if the defect or damage deprives the Employer of substantially the whole benefit of the Works or any major part of the Works, terminate the Contract as a whole, or in respect of such major part of the Works, termi-nate the Contract as a whole, or in respect of such major part which cannot be put to the intended use. Without prejudice to any other rights, under the Contract or otherwise, the Employer shall then be entitled to recover all sums paid for the Works or for such part (as the case may be), plus financing costs and the cost of dismantling the same, clearing the Site and returning Plant and Materials to the Contractor.*

OVERVIEW OF KEY FEATURES

11–016 • Where the Contractor fails to make good defects or damage within a reasonable time, then provided reasonable notice is given, the Engineer may fix the time within which the defects are to be remedied.
• If the Contractor fails to carry out required work within the time notified (or at all), then the Employer has a series of options available to him as identified in sub-clauses (a), (b) and (c).

COMMENTARY

11–017 The range of remedies provided for in the case of the Contractor's poor performance is extensive, although it may be rare for sub-cl. (c) termination, to be used, particularly on the partial basis which is permitted.

The need to notify the Contractor plays a critical role in sub-cl.11.4 and the Employer will be required to follow the procedures set out in sub-cl.11.4 with care before exercising the options at (a), (b) or (c).

It is suggested that, if the Employer purported to exercise the options in sub-cl.11.4 without giving the Contractor reasonable notice, then the Employer may not later be able to rely upon the provisions of sub-cl.11.4. Certainly, the Employer would risk arguments in relation to alleged failure to "instigate" (or reduce) his loss by following the procedure as to notification of the Contractor.

Note that under sub-cl.11.4(c), the Employer does not (as in the Old Red **11–018** Book FIDIC 4th edn) even need to refer the issue of the *purported* defect to the Engineer. If the Employer is wrong (e.g. if the defect has been remedied or does not deprive the Employer of substantially the whole benefit of the Works), then the Contractor may then be left with no other alternative than to start proceedings against the Employer for wrongful termination.

In the meantime, however, the Employer will be able to claim:

"all sums paid for the Works or for such part (as the case may be), plus financing costs and the cost of dismantling the same, clearing the Site and returning Plant and Materials to the Contractor".

Before taking advantage of this new but potentially significant benefit or draconian remedy (depending on your point of view), Employers should bear in mind that it can only be used when the defect in question deprives them "of substantially the whole benefit of the Works". A similar wording is used in sub-cl.9.4(b). This is a significant hurdle to overcome. For example, under English law, this is the wording used to determine whether or not there has been a repudiatory breach of contract.[8]

MDB HARMONISED EDITION

There is no change. **11–019**

11.5 REMOVAL OF DEFECTIVE WORK

If the defect or damage cannot be remedied expeditiously on the Site and the **11–020** *Employer gives consent, the Contractor may remove from the Site for the purposes of repair such items of Plant as are defective or damaged. This consent may require the Contractor to increase the amount of the Performance Security by the full replacement cost of these items, or to provide other appropriate security.*

[8] Per Lord Diplock – *Hong Kong Fir Shipping Co v Kawasaki Kisen Kaisha* [1962] 2 QB 26, 1 All E.R. 474.

OVERVIEW OF KEY FEATURES

11–021
- If defects or damage cannot be made good expeditiously *on site*, then the Contractor may remove such items of Plant as need to be repaired.
- The Employer's consent is required.
- The Employer's consent might require additional Performance Security.

COMMENTARY

11–022 This provision makes practical sense since it may often be far easier to address problems with Plant and equipment remote from the site. This requires the Employer to give its consent, and a question which may arise in practice is whether the Employer can at his absolute discretion insist that plant and machinery remain at site during repairs, or whether consent cannot unreasonably be withheld. The latter approach appears more consistent with the rest of the Contract, and with the additional safeguard offered in respect of security.

If without good reason the consent of the Employer for removal is withheld then such additional trouble and expense, as may arise from the execution of repair work on site is likely to be the object of debate between the parties. The correct approach will be a question of fact to be resolved in each case, although in a case where consent for removal is withheld by the Employer on an unreasonable basis, it is hard to see how the associated cost should be attributed to any default of the Contractor.

If the Contractor is required to increase the amount of the Performance Security then it should do so in accordance with the provisions of sub-cl.4.2.

MDB HARMONISED EDITION

11–023 There is no change here.

11.6 FURTHER TESTS

11–024 *If the work of remedying of any defect or damage may affect the performance of the Works, the Engineer may require the repetition of any of the tests described in the Contract. The requirement shall be made by notice within 28 days after the defect of damage is remedied.*

These tests shall be carried out in accordance with the terms applicable to the previous tests, except that they shall be carried out at the risk and cost of the Party liable, under Sub-Clause 11.2 [Cost of Remedying Defects], for the cost of the remedial work.

OVERVIEW OF KEY FEATURES

- If the remedying of any defect or damage could affect performance, then **11–025** the Engineer can require the repetition of any tests required under the Contract.
- The further testing shall be carried out in accordance with the Contract, but at the risk and cost of the Party responsible for the defect or damage.

COMMENTARY

Testing on completion is dealt with in cll 9 and 10. **11–026**
This clause envisages re-testing after the completion of remedial work in circumstances where the remedial works could affect the performance overall. It is thought that re-testing in circumstances described by the clause will be quite a usual occurrence. There are two important points to note here:

(i)   The testing regime or specification used for the re-testing should be that set out in the Contract, and not a different or more onerous regime.
(ii) The re-testing costs are to be borne by the party responsible for the remedial work itself under sub-cl.11.2.

MDB HARMONISED EDITION

There is no change. **11–027**

11.7 RIGHT OF ACCESS

Until the Performance Certificate has been issued, the Contractor shall have **11–028**
such right of access to the Works as is reasonably required in order to comply with this Clause, except as may be inconsistent with the Employer's reasonable security restrictions.

OVERVIEW OF KEY FEATURES

11–029
- The Contractor has a right of reasonable access in order to comply within his obligations under cl.11.
- The right to access is subject to the Employers' reasonable security restrictions.

COMMENTARY

11–030   The Contractor's right of access is often dealt with in contracts by the operation of necessary implied terms relating to non-prevention. The express provision in sub-cl.11.7 relating to Contractor's access for defects liability obligations is a useful clarification of what, in practice, could be an area of difficulty, since the Works (or Plant) are potentially in operation at this stage. In other words, the Employer is under an obligation to grant the Contractor access to remedy defects.

The Contractor is given a right of "reasonable access" in order to comply with his cl.11 obligations, save as is inconsistent with the Employer's "reasonable security restrictions". In both case, what is reasonable will depend on the actual circumstances on site at the time.

MDB HARMONISED EDITION

11–031   There is no change.

11.8 CONTRACTOR TO SEARCH

11–032   *The Contractor shall, if required by the Engineer, search for the cause of any defect, under the direction of the Engineer. Unless the defect is to be remedied at the cost of the Contractor under Sub-Clause 11.2 [Cost of Remedying Defects], the Cost of the search plus reasonable profit shall be agreed or determined by the Engineer in accordance with Sub-Clause 3.5 [Determinations] and shall be included in the Contract Price.*

OVERVIEW OF KEY FEATURES

- The Engineer can require the Contractor to search for the cause of any defect, under the direction of the Engineer. **11–033**
- The costs of search may be required to be paid by the Contractor under sub-cl.11.2.
- Alternatively, the Contractor is paid his costs and reasonable profit, to be determined by the Engineer under sub-cl.3.5.

COMMENTARY

It can often occur in practice that problems of operation of the Works (or **11–034**
Plant) do not indicate the immediate cause of difficulty, so their investigation is required. Difficulties can arise, in these circumstances, as to which party is to direct the investigations; which is to undertake the investigations; and what is to happen in relation to the costs of these operations. Here, the Contractor is required to look for defects, if so-directed by the Engineer.

Where the defect is to be remedied at the Contractor's cost (as per sub-cl.11.2) then he bears the expense, but otherwise, the Contractor receives his costs plus profit, either as agreed (if not) as determined by the Engineer.

MDB HARMONISED EDITION

The words *"reasonable profit"* have been replace by *"profit"* **11–035**

11.9 PERFORMANCE CERTIFICATE

Performance of the Contractor's obligations shall not be considered to have **11–036**
been completed until the Engineer has issued the Performance Certificate to the Contractor, stating the date on which the Contractor completed his obligations under the Contract.

The Engineer shall issue the Performance Certificate within 28 days after the latest of the expiry dates of the Defects Notification Periods, or as soon thereafter as the Contractor has supplied all the Contractor's Documents and completed and tested all the Works, including remedying any defects. A copy of the Performance Certificate shall be issued to the Employer.

Only the Performance Certificate shall be deemed to constitute acceptance of the Works.

OVERVIEW OF KEY FEATURES

11–037
- The issue of the Performance Certificate must occur before the obligations of the Contractor are regarded as having been completed.
- The Performance Certificate is to be issued by the Engineer within 28 days of the expiry of the latest of any Defects Notification Periods.
- A copy of the Performance Certificate is to be issued to the Employer.
- Only the Performance Certificate will constitute deemed acceptance of the Works.

COMMENTARY

11–038
The issue of the Performance Certificate is for practical purposes the conclusion of the defects liability machinery set out in cl.11 and is thus of considerable importance. The Performance Certificate is to be issued by the Engineer within 28 days of the latest of the expiry dates of the Defects Notification Periods ". . . or as soon after as the Contractor has supplied all the Contractor's Documents and completed and tested all the Works . . .". Once the certificate is issued, the Works shall be deemed to have been accepted.

The Contractor will note that the issue of the Performance Certificate is linked to the provision of the Contractor's Documents. The provision of documents can often be overlooked in practice and it is important for the Contractor to prepare for this obligation.

This sub-clause makes no mention as to whether the issuing of the Performance Certificate is final and conclusive as to the Contractor's satisfactory performance of the contract – in other words does it prevent the Employer from being able to sue the Contractor for defective work? There have been a number of English authorities on this point and the JCT was forced to re-draft one of its standard clauses following the 1994 decision of the Court of Appeal[9] that the issuing of a final certificate was conclusive evidence as to the quality of materials and standard of workmanship.

11–039
Given that the wording of sub-cl.11.9 does not include terms such as "final and binding", it is submitted that the issuing of the Performance Certificate will not serve as final and conclusive proof as to the Contractor's satisfactory performance under the Contract.

---

[9] *Crown Estates Commissioners v John Mowlem & Co Ltd* (1994) 70 BLR 1. See also, by way of example, *Attorney-General of Hong Kong v Wang Chong Construction* (1992) 8 Const L.J. 137 or *Matthew Hall Ortech v Tarmac Roadstone* (1997) 87 BLR 96.

As discussed below, any argument as to the potential conclusive effect of this sub-clause is also tempered by sub-cl.11.10 which deals with "unfulfilled obligations".

MDB HARMONISED EDITION

There is no change.   **11–040**

11.10 UNFULFILLED OBLIGATIONS

After the Performance Certificate has been issued, each Party shall remain   **11–041**
liable for the fulfilment of any obligation which remains unperformed at that
time. For the purposes of determining the nature and extent of unperformed
obligations, the Contract shall be deemed to remain in force.

OVERVIEW OF KEY FEATURES

- Each Party remains liable for the performance of unfulfilled obligations   **11–042**
  after the issue of the Performance Certificate.
- For the purposes of the performance of any unfulfilled obligations, the
  Contract is deemed to remain in force.

COMMENTARY

In a complex project, there will often be further outstanding obligations in   **11–043**
spite of which the Performance Certificate may be issued. In such a case, the
Contractor will not be able to rely upon the Performance Certificate as an
answer to outstanding obligations, since they remain to be performed by
virtue of sub-cl.11.10. The Engineer will wish, in deciding the time of issue
of the Performance Certificate, to be cautious about leaving too much
outstanding or "unperformed" within the meaning of sub-cl.11.10 since, in
contrast to the rest of cl.11, sub-cl.11.10 does not provide a mechanism for
dealing with these "unperformed obligations" in a particular manner or time
frame.

MDB HARMONISED EDITION

11–044   There is no change.

11.11 CLEARANCE OF SITE

11–045   *Upon receiving the Performance Certificate, the Contractor shall remove any remaining Contractor's Equipment, surplus material, wreckage, rubbish and Temporary Works from the Site.*

If all these items have not been removed within 28 days after the Employer receives a copy of the Performance Certificate, the Employer may sell or otherwise dispose of any remaining items. The Employer shall be entitled to be paid the costs incurred in connection with, or attributable to, such sale or disposal and restoring the Site.

Any balance of the moneys from the sale shall be paid to the Contractor. If these moneys are less than the Employer's costs, the Contractor shall pay the outstanding balance to the Employer.

OVERVIEW OF KEY FEATURES

11–046
- Upon receiving the Performance Certificate, the Contractor is obliged to clear the site.
- If any plant materials, wreckage or rubbish remains after 28 days of the Employer's receipt of its copy of the Performance Certificate, these can be sold or disposed of by the Employer.
- The Employer is entitled to recover from the Contractor the costs of any such disposal, with any surplus being paid to the Contractor.

COMMENTARY

11–047

It is unlikely in practice that the Contractor will leave any significant items of plant and materials on site for a period such as would fall within the operation of this sub-clause. The more likely scenario is that where the Contractor leaves debris and rubbish on site and this then has to be dealt with by the Employer. In such circumstances, the Employer is entitled to recover from the Contractor the *costs* of restoring the site – and that is the actual costs and not the "costs" as defined by sub-cl.1.1.4.3.

MDB HARMONISED EDITION

The words *"the Employer receives a copy"* have been replaced by *"receipt by*   **11–048**
the Contractor". Thus the 28-day deadline runs from when the Contractor
receives the Performance Certificate.

CLAUSE 12 – MEASUREMENT AND EVALUATION

12.1 WORKS TO BE MEASURED

The Works shall be measured, and valued for payment, in accordance with this Clause.   **12–001**

Whenever the Engineer requires any part of the Works to be measured, reasonable notice shall be given to the Contractor's Representative, who shall:

(a) promptly either attend or send another qualified representative to assist the Engineer in making the measurement, and
(b) supply any particulars requested by the Engineer.

If the Contractor fails to attend or send a representative, the measurement made by (or on behalf of) the Engineer shall be accepted as accurate.

Except as otherwise stated in the Contract, wherever any Permanent Works   **12–002**
are to be measured from records, these shall be prepared by the Engineer. The Contractor shall, as and when requested, attend to examine and agree the records with the Engineer, and shall sign the same when agreed. If the Contractor does not attend, the records shall be accepted as accurate.

If the Contractor examines and disagrees the records, and/or does not sign them as agreed, then the Contractor shall give notice to the Engineer of the respects in which the records are asserted to be inaccurate. After receiving this notice, the Engineer shall review the records and either confirm or vary them. If the Contractor does not so give notice to the Engineer within 14 days after being requested to examine the records, they shall be accepted as accurate.

OVERVIEW OF KEY FEATURES

- This sub-clause provides for the value of the work to be determined by   **12–003**
  means of measurement.
- Where measurement is required by the Engineer, reasonable notice is given to the Contractor's representative (on site) who shall either attend or provide particulars requested.
- This must be done promptly.
- If the Contractor does not attend, the Engineer's measurement shall be accepted as accurate.

- Measurement records are (save as otherwise provided) to be prepared by the Engineer and agreed by the Contractor, and shall be signed if agreed.
- If the Contractor does not give notice of any area of disagreement, the Engineer's records are accepted as accurate.

COMMENTARY

12–004 Measurement is a central feature of cl.12 and is the basis ultimately upon which payment to the Contractor is calculated. Sometimes called a "measure and value" type of contract, the arrangements in place in the FIDIC Form proceed on the basis that the Works are to be measured by the Engineer, and those quantities and measured amounts of work are then to be paid for alternatively at the rates and prices in the Contract, or else on the basis of adjusted rates, or entirely new rates (if there is no basis for using or altering Contract rates for the work).

As a matter of practicality, measurements and quantities are often capable of being agreed at site level or on the basis of records, and disputes in relation to these matters can be extremely time-consuming and expensive. Sub-clause 12.1 therefore includes a number of default positions, i.e. explanations of what is to occur if the Contractor does not attend. These are very important provisions since the Contractor will face great difficulties in advancing an alternative case on measurement if he has not complied with these.

Firstly, the Contractor must respond "promptly" to requests from the Engineer. The Contractor should ensure that his surveying team are alive to this requirement as this sub-clause provides little time to react, particularly bearing in mind the likely extensive nature of any valuation records. If the Contractor does not respond, then the Engineer's measurement shall be deemed to be accurate. Equally, if the Contractor does not take up the request to examine and agree the measurement records, they will be accepted as accurate. Finally, the Contractor should be aware that he only has 14 days to give notice of any disagreement.

12–005 The Particular Conditions provide some sensible guidance, suggesting that if any part of the Permanent Works are to be measured, as sub-cl.11.1 says, from records of its construction, then the type of records required should be specified in the tender documentation.

MDB HARMONISED EDITION

12–006 There is a new second sentence in sub-cl.12.1 which provides that the Contractor shall show in each application for payment be they interim or final (i.e. under sub-cll 14.3, 14.10 and 14.11):

. . . the quantities or other particulars detailing the amounts which he considers to be entitled under the Contract . . ..

This is an important change and clearly places an obligation upon the Contractor to identify quantities applied for on his Applications. The practical effect of the amendments remains to be seen and it may mark a shift towards the Engineer focusing upon the Contractor's Application as opposed to his own measurement.

In addition, the words *"and certify the payment of the undisputed part"* have **12–007** been added to the penultimate sentence of the final paragraph. This imposes a minor, but clearly important, additional obligation on the Engineer. It is minor because noting areas of agreement forms part of the Engineer's measurement role in any event, but important since the more items that can be certified and thus formally agreed and moved out of any arenas for potential dispute, the better.

It is to be noted, however, that there is nothing in the amendment which indicates that the measurement obligations of the Engineer have been in any way diminished.

12.2 METHOD OF MEASUREMENT

Except as otherwise stated in the Contract and notwithstanding local practice: **12–008**

(a)   measurement shall be made of the net actual quantity of each item of the Permanent Works, and

(b)   the method of measurement shall be in accordance with the Bill of Quantities or other applicable Schedules.

OVERVIEW OF KEY FEATURES

- The method of measurement shall (except where otherwise stated in the   **12–009** Contract) be the measurement of the net actual quantity of each item of the Permanent Works.
- The method of measurement adopted must accord with the Bill of Quantities (or equivalent Schedules).

COMMENTARY

12–010   Although the parties can make alternative provision for method of measurement in this Contract, the purpose of sub-cl.12.2 is to impose a standard method of measurement rather than allowing this important process to be determined, in default of agreement, by local custom and practice. It is therefore important the proposal method of measurement is actually defined in the Contract.

    "Schedules" are defined by sub-cl.1.1.1.7 as documents completed by the Contractor and submitted with the Letter of Tender. The Schedules may include Bills of Quantities or simple schedules of lists and prices. Sub-clause 1.1.1.10 rather unhelpfully adds little to this definition, merely referring to "Bills of Quantities" as being the "documents so named (if any) which are comprised in the Schedules".

MDB HARMONISED EDITION

12–011   There is no change here.

12.3 EVALUATION

12–012   *Except as otherwise stated in the Contract, the Engineer shall proceed in accordance with Sub-Clause 3.5 [Determinations] to agree or determine the Contract Price by evaluating each item of work, applying the measurement agreed or determined in accordance with the above Sub-Clauses 12.1 and 12.2 and the appropriate rate or price for the item.*

    For each item of work, the appropriate rate or price for the item shall be the rate or price specified for such item in the Contract or, if there is no such item, specified for similar work. However, a new rate or price shall be appropriate for an item of work if:

    (a) *(i)*   *the measured quantity of the item is changed by more than 10% from the quantity of this item in the Bill of Quantities or other Schedule,*

          (ii)   *this change in quantity multiplied by such specified rate for this item exceeds 0.01% of the Accepted Contract Amount,*

          (iii)   *this change in quantity directly changes the Cost per unit quantity of this item by more than 1%, and*

          (iv)   *this item is not specified in the Contract as a "fixed rate item";*

or

(b) *(i)* the work is instructed under Clause 13 [Variations and Adjustments],

(ii) no rate or price is specified in the Contract for this item, and

(iii) no specified rate or price is appropriate because the item of work is not of similar character, or is not executed under similar conditions, as any item in the Contract.

Each new rate or price shall be derived from any relevant rates or prices in the Contract, with reasonable adjustments to take account of the matters described in sub-paragraph (a) and/or (b) as applicable. If no rates or prices are relevant for the derivation of a new rate or price, it shall be derived from the reasonable Cost of executing the work, together with reasonable profit, taking account of any other relevant matters. **12–013**

Until such time as an appropriate rate or price is agreed or determined, the Engineer shall determine a provisional rate or price for the purposes of Interim Payment Certificates.

OVERVIEW OF KEY FEATURES

- Evaluation is to proceed on the basis of sub-cll 12.1 and 12.2 and the appropriate rate or price for that item of work. **12–014**
- If there is no rate or price for that item of work, then the rate or price for similar work is to be used.
- However, a new rate or price will be appropriate if the circumstances described in (a) or (b) are fulfilled.
- The new rates and prices are to be derived from any relevant rates or prices in the Contract, with reasonable adjustments having regard to the matters set out in (a) and/or (b).
- If no existing rates and prices are relevant for the purposes of the derivation of a new rate or price, then the new rate or price should be based upon reasonable cost and reasonable profit.

COMMENTARY

It is the Engineer who is to value the work and that valuation is to proceed on a measure and value basis. **12–015**

Under sub-cl.3.5, the Engineer must first try to achieve an agreed valuation. If that is not possible he must make his own determination. That determination must be, as defined in sub-cl.3.5, a fair one which is in accordance

259

with the Contract and which has taken due regard of all of the relevant circumstances.

The considered rule for such an evaluation is stated in the opening words of the second paragraph of sub-cl.12.3: the appropriate rate and price for an item of work shall be the rate or price for that work specified in the Contract. This sub-clause, perhaps in an attempt to limit claims, expressly limits the areas where new rates can apply. Pricing or measurement errors by the Contractor can often occur. So far as measurement errors are concerned, where the Contractor is to be paid a rate or price per unit measured, then the error will disappear.

12–016   Where there are "errors" in the pricing of the unit measured, then, save in the very exceptional case where there is some basis for rectification, the Contractor will be held to his rate or price. There is no basis for the adjustment of the rate or price on this ground alone: *Henry Boot Construction v Alstom*[1]. The focus in sub-cl.12.3 is the definition of the situations and circumstances in which a rate in the Contract will be inappropriate so that a new rate or price is to be found. It is to be noted that a variation may or may not be the trigger for this process.

The position in relation to *variations* is straightforward: the question is whether rate is, in effect, *appropriate*, for the varied work. The more difficult area, addressed by this sub-clause, is where quantities have increased or decreased without there being a variation.

MDB HARMONISED EDITION

12–017   The rates shown in sub-paras (a)(i) and (ii), which can trigger the use of rates other than those specified in the Contract, have been increased from 10 per cent and 0.01 per cent to 25 per cent and 0.25 per cent respectively. This seems to be a pro-Employer change as the increase in the threshold amount is of no benefit to the Contractor.

A new third paragraph has been added as follows:

> *Any item of work included in the Bill of Quantities for which no rate or price was specified shall be included in other rates and process in the Bills of Quantities and will not be paid for separately.*

12–018   This could have a potentially severe effect on the Contractor depending on the nature and cost of any item left unpriced in the Bill of Quantities as this addition clearly makes the Contractor responsible for any additional costs caused by the omission.

---

[1] [1999] BLR 123 (TCC); and 2000 BLR 247 (CA).

Finally, the words *"as soon as the concerned Works commences"* have been added to the end of the final paragraph. This sensible addition sets a time limit as to when the Engineer should set any provisional rates.

12.4 OMISSIONS

Whenever the omission of any work forms part (or all) of a Variation, the value   **12–019** *of which has not been agreed, if:*

(a)   *the Contractor will incur (or has incurred) cost which, if the work had not been omitted, would have been deemed to be covered by a sum forming part of the Accepted Contract Amount;*

(b)   *the omission of the work will result (or has resulted) in this sum not forming part of the Contract Price; and*

(c)   *this cost is not deemed to be included in the evaluation of any substituted work;*

then the Contractor shall give notice to the Engineer accordingly, with supporting particulars. Upon receiving this notice, the Engineer shall proceed in accordance with Sub-Clause 3.5 [Determinations] to agree or determine this cost, which shall be included in the Contract Price.

OVERVIEW OF KEY FEATURES

- The Contractor must give notice where a variation gives rise to an   **12–020** omission, or element of omission.
- If the omission produces abortive work for the Contractor or if the omission will result in a sum not forming part of the Contract Price, or if the cost is not part of replacement work, then upon receipt of the notices, the Engineer shall proceed to "agree or determine" the cost flowing from the omission.

COMMENTARY

It is open to the Engineer and the Contractor to agree the *value* of any omis-   **12–021** sion. However, if there is no agreement the Contractor must identify a *cost* which has been (or will be) incurred and which would have been reasonable under the Contract (absent the omission) but which is now not recoverable (in light of the omission) and will not be recovered under substituted work.

The use of the word cost should be noted. It is not "Cost". Accordingly the contract definition which excludes profit, to be found at sub-cl.1.1.4.3 will not apply and the Contractor should make provision accordingly in any claim which it might submit.

MDB HARMONISED EDITION

12–022   There is no change.

CLAUSE 13 – VARIATIONS AND ADJUSTMENTS

13.1 RIGHT TO VARY

Variations may be initiated by the Engineer at any time prior to issuing the Taking-Over Certificate for the Works, either by an instruction or by a request for the Contractor to submit a proposal.

13–001

The Contractor shall execute and be bound by each Variation, unless the Contractor promptly gives notice to the Engineer stating (with supporting particulars) that the Contractor cannot readily obtain the Goods required for the Variation. Upon receiving this notice, the Engineer shall cancel, confirm or vary the instruction.

Each Variation may include:

(a) changes to the quantities of any item of work included in the Contract (however, such changes do not necessarily constitute a Variation),

(b) changes to the quality and other characteristics of any item of work,

(c) changes to the levels, positions and/or dimensions of any part of the Works,

(d) omission of any work unless it is to be carried out by others,

(e) any additional work, Plant, Materials or services necessary for the Permanent Works, including any associated Tests on Completion, boreholes and other testing and exploratory work, or

(f) changes to the sequence or timing of the execution of the Works.

The Contractor shall not make any alteration and/or modification of the Permanent Works, unless and until the Engineer instructs or approves a Variation.

13–002

OVERVIEW OF KEY FEATURES

- Variations can be initiated at any time prior to the issue of the Taking-Over Certificate by the Engineer.

13–003

- The initiation of a variation is either by way of instruction or the request for a proposal from the Contractor.
- The Contractor is bound to perform unless he gives a notice stating that he cannot obtain the Goods required for the Variation.

- There is a wide definition of what a Variation may include.
- The Contractor may not of itself execute any change to the Works.

COMMENTARY

13–004   The contractual machinery for additional and/or varied work is a difficult and important area in the Engineering Contract. There is often a complex question involved in determining in the first place whether the work claimed by the Contractor is truly extra and/or varied, having regard to all of the Contract documents (and all of those things of which the Contractor is to be regarded as having had notice when he made the Contract and put in his tender). The mere fact that a Contractor has carried out extra work is not itself sufficient to give rise to a right to claim additional payment. On first principles, the Contractor must show an express (or implied) right to seek payment for this work.

A further difficult area is the question of written requests for additional work and/or written orders, and then the consequential issue of whether, and in what circumstances, the Contractor can recover payment for additional work without formal requests and/or written orders. It will be a question of construction in each case, but the requirements for requests to be in writing – or some stipulation as to a proper written order – may often be regarded as a condition precedent to the right to claim payments for extras.

Sub-clause 13.1 gives the Engineer the right to initiate a variation at any time prior to the issuing of a Taking-Over Certificate for the Works. However, it would appear that the Engineer is not permitted to initiate a variation during the Defects Notification Period. There is no restriction on this right to initiate a variation (subject of course to the delay and/or disruptive effects of such variation being borne by the Employer) and it is submitted that there is, with such terms, difficulty in an argument (perhaps based on an alleged implied term) that the procurement of work through variations ought to be carried out in a particular way by the Engineer and/or Employer.

13–005   The Contractor must carry out the variation if instructed. It has limited grounds to object. Even if the Contractor gives notice that it cannot readily obtain the goods required, the Engineer has the option of cancelling, varying, or confirming (and thereby ignoring the Contractor) the instructions.

The matters which can be made the subject-matter of an instructed variation under cl.13 are very wide indeed: it is to be noted, in particular, that sub-cl.(f) allows the Engineer to proceed by way of instructed variation in respect of changes to the sequence or timing of the execution of the Works. However, the power does not relate to the timing of the completion of the works. Acceleration and the bringing forward of the time of completion are covered by sub-cll 8.6 and 13.2.

The following words are very significant:

the Contractor shall not make any alteration and/or modification of the Permanent Works, unless and until the Engineer instructs or approves a Variation . . ..

Thus the variation procedure expressly requires that any change is initiated by the Employer (through the Engineer). This is an obligation of some significance as until the variation procedure (set out in sub-cl.13.3) has been complied with, sub-cl.13.1 provides that no alteration and/or modification in the Permanent Works shall be made by the Contractor. The question which arises in such a case is whether, if the Engineer were to seek to instruct a variation other than in accordance with the variation procedure, the Contractor should comply and make the alteration and/or modification. **13–006**

It is submitted that the sensible and prudent course for the Contractor is that he should insist upon compliance with the variation procedure prior to the carrying out of any alteration and/or modification work, particularly if he wishes to avoid a debate in relation to the recovery of the costs for such work. That said, there may be occasions where the need for a speedy or instantaneous decision is called for, and strictly this is not something covered by this sub-clause. Therefore the onus would be on the parties to agree[1] something amongst themselves to maintain progress.

This requirement could also cause difficulties if the Employer holds that there has been no change to the project. Whilst always dependent on the particular facts of the individual project, in reality such an argument would be difficult to maintain and might leave the Employer open to charges that it was acting unreasonably and/or in breach of its obligations of good faith (if any)[1a] or even that the effect of such an argument was to cause a wholesale breakdown of the change mechanism set out in this sub-clause.

As noted above, it appears that after the issuance of a Taking-Over Certificate, the Engineer can no longer instruct variations. Indeed, the only instructions that the Engineer can issue at this stage are those contemplated by cl.11. It is important, therefore, that Employers realize that the Works can no longer be varied after the issuance of the Taking-Over Certificate. If it is decided, after most of the project has been built, that additional facilities are required then presumably these additional works will need to be the subject of a separate side-agreement. **13–007**

---

[1] And to record that agreement.
[1a] See sub-cl.4.10 for fuller discussion on this point.

MDB HARMONISED EDITION

13–008 In the second paragraph, the MDB edition has introduced a new exception whereby the Contractor may not be bound to execute each variation and that is where:

> *Such Variation triggers a substantial change in the sequence or progress of the Works*

It will be for the Contractor to demonstrate in its notice why the change is a substantial one. Nevertheless the Engineer still has the right, having considered the objection, to ignore the Contractor and proceed to confirm the variation.

13.2 VALUE ENGINEERING

13–009 *The Contractor may, at any time, submit to the Engineer a written proposal which (in the Contractor's opinion) will, if adopted (i) accelerate completion, (ii) reduce the cost to the Employer of executing, maintaining or operating the Works, (iii) improve the efficiency or value to the Employer of the completed Works, or (iv) otherwise be of benefit to the Employer.*

The proposal shall be prepared at the cost of the Contractor and shall include the items listed in Sub-Clause 13.3 [Variation Procedure].

If a proposal, which is approved by the Engineer, includes a change in the design of part of the Permanent Works, then unless otherwise agreed by both Parties:

13–010 *(a)   the Contractor shall design this part,*
(b)   sub-paragraphs (a) to (d) of Sub-Clause 4.1 [Contractor's General Obligations] shall apply and
(c)   if this change results in a reduction in the contract value of this part, the Engineer shall proceed in accordance with Sub-Clause 3.5 [Determinations] to agree or determine a fee, which shall be included in the Contract Price. This fee shall be half (50%) of the difference between the following amounts:

> *(i)   such reduction in contract value, resulting from the change, excluding adjustments under Sub-Clause 13.7 [Adjustments for Changes in Legislation] and Sub-Clause 13.8 [Adjustments for Changes in Cost], and*
>
> *(ii)   the reduction (if any) in the value to the Employer of the varied works, taking account of any reductions in quality, anticipated life or operational efficiencies.*

However, if amount (i) is less than amount (ii), there shall not be a fee.

266

OVERVIEW OF KEY FEATURES

- The Contractor may make a value engineering proposal at any time. **13–011**
- The proposal is prepared at the cost of the Contractor and shall comply with the list of requirements set out in sub-cl.13.3 (Variations).
- If there is a design element in the proposal, then it is for the Contractor to ensure the design complies with sub-cl.4.1.
- If there is a resulting reduction in the Contract value, then the fee is calculated in accordance with the prescribed formula which clearly incentivises the making of savings.

COMMENTARY

Value engineering is an increasingly important part of the construction **13–012**
process, both in the United Kingdom and abroad. By mechanisms which display a wide range of complexity, parties to substantial engineering and construction contracts seek to derive mutual benefit, and to benefit jointly from the project as a whole. The aim is that each party is encouraged, and in particular the Contractor is encouraged, in the initiative in respect of the possibility of saving and/or gains.

In sub-cl.13.2 there is a complete freedom on the part of both the Contractor to submit value engineering proposals and on the Engineer to accept or reject those proposals. However, if the proposal is accepted, then the Contractor gets 50 per cent of the *net* benefit of the proposal to the Employer. There are some complexities involved in the concept of net benefit. Problems may arise in practice where the capital expenditure involved in a different piece of equipment is less but the costs of upkeep and maintenance may be higher.

Again, and as one commentator has pointed out,[2] value engineering proposals made by the Contractor (resulting in the Employer having to share net gains) could re-bound in the form of criticism of the Engineer's original proposals.

The final paragraph of this sub-clause sets out the procedure if the parties **13–013**
do not set-out what has been agreed in respect of any value engineering items that may be adopted. The Contractor should note that the fall-back position is that it is responsible for the design aspect of any such changes[3].

---

[2] Edward Corbett "FIDIC's New Rainbow 1st Edition – An Advance?" [2000] 1 CLR 253 at 264.
[3] See sub-cl.4.1 above for discussion of the Contractor's design obligations.

MDB HARMONISED EDITION

13–014   There is no change.

13.3 VARIATION PROCEDURE

13–015   *If the Engineer requests a proposal, prior to instructing a Variation, the Contractor shall respond in writing as soon as practicable, either by giving reasons why he cannot comply (if this is the case) or by submitting:*

   (a)   a description of the proposed work to be performed and a programme for it's execution,
   (b)   the Contractor's proposal for any necessary modifications to the programme according to Sub-Clause 8.3 [Programme] and to the Time for Completion, and
   (c)   the Contractor's proposal for evaluation of the Variation.

   The Engineer shall, as soon as practicable after receiving such proposal (under Sub-Clause 13.2 [Value Engineering] or otherwise), respond with approval, disapproval or comments. The Contractor shall not delay any work whilst awaiting a response.

13–016   *Each instruction to execute a Variation, with any requirements for the recording of Costs, shall be issued by the Engineer to the Contractor, who shall acknowledge receipt.*

   Each Variation shall be evaluated in accordance with Clause 12 [Measurement and Evaluation], unless the Engineer instructs or approves otherwise in accordance with this Clause.

OVERVIEW OF KEY FEATURES

13–017   • Prior to issuing a variation, the Engineer may issue a request for a proposal to the Contractor.
   • The Contractor must respond to that Request, either saying why he cannot comply or by providing a proposal for the Works; a programme proposal; and a proposed evaluation.
   • The Engineer is then to respond, either agreeing or setting out reasons for disagreement.
   • The Contractor is not to delay any work whilst awaiting a response.
   • The instruction is then to be issued by the Engineer and receipt acknowledged by the Contractor.

- Valuation of the Variation is to occur in accordance with the principles of cl.12.

COMMENTARY

The variation procedure set out in sub-cl.13.3 is relatively straightforward, **13–018** and places some emphasis upon a procedure whereby the Engineer, if he so chooses, requests proposals from the Contractor. Whilst this provision is optional, the emphasis upon a procedure based on Contractor's proposals has the practical advantage that the Engineer can use the Contractor's experience and knowledge of the Works to best advantage in deciding how to instruct the additional work. The sub-clause makes no reference to the costs of submitting proposals. These may be significant, if substantial design work is required. The proposals must be prepared as quickly as practicable. No definition of this time period is provided, but the timing will depend on the circumstances of the Project and the amount of documentation included.

The sub-clause expressly provides that the Contractor shall not delay any work whilst awaiting a response from the Engineer in relation to, for example, the Engineer's approval of the proposed work or method of working. Whilst the Engineer is supposed to respond as quickly as practicable, this provision may cause complications in practice and place an additional burden of risk upon the Contractor. Having made a proposal, and whilst awaiting the comments/approval of the Engineer, the Contractor is not permitted to delay the relevant works and in proceeding there must be a risk that the work undertaken will not be ultimately given the approval of the Engineer. This uncertainty means that this is an area where the FIDIC Form is less clear than it might be in providing a sensible commercial allocation of risk.

Although straightforward, it would be fair to say that the approval process is very formal, even if "approval" as such is not defined. Practically, as noted above in the discussion of sub-cl.13.1, changes on site are often agreed orally or through informal correspondence. In reality these more formal changes will still fall under the operation of sub-cl.13.3.

MDB HARMONISED EDITION

There is no change. **13–019**

13.4 PAYMENT IN APPLICABLE CURRENCIES

13–020 *If the Contract provides for payment of the Contract Price in more than one currency, then whenever an adjustment is agreed, approved or determined as stated above, the amount payable in each of the applicable currencies shall be specified. For this purpose, reference shall be made to the actual or expected currency proportions of the Cost of the varied work, and to the proportions of various currencies specified for payment of the Contract Price.*

OVERVIEW OF KEY FEATURES

13–021 Where different currencies are specified, the proportions of currency payments should be specified.

COMMENTARY

13–022 This sub-clause confirms that the currency provisions set out in the Contract will apply to variations.

Indeed, it should be noted that, under the current interpretation of s.48(4) of the Arbitration Act 1996 given by the House of Lords in the case of *Lesotho Highlands Development Authority v Impreglio SpA and Others*,[4] an arbitral tribunal sitting in England and Wales would commit an error of law if it decided to issue an award in a currency different from that provided for under the contract.

MDB HARMONISED EDITION

13–023 There is no change.

13.5 PROVISIONAL SUMS

13–024 *Each Provisional Sum shall only be used, in whole or in part, in accordance with the Engineer's instructions, and the Contract Price shall be adjusted*

[4] (2005) W.L.R. 129.

accordingly. The total sum paid to the Contractor shall include only such amounts, for the work, supplies or services to which the Provisional Sum relates, as the Engineer shall have instructed. For each Provisional Sum, the Engineer may instruct:

(a) *work to be executed (including Plant, Materials or services to be supplied) by the Contractor and valued under Sub-Clause 13.3 [Variation Procedure]; and/or*

(b) *Plant, Materials or services to be purchased by the Contractor, from a nominated Subcontractor (as defined in Clause 5 [Nominated Subcontractors]) or otherwise; and for which there shall be included in the Contract Price:*

(i) *the actual amounts paid (or due to be paid) by the Contractor, and* **13–025**
(ii) *a sum for overhead charges and profit, calculated as a percentage of these actual amounts by applying the relevant percentage rate (if any stated in the appropriate Schedule. If there is no such rate, the percentage rate stated in the Appendix to Tender shall be applied.*

The Contractor shall, when required by the Engineer, produce quotations, invoices, vouchers and accounts or receipts in substantiation.

OVERVIEW OF KEY FEATURES

- Provisional sums must be expended in accordance with instructions issued **13–026** from the Engineer.
- The Contract Price is then adjusted accordingly.
- The powers of the Engineer in relation to the expenditure of Provisional Sums are defined widely in sub-paras (a) and (b).
- The expenditure of the Contractor shall be substantiated as and when required by the Engineer.

COMMENTARY

Provisions Sums are defined in sub-cl.1.1.4.10 – a definition which makes it **13–027** clear that provisional sums relate solely to sub-cl.13.5. Provisional sums are only to be used when the Engineer so instructs. The sub-clause envisages a tight control on provisional sums, noting that the Engineer should instruct either the work to be carried out or the materials to be purchased. The Contractor should take care to keep appropriate records as the Engineer may well (and indeed should) require proof through receipted invoices of the costs expended.

MDB HARMONISED EDITION

13–028   The words "Contract Data" replace "Appendix to Tender".

13.6 DAYWORK

13–029   *For work of a minor or incidental nature, the Engineer may instruct that a Variation shall be executed on a daywork basis. The work shall then be valued in accordance with the Daywork Schedule included in the Contract, and the following procedure shall apply. If a Daywork Schedule is not included in the Contract, this Sub-Clause shall not apply.*

   Before ordering Goods for the work, the Contractor shall submit quotations to the Engineer. When applying for payment, the Contractor shall submit invoices, vouchers and accounts or receipts for any Goods.

   Except for any items for which the Daywork Schedule specifies that payment is not due, the Contractor shall deliver each day to the Engineer accurate statements in duplicate which shall include the following details of the resources used in executing the previous day's work:

13–030   *(a)   the names, occupations and time of Contractor's Personnel,*
   (b)   the identification, type and time of Contractor's Equipment and Temporary Works, and
   (c)   the quantities and types of Plant and Materials used.

One copy of each statement will, if correct, or when agreed, be signed by the Engineer and returned to the Contractor. The Contractor shall then submit priced statements of these resources to the Engineer, prior to their inclusion in the next Statement under Sub-Clause 14.3 [Application for Interim Payment Certificates].

OVERVIEW OF KEY FEATURES

13–031   Provided a Daywork Schedule is attached as part of the Contract:
   • The Engineer may instruct variations on a daywork basis if the work concerned is of a minor nature.

- Such daywork shall be valued in accordance with the Daywork Schedule.
- The Contractor must deliver daily statements of the work carried out, including details of the personnel and materials used.
- The daily statements must be counter-signed and agreed by the Engineer.

COMMENTARY

The nature of work carried out on a daywork basis can easily become uncontrolled and extended which can lead to unanticipated expense for the Employer. The purpose of sub-cl.13.6 is expressly to limit situations in which daywork can be carried out.[5]   **13–032**

One of the important points about sub-cl.13.6 is the opening words: "*for work of a minor or incidental nature . . .*". Accordingly, the power given to the Engineer to require that Works be carried out on a daywork basis is restricted to such type of work although, following the instruction of the carrying out of work on a daywork basis by the Engineer, it would seem to be very difficult for the Employer to deny payment on such a basis by reference to the argument that the work is *not* minor or incidental in its nature.

Accordingly, the Engineer will need to take great care in the way he uses the power to instruct Dayworks.

As a second point to note, the Contract specifically provides that ". . . *if a Dayworks Schedule is not included in the Contract, this sub-clause shall not apply . . .*". This provision has the practical advantage of ensuring that, where daywork is instructed, and the cost risks of doing so are relatively well-defined by reference to the Daywork Schedule to be found in the Contract. If the Daywork Schedule is not included then the Employer is likely to have no option but to agree dayworks at short notice and thus at a rate that is likely to be higher than it might otherwise have been.   **13–033**

The FIDIC Guide recommends that the Daywork Schedule includes time charge rates for individuals and equipment and the payment due for each category of materials, a term delivered by sub-cl.1.1.5.3.

The importance of ensuring the daily record statements are countersigned should not be underestimated, by both parties. In the case of *JDM Accord Ltd v Secretary of State for the Environment, Food and Rural Affairs,*[6] the Employer in the midst of the Foot and Mouth epidemic in the UK failed to verify, confirm or reject timesheets. Later, the Court held that it was the Employer who carried the evidential burden of showing that the timesheets were inaccurate. Whilst this is not a result which will always be repeated, it

---

[5] And also perhaps to limit protracted claims as the Contractor seeks recovery of what, it says, it is entitled to.
[6] [2004] EWHC 2 (TCC) 16 January, 2004.

clearly demonstrates the risks in failing to ensure your site procedures are set up to deal with tasks of this nature.

MDB HARMONISED EDITION

13–034   There is no change.

13.7 ADJUSTMENTS FOR CHANGES IN LEGISLATION

13–035   *The Contract Price shall be adjusted to take account of any increase or decrease in Cost resulting from a change in the Laws of the Country (including the introduction of new Laws and the repeal or modification of existing Laws) or in the judicial or official governmental interpretation of such Laws, made after the Base Date, which affect the Contractor in the performance of obligations under the Contract.*

If the Contractor suffers (or will suffer) delay and/or incurs (or will incur) additional Cost as a result of these changes in the Laws or in such interpretations, made after the Base Date, the Contractor shall give notice to the Engineer and shall be entitled subject to Sub-Clause 20.1 [Contractor's Claim] to:

(a)   an extension of time for any such delay, if completion is or will be delayed, under Sub-Clause 8.4 [Extension of Time for Completion], and
(b)   payment of any such Cost, which shall be included in the Contract Price.

After receiving this notice, the Engineer shall proceed in accordance with Sub-Clause 3.5 [Determinations] to agree or determine these matters.

OVERVIEW OF KEY FEATURES

13–036
- Where there is a change in the law of the country which affects the performance of the Contractor the Contract Price shall be adjusted.
- If the Contractor considers that it will suffer delay or increased cost it must give notice to the Engineer within 28 days as required by sub-cl.20.1.
- If the Engineer so determines, the Contractor may be entitled to an extension of time and payment of cost.

COMMENTARY

This sub-clause relates to the impact of changes to the law of the country that **13–037** the site (or the majority of that site) is located in.[7] In other words it deals with changes such as in taxation legislation rather than those caused by economic factors such as inflation. The type of change envisaged might include changes in labour laws which might affect clause 6 or perhaps the introduction of trade embargos or sanctions.

This is a relatively sophisticated provision which deals in greater detail than other standard forms with legislative changes during the course of the Contract. Whilst this is not a new concept, it has been extended. Sub-clause 70.2 of the Old Red Book FIDIC 4th edn, provided that account would be taken of any subsequent legislation, however it only referred to cost and not time.

The sub-clause is thus in two parts. In the first, the aim of the sub-clause is to insulate both the Contractor and the Employer (depending on whether the Contract Price goes up or down) from the impact of any changes in the law which take place after the base date – defined by the date 28 days prior to the latest date for submission of the Tender.[8] The second part provides that the Contractor may be entitled to an extension of time and/or additional costs as a consequence of any legislative changes. It would be difficult for the Employer to claim a reduction in the time for completion and no attempt to give any such right has been provided here.

The sub-clause makes no reference to whether the Contractor is able to **13–038** make a claim for additional costs or time as a consequence of a change in the law which he could have reasonably foreseen upon entering into the contract. This is notwithstanding that sub-cl.1.13 makes it clear that the Contractor shall comply with applicable laws and pay taxes and duties. As it does not, it is submitted that, although contrary to the discussions about unforeseeable above[9], the question of forseeability does not apply here. This is in particular when sub-cl.8.5 (delays caused by authorities) specifically imposes at subs.(c) a requirement that any delay caused by following legal procedures must have been unforeseeable. Thus, especially as legislation is usually subject to a lengthy lead-in period, the prudent Employer might want to give consideration to whether there are any legislative changes which might be in the pipeline and, if appropriate, require that due allowance of these potential changes is made by the Contractor in its tender.

---

[7] See sub-cll 1.1.6.2 and 1.1.6.7.
[8] See sub-cll 1.1.3.1.
[9] See sub-cll 1.1.6.8 and 4.12.

MDB HARMONISED EDITION

13–039   The following has been added to the end of the sub-clause:

> *"Notwithstanding the foregoing, the Contractor shall not be entitled to an extension of time if the relevant delay has already been taken into account in the determination of a previous extension of time and such Costs shall not be separately paid if the same shall already have been taken into account in the indexing of any inputs to the table of adjustment data in accordance with the provisions of Sub-Clause 13.8 [Adjustments for Changes in Cost]."*

This has been introduced to ensure that there is no possibility of the Contractor being able to duplicate either any claim for additional time or money as a consequence of this sub-clause, notwithstanding the obvious difficulties in any such attempt being made in the first place.

13.8 ADJUSTMENTS FOR CHANGES IN COST

13–040   *In this Sub-Clause, "table of adjustment date" means the completed table of adjustment data included in the Appendix to Tender. If there is no such table of adjustment data, this Sub-Clause shall not apply.*

If this Sub-Clause applies, the amounts payable to the Contractor shall be adjusted for rises or falls in the cost of labour, Goods and other inputs to the Works, by the addition or deduction of the amounts determined by the formulae prescribed in this Sub-Clause. To the extent that full compensation for any rise or fall in Costs is not covered by the provisions of this or other Clauses, the Accepted Contract Amount shall be deemed to have included amounts to cover the contingency of other rises and falls in costs.

The adjustment to be applied to the amount otherwise payable to the Contractor, as valued in accordance with the appropriate Schedule and certified in Payment Certificates, shall be determined from formulae for each of the currencies in which the Contract Price is payable. No adjustment is to be applied to work valued on the basis of Cost or current prices. The formulae shall be of the following general type:

$$Pn = a + b\,\frac{Ln}{Lo} + c\,\frac{En}{Eo} + d\,\frac{Mn}{Mo} + \ldots\ldots$$

13–041   *Where:*

"Pn" is the adjustment multiplier to be applied to the estimated contract value in the relevant currency of the work carried out in period "n", this period being a month unless otherwise stated in the Appendix to Tender;

"a" is a fixed coefficient, stated in the relevant table of adjustment data, representing the non-adjustable portion in contractual payments;

"b", "c", "d", . . . are coefficients representing the estimated proportion of each cost element related to the execution of the Works, as stated in the relevant table of adjustment data; such tabulated cost elements may be indicative of resources such as labour, equipment and materials;

"Ln", "En", "Mn", . . . are current cost indices or reference prices for period "n", expressed in the relevant currency of payment, each of which is applicable to the relevant tabulated cost element on the date 49 days prior to the last day of the period (to which the particular Payment Certificate relates); and   **13–042**

"Lo", "Eo", "Mo", . . . are the base indices or reference prices, expressed in the relevant currency of payment, each of which is applicable to the relevant tabulated cost element on the Base Date.

The cost indices or reference prices stated in the table of adjustment data shall be used. If their source is in doubt, it shall be determined by the Engineer. For this purpose, reference shall be made to the values of the indices at stated dates (quoted in the fourth and fifth columns respectively of the table) for the purposes of clarification of the source; although these dates (and thus these values) may not correspond to the base cost indices.

In cases where the "currency of index" (stated in the table) is not the relevant currency of payment, each index shall be converted into the relevant currency of payment at the selling rate, established by the central bank of the Country, of this relevant currency on the above date for which the index is required to be applicable.   **13–043**

Until such time as each current cost index is available, the Engineer shall determine a provisional index for the issue of Interim Payment Certificates. When a current cost index is available, the adjustment shall be recalculated accordingly.

If the Contractor fails to complete the Works within the Time for Completion, adjustment of prices thereafter shall be made using either (i) each index or price applicable on the date 49 days prior to the expiry of the Time for Completion of the Works, or (ii) the current index or price: whichever is more favourable to the Employer.

The weightings (coefficients) for each of the factors of cost stated in the table(s) of adjustment data shall only be adjusted if they have been rendered unreasonable, unbalanced or inapplicable, as a result of Variations.   **13–044**

277

OVERVIEW OF KEY FEATURES

13–045
- This sub-clause only applies if the adjustment data table is part of the Contract.
- If this sub-clause applies, the formulae set out must be used to adjust changes in the cost of labour and materials.

COMMENTARY

13–046 The detailed provisions of sub-cl.13.8 are in large parts typical of those to be found in complex, long-term construction engineering contracts in relation to price escalation. Here, it is intended that where this sub-clause does apply, it is the Employer who bears the risk of any cost increases. However the parties should note that the particular feature of sub-cl.13.8 is that it proceeds only by reference to the "*table of adjustment data*" which is included in the Appendix to the Tender – without which the entire sub-clause does not apply.

MDB HARMONISED EDITION

13–047 There are a handful of small changes.

The words "included in the Appendix to Tender" in the first paragraph have been replaced by "for local and foreign currencies in the Schedules".

Elsewhere, the words "Appendix to Tender" have been replaced by "Contract Date".

Finally the words in brackets in the fourth and fifth paragraphs, which refer to the table, have been deleted. However the table itself, it should be noted, remains.

CLAUSE 14 – CONTRACT PRICE AND PAYMENT

14.1 THE CONTRACT PRICE

(a)   *the Contract Price shall be agreed or determined under Sub-Clause 12.3*   **14–001**
[Evaluation] and be subject to adjustments in accordance with the Contract;

(b)   *the Contractor shall pay all taxes, duties and fees required to be paid by him under the Contract, and the Contract Price shall not be adjusted for any of these costs except as stated in Sub-Clause 13.7 [Adjustment for Changes in Legislation];*

(c)   *any quantities which may be set out in the Bill of Quantities or other Schedule are estimated quantities and are not to be taken as the actual and correct quantities:*

   (i)   *of the Works which the Contractor is required to execute, or*   **14–002**
   (ii)   *for the purposes of Clause 12 [Measurement and Evaluation]; and*

(d)   *the Contractor shall submit to the Engineer, within 28 days after the Commencement Date, a proposed breakdown of each lump sum price in the Schedules. The Engineer may take account of the breakdown when preparing Payment Certificates, but shall not be bound by it.*

OVERVIEW OF KEY FEATURES

- The Contract Price shall be evaluated by the Engineer in accordance with   **14–003**
  the provisions of sub-cl.12.3.
- It is the responsibility of the Contractor to pay all taxes and other fees as required.
- Any quantity set out in the Bill of Quantities or other schedule is an estimated figure only.
- It is the responsibility of the Contractor to submit, within 28 days after the Commencement Date, a non-binding breakdown of each lump sum price shown in any schedule.

COMMENTARY

14–004   The amount the Contractor is going to be paid, and the timing of that payment is of fundamental importance to both Contractor and Employer alike. The manner in which the payment is made is traditionally dependent on the precise wording of the contract.[1] Under the code of Hammurabi[2] the rule was as follows:

> *If a builder build a house for some one and complete it, he shall give him a fee of two shekels in money for each sar of surface.*

Thus, the amount to be paid was clear and given that the punishment for violating most of the provisions of the Code was death, it might be presumed that most builders were paid, provided the house was constructed properly. However, the rule does not say when the payment has to be made.

Clause 14 provides the basis for the FIDIC payment regime. It duly sets out when the amounts due are to be paid out by the Employer. The timing of payment is primarily a commercial matter and it is important both parties are content with the proposed procedures.

14–005   Under sub-cl.1.1.4.2, the Contract Price is stated to be the "price defined in sub-cl.14.1". This is slightly misleading as this sub-clause immediately cross-refers to the mechanism set out in sub-cl.12.3 whereby the Engineer will either agree or determine the Contract Price.

Sub-paragraph (b) reconfirms the requirement detailed in sub-cl.1.13 that the Contractor must comply with the law of the contract and pay any taxes or other duties. Where sub-para.(b) does not apply, the Particular Conditions suggest example sub-clauses dealing with the payment of duties or taxes. Obviously, it is the responsibility of the Contractor to ensure that he fully understands any tax and excise laws of the country where the project is based prior to the commencement of the Contract, ideally prior to submitting the tender. That said, the Contractor has some protection as from sub-cl.13.7 it appears that if the tax rates go up (or as occasionally happens go down) the Contract Price will be adjusted accordingly.

The final sub-paragraph requires the Contractor to submit a breakdown of every lump sum price. This needs to be done quite promptly, within 28 days of the commencement date. However this will serve as no more than a guide to the Engineer as the sub paragraph expressly makes clear that he is not bound by it.

As noted above, the FIDIC form is a "measure and value" contract. Thus a number of the measurement and payment sub-clauses in the standard

---

[1] Although in the UK, s.109 of the 1998 Housing Grant Construction & Regeneration Act, now gives most contractors the right to payment by instalments.
[2] King Hammurabi ruled the kingdom of Babylon from 1792 to 1750 BC.

FIDIC form, for example sub-cl.14.1(a), would be inappropriate for contracts let on a lump sum basis. With lump sum contracts, the contract sum is fixed, subject to the correction of any errors and adjustment to the scope of the works by way of a variation or change order. With the measure and value contract, the Contractor is paid the actual value of the works carried out – the valuation being based on the Bill of Quantities or if applicable the Schedules.

There is small scope for confusion as it could be argued that strictly re-measurement contracts are the equivalent of lump sum contracts in that the rates for the work are fixed and thus the individual rates for each element in the bill of quantities or other schedules are also individual lump sums in their own rights. In other words, while the contractor is to be paid for the items as eventually carried out and measured, the contractor will be paid the rate upon which his original tender was based. That is not the case here as sub-para.(c) confirms that Bills of Quantities are estimated only and will thus remain to be calculated. In fact, if the quantities change substantially, then arguments might be raised that the rate should be varied because of the substantial change in the quantities, resulting in a change to the nature of the works.

14–006

If the lump sum form is to be adopted, the Particular Conditions recommend the following amendments:

Delete clause 12.
Delete the last sentence of sub-clause 13.3 and substitute:

"Upon instructing or approving a Variation, the Engineer shall proceed in accordance with sub-clause 3.5 to agree or determine adjustments to the Contract Price and to the Schedule of Payments under Sub-clause 14.4. These adjustments shall include reasonable profit, and shall take account of the Contractor's submissions under sub-clause 13.2 if applicable."

Delete sub-paragraph (8) of sub-clause 14.1 and substitute:

(a)   The Contract Price shall be the lump sum accepted contract amount and shall be subject to adjustments in accordance with the Contract.

MDB HARMONISED VERSION

The following sub-cl. has been added:

14–007

(e) Notwithstanding the provisions of sub paragraph (b), Contractor's Equipment, including essential spare parts therefore, imported by the Contractor for the sole purpose of executing the Contract shall be exempt from the payment of import duties and taxes upon importation.

This introduces a potentially significant exception to the Contractor's liability to pay taxes and other duties.

14.2 ADVANCE PAYMENT

14–008   *The Employer shall make an advance payment, as an interest-free loan for mobilisation, when the Contractor submits a guarantee in accordance with this Sub-Clause. The total advance payment, the number and timing of instalments (if more than one), and the applicable currencies and proportions, shall be as stated in the Appendix to Tender.*

Unless and until the Employer receives this guarantee, or if the total advance payment is not stated in the Appendix to Tender, this Sub-Clause shall not apply.

The Engineer shall issue an Interim Payment Certificate for the first instalment after receiving a Statement (under Sub-Clause 15.3 [Application for Interim Payment Certificates]) and after the Employer receives (i) the Performance Security in accordance with Sub-Clause 4.2 [Performance Security] and (ii) a guarantee in amounts and currencies equal to the advance payment. This guarantee shall be issued by an entity and from within a country (or other jurisdiction) approved by the Employer, and shall be in the form annexed to the Particular Conditions or in another form approved by the Employer.

14–009   *The Contractor shall ensure that the guarantee is valid and enforceable until the advance payment has been repaid, but its amount may be progressively reduced by the amount repaid by the Contractor as indicated in the Payment Certificates. If the terms of the guarantee specify its expiry date, and advance payment has not been repaid by the date 28 days prior to the expiry date, the Contractor shall extend the validity of the guarantee until the advance payment has been repaid.*

The advance payment shall be repaid through percentage deductions in Payment Certificates. Unless other percentages are stated in the Appendix to Tender:

(a)   deductions shall commence in the Payment Certificate in which the total of all certified interim payments (excluding the advance payment and deductions and repayments of retention) exceeds ten per cent (10%) of the Accepted Contract Amount less Provisional Sums; and

(b)   deductions shall be made at the amortisation rate of one quarter (25%) of the amount of each Payment Certificate (excluding the advance payment and deductions and repayments of retention) in the currencies and proportions of the advance payment, until such time as the advance payment has been repaid.

If the advance payment has not been repaid prior to the issue of the Taking-Over Certificate for the Works or prior to termination under Clause 15 [Termination by Employer], Clause 16 [Suspension and Termination by Contractor] or Clause 19 [Force Majeure] (as the case may be), the whole of the balance then outstanding shall immediately become due and payable the Contractor to the Employer. **14–010**

OVERVIEW OF KEY FEATURES

- Provided the Contractor submits a guarantee, which must remain valid **14–011** until the lump sum is repaid to obtain the advance payment, the Employer shall make an advance payment for mobilisation.
- The advance payment is an interest-free loan.
- The total advance payment shall be stated in the Appendix to Tender.
- If the advance payment has not been repaid 28 days prior to the expiry date, the Contractor must extend the validity of the guarantee.
- The advance payment is re-paid through percentage deductions in the payment certificates, which commence when the total of certified interim payments exceeds 10 per cent of the accepted contract amount less any provisional sums.

COMMENTARY

The right to an advance payment is not automatic. The key to this sub-clause **14–012** is that if the proposed advance payment is not set out in the Appendix to Tender, then it will not apply. The suggestion of an advance payment accords with the sentiments set out in the Particular Conditions which when commenting upon sub-cl.14.1 recommend that Employers take the following into account:

> *When writing the Particular Conditions consideration should be given to the amount and timing of payment(s) to the Contractor. A positive cash flow is clearly of benefit to the Contractor, and tenderers will take account of the interim payment procedures when preparing their tenders.*

The advance payment here is stated to be interest free and its purpose is expressly stated to assist with mobilisation. However the Employer is under no obligation to make any advance payment, until the Contractor has submitted a guarantee which conforms with the requirements of this sub-clause. Thus the Employer knows that its loan will be protected.

If it is intended to pay the advance payment in instalments, then the number and timing of the payments of the advance payments should also be stated in the Appendix to Tender. The timing of the payments is governed by sub-cl.14.7(a). The first instalment is to be paid within 42 days of the date of issue of the Letter of Acceptance or within 21 days after receipt of the documents and guarantees referred to in sub-cll 4.2 and 14.2. Equally, the scheme of the proposed payments is something which should be made clear in the advance payment guarantee.

14–013   The sub-clause also sets out the provisions for re-payment. Re-payment commences, unless agreed otherwise, when the certified sums reach 10 per cent of the Accepted Contract Amount and the Contractor should factor in the impact of the timing and in particular consider the cash flow implications of the repayments.

Sub-clause 14.5 provides another optional form of advance payment in relation to Plant and Materials.

MDB HARMONISED EDITION

14–014   There have been a number of minor changes. The words "Appendix to Tender" throughout this sub-clause have been replaced by "Contract Data".

The words "and cash flow support" have been added to follow the words "mobilisation" in the first paragraph, thereby reinforcing in contractual terms, one of the benefits to the Contractor of the advance payment.

In the third paragraph, the first part now reads:

> *The Engineer shall <u>deliver to the Employer and to the Contractor</u> an Interim Payment Certificate for the <u>advance payment</u> first instalment after receiving a statement. . .*

14–015   These words replace the words "issue".

The word "may" in the third line of the fourth paragraph has been replaced by "shall". Therefore, there is now a requirement that the amount of the loan is progressively reduced through the payment certificates.

The fifth paragraph now begins:

> *<u>Unless stated otherwise in the Contract Data</u>, the advance payment shall be repaid through percentage deductions <u>from the interim payments determined by the Engineer in accordance with Sub-Clause 14.6 [issue of interim payment certificates]</u>, as follows:*

In a move which favours the Contractor, the threshold value of when deductions/repayments shall commence to be found in sub paragraph (a) has been increased from 10 per cent to 30 per cent.

In sub-para.(b), the reference to payment certificate is now a reference to an interim payment certificate and the reference to "repayments of retention" has been replaced by "for its repayment as well as deductions for retention money".

14–016

Sub-paragraph (b) now includes a more specific time limit for repayment:

Provided that the advance payment shall be completely repaid prior to the time when 90% (ninety percent) of the Accepted Contract Amount less Provisional Sums has been certified for payment.

Finally, the following has been added to the end of the final paragraph:

. . .in case of termination under Clause 15 [termination by Employer] and Sub-Clause 19.6 [optional termination, payment and release].

14.3 APPLICATION FOR INTERIM PAYMENT CERTIFICATES

The Contractor shall submit a Statement in six copies to the Engineer after the end of each month, in a form approved by the Engineer, showing in detail the amounts to which the Contractor considers himself to be entitled, together with supporting documents which shall include the report on the progress during this month in accordance with Sub-Clause 4.21 [Progress Reports].

14–017

The Statement shall include the following items, as applicable, which shall be expressed in the various currencies in which the Contract Price is payable, in the sequence listed:

(a) *the estimated contract value of the Works executed and the Contractor's Documents produced up to the end of the month (including Variations but excluding items described in sub-paragraphs (b) to (g) below);*

(b) *any amounts to be added and deducted for changes in legislation and changes in cost, in accordance with Sub-Clause 13.7 [Adjustments for Changes in Legislations] and Sub-Clause 13.8 [Adjustments for Changes in Cost];*

14–018

(c) *any amount to be deducted for retention, calculated by applying the percentage of retention stated in the Appendix to Tender to the total of the above amounts, until the amount so retained by the Employer reaches the limit of Retention Money (if any) stated in the Appendix to Tender;*

(d) *any amounts to be added and deducted for the advance payment and repayments in accordance with Sub-Clause 14.2 [Advance Payment];*

(e) *any amounts to be added and deducted for Plant and Materials in accordance with Sub-Clause 14.5 [Plant and Materials intended for the Works];*

14–019

(f)   *any other additions or deductions which may have become due under the Contract or otherwise, including those under Clause 20 [Claims, Disputes and Arbitration]; and*

(g)   *the deduction of amounts certified in all previous Payment Certificates.*

OVERVIEW OF KEY FEATURES

14–020 • The Contractor must submit on a monthly basis, six copies of its application for interim payment.
- The application for interim payment must include the progress report prepared in accordance with sub-cl.4.21.
- It is a further requirement that the application includes the detailed supporting information set out at sub-paras (a)–(g).

COMMENTARY

14–021 This sub-clause provides for a monthly payment cycle, which is initiated by the Contractor submitting to the Engineer detail of the amount it considers it is entitled to be paid. This interim payment will either be calculated by reference to the estimated value of the executed works (sub-cl.14.3(a)) or by the Schedules of Payment, if they form part of the Contract (sub-cl.14.4).

As noted above, the requirement imposed by sub-cl.4.21 to provide detailed monthly progress reports amounts to a pre-condition of payment. This link between payment and the delivery of a monthly report has presumably been inserted because of the particular importance to the Employer of being kept up-to-date with progress.

By virtue of sub-cl.14.7(b), the time within which the Employer must pay the amount certified does not start to run until the Engineer has received the Contractor's statement and supporting documents. Therefore the Contractor should take steps to ensure that there is no room for doubt as to the date the application is delivered. The application must be submitted in accordance with the details required by sub-paras (a)–(g). These are:

14–022 (a)   Current estimated contract value.
(b)   Any adjustments due to legislative changes or cost fluctuations.
(c)   The retention amount.
(d)   The advance payment.
(e)   Amounts for Plant and Materials.
(f)   Amount of Contractor claims or deductions for Employer claims.
(g)   The amount previously certified.

It is therefore important that the Contractor is in a position to provide all that is required by this sub-clause on a regular basis. Any delay to the application will delay payment.

MDB HARMONISED EDITION

The words "Contract Data" replace "Appendix to Tender". 14–023
    The words "for the advance payment and (if more than one instalment) and to be deducted for its" replace "and deducted for the advance payment and" in sub-para.14.3. These merely serve to clarify the existing wording.

14.4 SCHEDULE OF PAYMENTS

If the Contract includes a schedule of payments specifying the instalments in 14–024
which the Contract Price will be paid, then unless otherwise stated in this schedule:

(a)   *the instalments quoted in this schedule of payments shall be the estimated contract values for the purposes of sub-paragraph (a) of Sub-Clause 14.3 [Application for Interim Payment Certificates];*
(b)   *Sub-Clause 14.5 [Plant and Materials intended for the Works] shall not apply; and*
(c)   *if these instalments are not defined by reference to the actual progress achieved in executing the Works, and if actual progress is found to be less than that on which this schedule of payments was based, then the Engineer may proceed in accordance with Sub-Clause 3.5 [Determinations] to agree or determine revised instalments, which shall take account of the extent to which progress is less than that on which the instalments were previously based.*

If the Contract does not include a schedule of payments, the Contractor shall 14–025
submit non-binding estimates of the payments which he expects to become due during each quarterly period. The first estimate shall be submitted within 42 days after the Commencement Date. Revised estimates shall be submitted at quarterly intervals, until the Taking-Over Certificate has been issued for the Works.

OVERVIEW OF KEY FEATURES

14–026
- The Contract may include a Schedule of Payments detailing provisions for payment by instalments.
- If the schedule is included, the instalments must be estimated contract values defined by reference to actual progress.
- If the Contract does not include a schedule, the Contractor shall submit non-binding estimates of the payments which he expects to become due.
- The first estimate must be submitted within 42 days after the Commencement Date.

COMMENTARY

14–027   It is quite likely that the majority of this sub-clause will not be required, as it only takes effect if the contract provides that interim payments are to be made in accordance with a specific Schedule of Payments. The contractual scheme as set out at sub-cl.14.3 is for payments to be monthly.

The final paragraph will however apply regardless. It requires the Contractor to provide a quarterly estimate of the amount of payment it anticipates. This is not binding and it may at first be thought to be rather unnecessary especially as there is (and can be) no sanction for the provision of inaccurate estimates. However it is not new and replaces the requirement for cash-flow estimates to be found at sub-cl.14.3 of the Old Red Book, FIDIC 4th edn. It will serve as a valuable guide as to the forthcoming expenditure and so will be of some assistance to those monitoring the cash-flow of both the Employer and Contractor.

MDB HARMONISED EDITION

14–028   The words "or more" have been added after the words "less" to sub-para.(c).

14.5 PLANT AND MATERIALS INTENDED FOR THE WORKS

14–029   *If this Sub-Clause applies, Interim Payment Certificates shall include, under sub-paragraph (e) of Sub-Clause 14.3, (i) an amount for Plant and Materials which have been sent to the Site for incorporation in the Permanent Works, and*

(ii) a reduction when the contract value of such Plant and Materials is included as part of the Permanent Works under sub-paragraph (a) of Sub-Clause 14.3 [Application for Interim Payment Certificates].

If the lists referred to in sub-paragraphs (b)(i) or (c)(i) below are not included in the Appendix to Tender, this Sub-Cause shall not apply.

The Engineer shall determine and certify each addition if the following conditions are satisfied:

(a)   The Contractor has:

(i)   kept satisfactory records (including the orders, receipts, Costs and use of Plant and Materials) which are available for inspection, and

(ii)   submitted a statement of the Cost of acquiring and delivering the Plant and Materials to the Site, supported by satisfactory evidence;

and either:

14–030

(b)   the relevant Plant and Materials:

(i)   are those listed in the Appendix to Tender for payment when shipped,

(ii)   have been shipped to the Country, en route to the Site, in accordance with the Contract; and

(iii) are described in a clean shipped bill of lading or other evidence of shipment, which has been submitted to the Engineer together with evidence of payment of freight and insurance, any other documents reasonably required, and a bank guarantee in a form and issued by an entity approved by the Employer in amounts and currencies equal to the amount due under this Sub-Clause: this guarantee may be in a similar form to the form referred to in Sub-Clause 14.2 [Advance Payment] and shall be valid until the Plant and Materials are properly stored on Site and protected against loss, dammar or deterioration;

or

14–031

(c)   the relevant Plant and Materials:

(i)   are those listed in the Appendix to Tender for payment when delivered to the Site, and

(ii)   have been delivered to and are properly stored on the Site, are protected against loss, damage or deterioration, and appear to be in accordance with the Contract.

The additional amount to be certified shall be equivalent of eighty percent of the Engineers determination of the cost of the Plant and Materials (including delivery to Site), taking account of the documents mentioned in this Sub-Clause and of the contract value of the Plant and Materials.

The currencies for this additional amount shall be the same as those in which payment will become due when the contract value is included under sub-paragraph (a) of Sub-Clause 14.3 [Application for Interim Payment Certificates]. At that time, the Payment Certificate shall include the applicable reduction which shall be equivalent to, and in the same currencies and proportions as, this additional amount for the relevant Plant and Materials.

OVERVIEW OF KEY FEATURES

14–032
- This sub-clause will only apply if lists of relevant Plant and Materials as set out in sub-paras (b) and (c) appear in the Appendix to Tender.
- The Engineer will only certify additions for Plants and Materials for payment if the Contractor's records are adequate.
- The additional amount to be certified will be up to 80 per cent of the Engineer's determination of the cost of the additional plant and materials.
- Payment will be made in the same currency as for the rest of the contract.

COMMENTARY

14–033 This is another optional sub-clause in that it will not apply if a Schedule of Payments has been adopted, as provided for in sub-cl.14.4(b), or if the lists set out in sub-para.(b) and (c) are not set out in the Appendix to Tender. Its purpose is to deal with payment for plant and materials which are being delivered to site. It acts as a further form of advance payment, in the sum of 80 per cent of the Engineer's valuation of the Plant and Materials in question. Thus, this is another sub-clause designed to assist the Contractor in a positive cash-flow by keeping his financing costs low. As with sub-cl.14.2, if an advance payment is to be made, a guarantee must be provided by the Contractor.

If this sub-clause is to operate, the Contractor must take care to keep the necessary records in order to satisfy the requirements listed in the sub-paragraph. It is for the Engineer to determine whether or not those requirements have been satisfied.

MDB HARMONISED EDITION

14–034 The words "Appendix to Tender" have been replaced by "Schedules".

14.6 ISSUE OF INTERIM PAYMENT CERTIFICATES

No amount will be certified or paid until the Employer has received and **14–035**
*approved the Performance Security. Thereafter, the Engineer shall, within 28
days after receiving a Statement and supporting documents, issue to the
Employer an Interim Payment Certificate which shall state the amount which
the Engineer fairly determines to be due, with supporting particulars.*

*However, prior to issuing the Taking-Over Certificate for the Works, the
Engineer shall not be bound issue an Interim Payment Certificate in an amount
which would (after retention and other deductions) be less than the minimum
amount of Interim Payment Certificates (if any) stated in the Appendix to
Tender. In this event, the Engineer shall give notice to the Contractor
accordingly.*

*An Interim Payment Certificate shall not be withheld for any other reason,
although:*

*(a)   If any thing supplied or work done by the Contractor is not in accordance
with the Contract, the cost of rectification or replacement may be withheld until rectification or replacement may be withheld until rectification
or replacement has been completed; and/or*

(b)   if the Contractor was or is failing to perform any work or obligation in **14–036**
*accordance with the Contract, and had been so notified by the Engineer,
the value of this work or obligation may be withheld until the work or
obligation has been performed.*

*The Engineer may in any Payment Certificate make any correction or modifi-
cation that should properly be made to any previous Payment Certificate. A
Payment Certificate shall not be deemed to indicate the Engineer's acceptance,
approval, consent or satisfaction.*

OVERVIEW OF KEY FEATURES

- The Engineer must make a fair determination of the amount due to the   **14–037**
  Contractor.
- The interim payment certificate can be withheld in two circumstances
  only:

  (i)   if the Employer has not received and approved the Performance
         Security; or

  (ii)   if, prior to issuing the taking over certificate, the amount to be
         certified would be less than the minimum amount of the interim
         payment certificate stated in the Appendix to Tender.

- Where work is carried out which is not in accordance with the Contract, the costs of rectification work may be withheld until the work is carried out.
- If the Engineer notifies the Contractor that it is failing to perform any obligation in accordance with the Contract, then the value of this work may be withheld until the work has been carried out.
- Payment certificates will not be deemed to indicate the Engineer's approval of any work that has been carried out.

COMMENTARY

14-038 Sub-clause 14.6 starts off by confirming the requirement expressed in sub-cl.4.2 that the Contractor must provide Performance Security before the Employer is under any obligation to make payment. The sub-clause thereafter concentrates on the question of payments and, more significantly, the right, if any, for payment to be withheld.

The sub-clause confirms that the Engineer must determine how much the Contractor is due "fairly". If the Contractor disputes the Engineer's determination then it must look to the provisions of cl.20 to resolve that dispute. This determination must be carried out within 28 days of receiving the Contractor's application and the result will be set out in the Interim Payment Certificate. If the Engineer fails to do this, sub-cl.16.1 gives the Contractor the option of giving notice of an intention to suspend the Works. If the Appendix to Tender provides a minimum value for an Interim Certificate and the Engineer's determination is less than that sum, then the Engineer is under no obligation to issue a Certificate.

Under sub-cl.14.7, the Employer "shall pay" to the Contractor the amount certified. The Employer is therefore bound by that certification. As stated above, sub-cl.2.5 was introduced in order to provide the Employer with a mechanism to make claims against the Contractor and thereby try and prevent the Employer from summarily withholding payment. If the Engineer determines that the Employer has a valid claim and is entitled to payment for that claim, then the amount will be deducted from the next interim certificate issued after the determination has been made.

14-039 That said, the sub-clause in sub-paras (a) and (b) sets out the limited circumstances where payments may be withheld. The Employer should note that the exception in sub-para.(b) is dependent on the appropriate notice being given by the Engineer. These are slightly confusing in that they refer to failures to perform which are the very items an Employer would expect to withhold payment for but which at the same time are the very items which sub-cl.2.5 had been set up to deal with. In reality these should not in any event be a separate withholding issue. If work has been carried out which is contrary to the Contract, it will presumably be valued at a reduced price or, even nil, by the Engineer.

The importance of the last paragraph should not be neglected. Its purpose serves both to prevent the Employer from asserting a right to withhold on the basis of an interim certificate and the Contractor from asserting that the interim certificate represents approval for work which has been carried out.

MDB HARMONISED EDITION

As usual the words "Appendix to Tender" have been replaced by "Contract Data".   **14–040**

The word "deliver" replaces "issue" in the first paragraph, putting a positive obligation on the Engineer to ensure the certificate reaches the Employer and also, (in another addition), the Contractor. The following has also been added to the end of the same paragraph:

for any reduction or withholding made by the Engineer on the Statement if any.

14.7 PAYMENT

The Employer shall pay to the Contractor: **14–041**

(a) *the first instalment of the advance payment within 42 days after issuing the Letter of Acceptance or within 21 days after receiving the documents in accordance with Sub-Clause 4.2 [Performance Security] and Sub-Clause 14.2 [Advance Payment], whichever is later;*

(b) *the amount certified in each Interim Payment Certificate within 56 days after the Engineer receives the Statement and supporting documents; and*

(c) *the amount certified in the Final Payment Certificate within 56 days after the Employer receives this Payment Certificate.*

Payment of the amount due in each currency shall be made into the bank account, nominated by the Contractor, in the payment country (for this currency) specified in the Contract.

OVERVIEW OF KEY FEATURES

14–042
- The Employer must make payment to the Contractor in accordance with the timescales set out in sub-paras (a)–(c).
- Payment is to be made into the bank account as chosen by the Contractor, as specified in the Contract.

COMMENTARY

14–043   This is a key sub-clause which sets out the dates by which the Employer should pay the Contractor.

During the project, the time for payment runs not from the date of the Interim Certificate but from the date the Engineer receives the Contractor's Statement applying for payment. However the Employer will not know exactly how much it has to pay until the Engineer has issued its certificate. As noted above in sub-cl.14.6, that payment must be in full for the amount shown in the Engineer's certificate. The payment must be made into the Contractor's nominated bank account.

With an interim payment, the Employer must pay the sum certified by the Engineer within 56 days of the date the Contractor's application. With the Final Payment, if the amount of the payment is agreed, the position is this:

(i)   The Contractor receives the Performance Certificate (see sub-cl.11.9).

(ii)   The Contractor then has 56 days to submit the draft Final Account (see sub-cl.14.11).

(iii)   If the Final Payment Certificate is agreed, the Engineer has 28 days to issue a Final Payment Certificate (see sub-cl.14.13).

(iv)   The Employer then has a further 56 days to make payment (see sub-cl.14.7(c)).

14–044   If the amount of payment is not agreed, then following sub-cl.14.11, the disagreement will be resolved in accordance with the principles of sub-cll 3.5, 20.4 and 5.

If the Employer fails to pay either on time or at all, then the Contractor has a right to interest under sub-cl.14.8, to give notice that it intended to suspend performance of its work under sub-cl.16.1 and ultimately to terminate under sub-cl.16.2.

MDB HARMONISED EDITION

The following has been added to sub-para.(b):   **14–045**

or, at a time when the Bank's loan or credit (from which part of the payments to the Contractor is being made) is suspended, the amount shown on any statement submitted by the Contractor within 14 days after such statement is submitted, any discrepancy being rectified in the next payment to the Contractor;

The following has been added to sub-para.(c):

or, at a time when the Bank's loan or credit (from which part of the payments to the Contractor is being made) is suspended, the undisputed amount shown in the Final Statement within 56 days after the date of notification of the suspension in accordance with Sub-Clause 16.2 [Termination by Contractor].

Both these additions serve to extend the time within which the Employer has to make payment in the limited circumstances set out, namely when the loan credit is suspended.

14.8 DELAYED PAYMENT

If the Contractor does not receive payment in accordance with Sub-Clause 14.7 **14–046**
[Payment], the Contractor shall be entitled to receive financing charges compounded monthly on the amount unpaid during the period of delay. This period shall be deemed to commence on the date for payment specified in Sub-Clause 14.7 [Payment], irrespective (in the case of its sub-paragraph (b) of the date on which any Interim Payment Certificate is issued.

Unless otherwise stated in the Particular Conditions, these financing charges shall be calculated at the annual rate of three percentage points above the discount rate of the central bank in the country of the currency of payment, and shall be paid such currency.

The Contractor shall be entitled to this payment without form notice of certification, and without prejudice to any other right or remedy.

OVERVIEW OF KEY FEATURES

14–047 • If the Employer does not make payment in accordance with sub-clause 14.7, the Contractor shall automatically be entitled to finance charges on the unpaid amount.
• The finance charges shall be calculated at a rate of 3 per cent above the base rate.

COMMENTARY

14–048 The purpose of this sub-clause is to compensate the Contractor if payment is made late. If payment is not made within 56 days of the submission of the Contractor's Statement applying for payment, then the Contractor has a right to interest at a rate of 3 per cent above the interest rate of the central bank of the country whose currency is used to make payment under the contract. Interest runs from the date the payment was due. This right to interest is automatic and the Contractor is not required to give notice of the entitlement, although practically it might be sensible.

In England, the question of interest for late payment has been settled by the Late Payment of Commercial Debts (Interest) Act 1998 which provides for payment of interest at a rate of 8 per cent above the Bank of England base rate.[3] This would apply to English contracts if sub-cl.14.8 was deleted. The interest rate may also apply, if the sub-clause remains, as s.9 of the Act notes that if the parties do agree an alternative contractual interest provision, it must amount to a "substantial remedy". A remedy will be substantial if it is sufficient to be a deterrent to making late payment, it compensates for losses due to late payment (i.e Bank interest) and it is reasonable to allow the contractual compensation to replace the statutory right to interest. Therefore it is submitted that when the Act applies an argument that only 3 per cent above base rate is not a substantial remedy might succeed.

Prior to the Act, interest on a debt had only been available in three limited circumstances. First, at common law, interest could only be claimed as special damages[4]. The interest would be claimed as either 'lost interest' (i.e. compensation to the creditor for interest the money would have earned in his bank account had the payment not been received late) or reimbursement for 'finance charges' that a creditor has had to pay, for example, on an overdraft or loan repayment due to late funds. In the case of *Amec Process and Energy*

---

[3] Indeed similar legislation should apply throughout the EU, thereby complying with EC Directive 2000/35.
[4] *Wadsworth v Lydall* [1981] 2 All E.R. 401 CA.

Ltd v Stork Engineers & Contractors BV (No.2)[5] the court developed the second limb of *Hadley v Baxendale*[6] to allow for a claim for lost interest in an action for breach of contract. In a construction context the Court has interpreted certain contract terms to imply a right to reimbursement for interest paid as a 'financial charge' due to late payment. In the case of *Minter v WHTSO*,[7] Stephenson L.J. said:

> *I do not think that today we should allow medieval abhorrence to usury to make us shrink from implying a promise to pay interest in a contract if by refusing to imply it we thereby deprive a party of what the contract appears on its natural interpretation to give him...*

Second, interest on a debt sum can be claimed in proceedings.[8] Such a claim for interest is lost, however, if not included in the creditor's pleadings.   **14–049**

Finally, interest is payable on a debt if an express contractual term provides for it. This is the case here, provided the sub-clause is not deleted from the Contract. However, the right to interest for delayed payment only applies to the circumstances set out in sub-cl.14.8. For example, it does not necessarily apply to determinations made by the Engineer or decisions of the DAB. Clause 8f of the DAB procedural rules only provides for the "payment of financing charges in accordance with the Contract".

Financing charges is not a defined term. In the *Minter* case, the Court of Appeal ruled that "direct loss and/or expense" could include interest paid on capital which had to be borrowed, as a consequence of the events which caused the loss and expense to be incurred. However, the period of entitlement only ran from the time the loss and/or expense was incurred and the date of the Contractor's application.

The FIDIC Guide notes that as an alternative to interest, the actual   **14–050** financing costs could be paid. However, this suggestion, though apparently on the face of it equitable, does not satisfy the principles behind and requirements of the EU late payment legislation.

---

[5] [2002]All E.R. 42.
[6] (1854) 9 Ex 341. The creditor will need to satisfy the test that "the special circumstances were communicated by the plaintiffs to the defendants, and thus known to both parties, the damages resulting from the breach of such a contract, which they would reasonably contemplate, would be the amount of injury which would ordinarily follow from a breach of contract under these special circumstances so known and communicated."
[7] *F.G. Minter v WHTSO* (1980) 13 BLR 1.
[8] In England under s.35A Supreme Court Act 1981 or s.69 County Court Act 1984 in the High Court and County Court respectively, or s.49 of the 1996 Arbitration Act.

MDB HARMONISED EDITION

14–051 The words "or if not available, the interbank offered rate" have been added to the second paragraph to follow "currency of payment".

14.9 PAYMENT OF RETENTION MONEY

14–052 *When the Taking-Over Certificate has been issued for the Works, the first half of the Retention Money shall be certified by the Engineer for payment to the Contractor. If a Taking-Over Certificate is issued for a Section or part of the Works, a proportion of the Retention Money shall be certified and paid. This proportion shall be two-fifths (40%) of the proportion calculated by dividing the estimated contract value of the Section or part, by the estimated final Contract Price.*

Promptly after the latest of the expiry dates of the Defects Notification Periods, the outstanding balance of the Retention Money shall be certified by the Engineer for a payment to the Contractor. If a Taking-Over Certificate was issued for a Section, a proportion of the second half of the Retention Money shall be certified and paid promptly after the expiry date of the Defects Notification Period for the Section. This proportion shall be two-fifths (40%) of the proportion calculated by dividing the estimated contract value of the Section by the estimated final Contract Price.

However, if any work remains to be executed under Clause 11 [Defects Liability], the Engineer shall be entitled to withhold certification of the estimated cost of this work until has been executed.

14–053 *When calculating these proportions, no account shall be taken of any adjustments under Sub-Clause 13.7 [Adjustments for Changes in Legislation] and Sub-Clause 13.8 [Adjustments for Changes in Cost].*

OVERVIEW OF KEY FEATURES

14–054
- The first half of the retention money shall be certified for payment to the Contractor after the issue of the Taking-Over Certificate.
- If the Taking Over Certificate is for a section or part of the work, 40 per cent of the value of that section shall be released by way of retention.
- The outstanding balance for retention money shall be certified for payment promptly after the expiry of the defects notification period.
- If any work remains to be executed, the Engineer shall be entitled to withhold certification of the estimated cost of this work.

- When calculating a retention, no account is made for sub-clauses 13.7 [adjustment for changes in legislation] and 13.8 [adjustment for changes in cost].

COMMENTARY

The relevance of the retention is often contentious. The aim of the retention is to either act as an incentive to the Contractor to avoid or eliminate defects or to provide reassurance to the Employer that funds exist to remedy any defects. With smaller contractors, the imposition of a retention can act as a considerable burden on cash-flow. In addition on some occasions parties are reluctant to release the retention as the contract demands.[9]

14–055

The scheme of the FIDIC form is straightforward. The first half of the retention is to be repaid when the Taking Over Certificate is issued in accordance with sub-cl.10.1. The second half is to be repaid on the expiry of the Defects Notification Period(s), in accordance with cl.11.

The sub-clause here does not say when the first half of the retention must be paid by the Employer. However sub-cl.14.3(c) requires that the Contractor include details of retention amounts in its interim payment applications. Thus, the retention will not be repaid until the Contractor makes the appropriate claim in its next payment application. The situation with the repayment of the second half is different, and the Engineer is required to certify the payment "promptly".

The Particular Conditions provide for an alternative option, which is the release of retention monies in return for a suitable guarantee. An example form is attached at Annex F. The Particular Conditions offer the following alternative sub-clause:

14–056

> When the Retention Money has reached three-fifths (60%) of the limit of retention money stated in the Appendix to Tender, the Engineer shall certify and the Employer shall make payment of half (50%) of the limit of Retention Money to the Contractor if he obtains a guarantee, in a form and provided by an entity approved by the Employer, in amounts and currencies equal to the payment.
>
> The Contractor shall ensure that the guarantee is valid and enforceable until the Contractor has executed and completed the Works and remedied any defects, as specified for the Performance Security in Sub-Clause 4.2, then shall be returned to the Contractor accordingly. This release of retention shall be in lieu on the release of the second half of the Retention Money under the second paragraph of Sub-Clause 14.9.

[9] In the UK, the Inland Revenue allows companies to write-off retentions which are not repaid.

MDB HARMONISED EDITION

14–057 The percentage to be repaid, referred to in the first paragraph has increased from 40 per cent to 50 per cent.

The MDB conditions also adopt the idea of the retention guarantee and the following lengthy paragraph has been added to the end of this sub-clause:

> *Unless otherwise stated in the particular conditions, when the taking-over Certificate has been issued for the Works and the first half of the Retention Money has been certified for payment by the Engineer, the Contractor shall be entitled to substitute a guarantee in the form next to the Particular Conditions or in another approved by the Employer and provided by an entity approved by the Employer, for the second half of the retention money. The Contractor shall ensure that the guarantee is in the amounts and currencies of the second half of the retention money and is valid and enforceful until the Contractor has executed and completed the works and remedied any defects, as specified for the performance security in sub-clause 4.2. On receipt by the Employer of the required guarantee, the Engineer shall certify and the Employer shall pay the second half of the retention money. The release of the second half of the retention money against their guarantee shall then be in lieu of the release under the second paragraph of this sub-clause. The Employer shall return the guarantee to the Contractor within 21 days after receiving a copy of the performance certificate.*

14–058 > *If the performance security required under sub-clause 4.2 is in the form of a demand guarantee, and the amount guaranteed under it when the taking over certificate is issued is more than half of the retention money, then the retention money guarantee will not be required. If the amount guaranteed under the performance security when the taking-over certificate is issued is less than half of the retention money, the retention money guarantee will only be required for the difference between half of the retention money and the amount guaranteed under the performance security.*

Thus this addition attempts to deal with the problem that is often encountered whereby Contractors suffer difficulty in achieving the repayment of the retention by setting introducing the retention bond.

14.10 STATEMENT AT COMPLETION

14–059 *Within 84 days after receiving the Taking-Over Certificate for the Works, the Contractor shall submit to the Engineer six copies of a Statement at completion* *with supporting documents, in accordance with Sub-Clause 14.3 [Application for Interim Payment Certificates], showing:*

(a) the value of all work done in accordance with the Contract up to the date stated in the Taking-Over Certificate for the Works,

(b) any further sums which the Contractor considers to be due, and

(c) an estimate of any other amounts which the Contractor considers will become due to him under the Contract. Estimated amounts shall be shown separately in this Statement at completion.

The Engineer shall then certify in accordance with Sub-Clause 14.6 [Issue of Interim Payment Certificates].

OVERVIEW OF KEY FEATURES

- The Contractor shall submit to the Engineer six copies of its completion statement within 84 days after receiving the Taking Over Certificate. **14–060**
- The Statement at completion shall show the value of all work done, any further sums considered due and estimates of any other amounts which may be due.
- The Engineer shall certify in accordance with the principles of sub-cl.14.6 i.e. fairly.

COMMENTARY

The Contractor's completion statement must be prepared in the same way **14–061** and with the same detail as required by sub-para.14.3 for the Statement applying for interim payment. Therefore the preparation of the completion statement should not cause the Contractor too many difficulties. Obviously the Contractor does not have to wait 84 days to submit the application. The sooner it is submitted, the sooner it can be dealt with.

When preparing the statement the Contractor should have regard to sub-cl.14.14 which deals with the cessation of the Employer's liability. This is part of the reason why sub-cll (b) and (c) require that the Contractor provide an estimate of any future amounts that it considers may be due.

In accordance with sub-cl.14.6, the Engineer has 28 days to certify any sums due and that determination must be carried out fairly. Presumably, although it is not stated, as with any previous interim payment, the Employer has 56 days from the date the Completion Statement was delivered to make payment.

MDB HARMONISED EDITION

14–062   There is no change.

14.11 APPLICATION FOR FINAL PAYMENT CERTIFICATE

14–063   *Within 56 days after receiving the Performance Certificate, the Contractor shall submit, to the Engineer, six copies of a draft final statement with supporting documents showing in detail in a form approved by the Engineer:*

   (a)   the value of all work done in accordance with the Contract, and
   (b)   any further sums which the Contractor considers to be due to him under the Contract or otherwise.

If the Engineer disagrees with or cannot verify any part of the draft final statement, the Contractor shall submit such further information as the Engineer may reasonably require and shall make such changes in the draft as may be agreed between them. The Contractor shall then prepare and submit to the Engineer the final statement as agreed. This agreed statement is referred to in these Conditions as the "Final Statement".

14–064   *However if following discussions between the Engineer and the Contractor and any changes to the draft final statement which are agreed, it become evidence that a dispute exists, the Engineer shall deliver to the Employer (with a copy to the Contractor) an Interim Payment Certificate for the agreed parts of the draft final statement. Thereafter, if the dispute is finally resolved under Sub-Clause 20.4 [Obtaining Dispute Adjudication Board's Decision] or Sub-Clause 20.5 [Amicable Settlement], the Contractor shall then prepare and submit to the Employer (with a copy to the Engineer) a Final Statement.*

OVERVIEW OF KEY FEATURES

14–065   • The Contractor is to submit six copies of the draft final statement within 56 days after receiving the Performance Certificate.
   • The draft final statement should show the value of all work done and the value of any further sums which the Contractor considers due.
   • If the Engineer disagrees with this, the Contractor shall submit any further information as may be reasonably required. After discussion, any disagreement must be resolved according to the principles of sub-cll 3.5 and 20.

- If a dispute exists, the Engineer must send to the Employer and Contractor an interim payment certificate for any amount that is agreed.
- If the dispute is agreed, the Contractor must submit a Final Statement.

COMMENTARY

This is an important sub-clause which sets out the procedure the parties must follow in order to settle issues of payment. As with all payment procedures, the timetable is initiated by the Contractor. **14–066**

The scheme of the Contract is to encourage the parties to agree the final payment amount. Provision is made for the Contractor to submit further information and for discussions to take place. However, if a dispute remains then any amount agreed must be certified and the Contractor must submit the disputed elements to the Dispute Adjudication Board per cl.20. If the dispute is finally resolved by the DAB, then the Contractor must initiate the final payment procedures again, by submitting a Final Statement.

MDB HARMONISED EDITION

The following words have been added to the middle of the second paragraph "within 28 days from receipt of said draft". Thus, a time limit has been imposed for the submission by the Contractor of further information. Note that no equivalent time limit has been imposed on the Engineer to consider that information. **14–067**

The reference to Dispute Board has dropped the word "Adjudication".

14.12 DISCHARGE

When submitting the Final Statement, the Contractor shall submit a written discharge which confirms that the total of the Final Statement represent full and final settlement of all moneys due to the Contractor under or in connection with the Contract. This discharge may state that it becomes effective when the Contractor has received the Performance Security and the outstanding balance of this total, in which event the discharge shall be effective on such date. **14–068**

CLAUSE 14 – CONTRACT PRICE AND PAYMENT

OVERVIEW OF KEY FEATURES

14–069 • The Contractor must submit with the Final Statement a written discharge confirming that the total of the final statement represents the full and final settlement of all monies due to it.
• This discharge becomes effective when the Contractor has received the Performance Security and payment of any outstanding balance.

COMMENTARY

14–070 The Contractor must take care when submitting its Final Statement as this sub-clause makes it clear that it covers all monies due to the Contractor, including claims. Thus, if the Contractor considers that it still has claims, then the Final Statement and written discharge clearly cannot be submitted. This is why sub-cl.14.11 requires that the Final Statement procedure is initiated after the DAB has made a ruling on any disputed claims in relation to final payment. However, if the dispute still remains and is referred to arbitration it may well be that the requirement for the written discharge and indeed ultimately the Final Statement will fall away.

The FIDIC Guide suggests the following form of discharge:

We hereby confirm, in terms of sub-clause 14.12 of the Conditions of Contract, that the total the attached Final Statement, namely . . ., represents to full an final settlement of all monies due to us under or in connection with the Contract. This discharge shall only be effective when we have received to Performance Security and the outstanding balance of this total of the attached Final Statement.

MDB HARMONISED EDITION

14–071 There is no change.

14.13 ISSUE OF FINAL PAYMENT CERTIFICATE

14–072 *Within 28 days after receiving the Final Statement and written discharge in accordance with Sub-Clause 14.11 [Application for Final Payment Certificate]* *and Sub-Clause 14.12 [Discharge], the Engineer shall issue, to the Employer, the Final Payment Certificate which shall state:*

(a)   the amount which is finally due, and

(b)   after giving credit to the Employer for all amounts previously paid by the Employer and for all sums to which the Employer is entitled, the balance (if any) due from the Employer to the Contractor or from the Contractor to the Employer, as the case may be.

If the Contractor has not applied for a Final Payment Certificate in accordance with Sub-Clause 14.11 [Application for Final Payment Certificate] and Sub-Clause 14.12 [Discharge], the Engineer shall request the Contractor to do so. If the Contractor fails to submit an application within a period of 28 days, the Engineer shall issue the Final Payment Certificate for such amount as he fairly determines to be due.

OVERVIEW OF KEY FEATURES

- The Final Payment Certificate is to be issued within 28 days after receiving the final statement and written discharge.   **14–073**
- The Final Payment Certificate is issued to the Employer by the Engineer.
- The Final Payment Certificate states the amount finally due and the balance, if any, due to the Contractor.
- If the Contractor does not apply for a Final Payment Certificate, the Engineer must request one.
- If the application is still not made within 28 days, the Engineer must issue a final certificate having determined the sum it believes to be due.

COMMENTARY

This sub-clause envisages that the Final Payment Certificate cannot be issued   **14–074**
until the Contractor has submitted its Final Statement and written discharge. Under sub-cl.14.7, the Employer must make payment of any sum certified in the Final Payment Certificate within 56 days. However, if the Contractor fails to submit the statement, even in response to a request from the Engineer, then the sub-clause provides for the Engineer to issue the Final Payment Certificate of its own volition. If the Engineer does this, he must make his determination "fairly". As a likely reason why the Contractor has not issued a Final Payment Certificate will be the existence of claims (see sub-cl.14.2 below), the Engineer will, in essence, be required to determine the value of those claims.

MDB HARMONISED EDITION

14–075 There has been a small addition to the first paragraph which makes it clear that the Engineer must deliver *"to both the Employer and the Contractor, the Final Payment Certificate"*. In addition the word "written" has been deleted, although it is considered that the form of discharge must still be in writing.

The words "fairly determines" have been added to sub-para.(a) simply confirming how the Engineer is to proceed.

14.14 CESSATION OF EMPLOYER'S LIABILITY

14–076 *The Employer shall not be liable to the Contractor for any matter or thing under or in connection with the Contract or execution of the Works, except to the extent that the Contractor shall have included an amount expressly for it:*

(a) *in the Final Statement and also*

(b) *(except for matters or things arising after the issue of the Taking-Over Certificate for the Works) in the Statement at completion described in Sub-Clause 14.10 [Statement at Completion].*

However, this Sub-Clause shall not limit the Employer's liability under his indemnification obligations, or the Employer's liability in any case of fraud, deliberate default or reckless misconduct by the Employer.

OVERVIEW OF KEY FEATURES

14–077 • The Employer will only be (potentially) liable for matters which the Contractor has made specific provision in the Final Statement.
  • The only exceptions to this relate to the Employer's indemnity obligations or in the cases of deliberate default, reckless misconduct or fraud by the Employer.

COMMENTARY

14–078 This is an important sub-clause. It is not new and can be found at sub-cl.60.9 of the Old Red Book FIDIC 4th edn. It refers to the cessation of the Employer's liability and not that of the Contractor.

This sub-clause reinforces the importance of the Contractor making clear in either the Final Statement or Statement at completion if it disputes any valuation or anything else relating to the project. As part of what is in effect a reservation of rights, the Contractor must put forward a value of any claim. The clear intention of this sub-clause is that if the Contractor fails to notify any claim as required here, then the right to raise that item in the future will be barred. The Contractor should note that it also appears that the notification requirement here is in addition to any notice requirements under cl.20.

The key questions are, whether or not this clause is binding and whether or not it acts as a bar to any claim. The words of the sub-clause seem clear and certainly Contractors should proceed on the basis that it is binding and should try and raise any unsettled issues during the period provided for the agreement of the Final Statement. However, it is submitted that this sub-clause can only serve to exclude items about which the Contractor was aware, or ought to have been aware, at the relevant time. That this is the case, is hinted at by sub-para.(b) which refers to matters arising after the issue of the Taking-Over Certificate for the Works. In addition, there are no clear words which would serve to extend this sub-clause to include future (in the sense of unknown) liabilities.

14–079 This sub-clause will not apply in cases of fraud or "deliberate default" or "reckless misconduct" by the Employer. Sub-clause 17.6 which deals with limitation of liability includes a similar exception. "Deliberate default" means more than mere negligence; there must have been an intention to carry out the behaviour which lead to the act of default. Likewise, although "reckless misconduct" does not appear to have been judicially defined, an understanding of their meaning can be found from the words of Webster J. who defined "wilful misconduct" as meaning "deliberately doing something which is wrong, knowing it to be wrong or with reckless indifference as to whether it is wrong or not".[10]

Finally, this sub-clause does not relieve the Employer of its contractual obligations to indemnify the Contractor in certain circumstances. These include the requirements of sub-cl.1.13 [Compliance with Laws], 4.2 [Performance Security], 5.2 [Objection to Nomination], and 17.1 and 17.7 [Indemnities and Limits on Indemnities respectively].

MDB HARMONISED EDITION

There is no change. **14–080**

---

[10] *Graham v Teesdale* (1981) 81 LGR 117.

14.15 CURRENCIES OF PAYMENT

14–081   *The Contract Price shall be paid in the currency or currencies named in the Appendix to Tender. Unless otherwise stated in the Particular Conditions, if more than one currency is so named, payments shall be made as follows:*

    (a)   if the Accepted Contract Amount was expressed in Local Currency only:

        (i)   the proportions or amounts of the Local and Foreign Currencies, and the fixed rates of exchange to be used for calculating the payments, shall be as stated in the Appendix to Tender, except as otherwise agreed by both Parties;

14–082   *(ii)   payments and deductions under Sub-Clause 13.5 [Provisional Sums] and Sub-Clause 13.7 [Adjustments for Changes in legislation] shall be made in the applicable currencies and proportions; and*

        (iii)   other payment and deductions under sub-paragraphs (a) to (d) of Sub-Clause 14.3 [Application for Interim Payment Certificates] shall be made in the currencies and proportions specified in sub-paragraph (a)(i) above;

    (b)   payment of the damages specified in the Appendix to Tender shall be made in the currencies and proportions specified in the Appendix to Tender;

14–083   *(c)   other payments to the Employer by the Contractor shall be made in the currency in which the sum was expended by the Employer, or in such currency as may be agreed by both Parties;*

    (d)   if any amount payable by the Contractor to the Employer in a particular currency exceeds the sum payable by the Employer to the Contractor in that currency, the Employer may recover the balance of this amount from the sums otherwise payable to the Contractor in other currencies; and

    (e)   if no rates of exchange are stated in the Appendix to Tender, they shall be those prevailing on the Base Date and determined by the central bank of the Country.

OVERVIEW OF KEY FEATURES

14–084   • The intention of the Contract is that the contract price should be paid in the currency named in the Appendix to Tender.
    • If no such currency is named then the default provisions of sub-paras (a) to (e) will apply.

COMMENTARY

This is a simple clause defining the currency in which payment is to be made.   **14–085**

MDB HARMONISED EDITION

In the first paragraph, the words "Appendix to Tender" have been replaced by   **14–086**
"Schedule of Payment currencies". The words "unless otherwise stated in the
particular conditions" have been deleted from the first paragraph.

In sub-para.(b) the first reference to "Appendix to Tender" has been
replaced by "Contract Data", the second has been replaced by "schedule of
payment currencies". In sub-para.(e) the words "Appendix to Tender" has
been replaced by "schedule of payment currencies".

14.16 FINANCING ARRANGEMENTS

The Particular Conditions (in the 1999 edn only) include a detailed   **14–087**
discussion on financing arrangements and include sample sub-clauses
which financing institutions may require. This does not appear in the
MDB Harmonised Edition.

The Particular Conditions recognise that there might be a need to secure
finance from development banks, credit agencies or aid agencies. Whilst the
particular conditions recognise that the exact wording of any financing
arrangement will depend on the particular requirements of the individual
international financing institution concerned, the Particular Conditions list
the following type of topics which may be of concern to both borrower and
lender:

(a)   prohibition from discrimination against shipping companies of any one
      company;
(b)   ensuring that the Contract is subject to a widely-accepted neutral law;
(c)   provision for arbitration under recognised international rules and at a
      neutral location;
(d)   giving the Contractor the right to suspend or terminate in the event of
      default under the financing arrangements;
(e)   restricting the right to reject Plant;
(f)   specifying the payments due in the event of termination;
(g)   specifying that the Contract does not become effective until certain
      conditions precedent have been satisfied, including pre-disbursements
      for the financial arrangements; and

(h) obliging the Employer to make payments from its own resources if, for any reason, the funds under the financing arrangements are insufficient to meet the payments due to the Contractor, whether due a default under the financing arrangements or otherwise.

CLAUSE 15 – TERMINATION BY EMPLOYER

15.1 NOTICE TO CORRECT

If the Contractor fails to carry out any obligation under the Contract, the Engineer may by notice require the Contractor to make good the failure and to remedy it within a specified reasonable time.   **15–001**

OVERVIEW OF KEY FEATURES

The Engineer is empowered to issue a notice requesting that the Contractor remedy any default under the Contract.   **15–002**

COMMENTARY

This is a short, simple sub-clause which gives the Engineer the right to issue a notice to the Contractor requesting that it rectifies poor performance under the Contract. Under sub-cl.1.3, the notice must be in writing. Although the giving of a notice potentially has a very significant effect, the requirement on the Engineer is an optional one. He does not have to issue such a notice.   **15–003**

Nevertheless as the notice (and the failure to comply with any such notice) is one of the pre-cursors to termination, the FIDIC Guide recommends that the notice should:

(i)   state that it is a notice under sub-cl.15.1;
(ii)   describe clearly the nature of the failure; and
(iii)   specify what constitutes a reasonable time to rectify the failure.

All three recommendations are sensible. The reason for including an actual reference to sub-cl.15.1 is to avoid disputes as to the nature of notice. For example, the Engineer can issue instructions under sub-cl.3.3 requiring the remedying of defects or serve a notice under sub-cl.7.5 rejecting works carried out by the Contractor. The second item adds similar clarity, whilst the third is already a requirement of the sub-clause.   **15–004**

The sub-clause refers to a failure to carry out "any" obligation under the Contract. Given the fact that minor defects by themselves would be unlikely to amount to a sufficient breach of contract to justify termination, there might be thought to be a question mark over whether the failure to remedy a minor defect will empower the Employer to exercise its right to terminate under sub-cl.15.2.[1] However, it is not the nature of defect itself which becomes relevant, but the failure of the Contractor to remedy the defect when requested to do so.

At common law the non-compliance with such a term after receipt of notice may amount to repudiation regardless of the contractual effect set-out in sub-cl.15.2.[2] In the *Hong Kong Fir* case,[3] the charterer argued that it had had the right to terminate the charterparty on the basis that the incompetence of the engine room staff was in breach of an implied term that the ship would be seaworthy. This claim was rejected. However Sellers L.J. suggested an alternative way in which the claim could have been made:

15–005   *It would be unthinkable that all the relatively trivial matters which have been held to be unseaworthiness could be regarded as conditions of the contract or conditions precedent to a charterer's liability and justify in themselves a cancellation or refusal to perform on the part of the charter. If, in the present case, the inadequacy and incompetence of the engine-room staff had been known to them, the charterers could have complained of the failure by the owners to deliver the vessel at Liverpool in accordance with clause 1 of the chaterparty. . . have given the shipowners a week in which to bring the engine room staff into suitable strength and competency for the vessel's 'ordinary cargo service'. If the shipowners had refused or failed so to do, their conduct and not the unseaworthiness would have amounted to a repudiation of the charterparty and entitled the charterers to accept it and treat the contract as at an end.*

This is exactly the type of situation the sub-clause is attempting to deal with. It is not the defect itself but the conduct in failing to remedy the defect that is important.

MDB HARMONISED EDITION

15–006   There is no change.

---

[1] The legal principle "de minimis non curat lex" – the law does not concern itself with trifles – applies both at common law and to most civil codes.
[2] That said, where the Contract contains express determination provisions, a party would be well-advised to rely upon this, if possible. See sub-cl.15.2 below.
[3] *Hong Kong Fir Shipping Co v Kawasaki Kisen Kaisha* [1962] 2 QB 26. See sub-cl.11.4 above.

15.2 TERMINATION BY EMPLOYER

The Employer shall be *entitled to terminate the Contract if the Contractor:*　**15–007**

(a)　fails to comply with Sub-Clause 4.2 [Performance Security] or with a notice under Sub-Clause 15.1 [Notice to Correct],

(b)　abandons the Works or otherwise plainly demonstrates the intention not to continue performance of his obligations under the Contract,

(c)　without reasonable excuse fails:

　　(i)　to proceed with the Works in accordance with Clause 8 [Commencement, Delays and Suspension], or

　　(ii)　to comply with a notice issued under Sub-Clause 7.5 [Rejection] or Sub-Clause 7.6 [Remedial Work], within 28 days after receiving it.

(d)　Subcontracts the whole of the Works or assigns the Contract without the　**15–008**
required agreement,

(e)　becomes bankrupt or insolvent, goes into liquidation, has a receiving or administration order made against him, compounds with his creditors, or carries on business under a receiver, trustee or manager for the benefit of his creditors, or if any act is done or event occurs which (under applicable Laws) has a similar effect to any of these acts or events, or

(f)　gives or offers to give (directly or indirectly) to any person any bribe, gift, gratuity commission or other thing of value, as an inducement or reward:

　　(i)　for doing or forbearing to do any action in relation to the Contract, or

　　(ii)　for showing or forbearing to show favour or disfavour to any person in relation to the Contract,

or if any of the Contractor's Personnel, agents or Subcontractors gives or offers to give (directly or indirectly) to any person any such inducement or reward as is described in this sub-paragraph (f). However, lawful inducements and rewards to Contractor's Personnel shall not entitle termination.

In any of these events or circumstances, the Employer may, upon giving 14 days'　**15–009**
notice to the Contractor, terminate the Contract and expel the Contractor from the Site. However, in the case of sub-paragraph (e) or (f), the Employer may by notice terminate the Contract immediately.

　The Employers election to terminate the Contract shall not prejudice any other rights of the Employer, under the Contract or otherwise.

　The Contractor shall then leave the Site and deliver any required Goods, all Contractor's Documents, and other design documents made by or from him, to the Engineer. However, the Contractor shall use his best efforts to comply immediately with any reasonable instructions included in the notice (i) for the assignment of any subcontract, and (ii) for the protection of life or property or for the safety of the Works.

15–010 *After termination, the Employer may complete the Works and/or arrange for any other entities to do so. The Employer and these entities may then use any Goods, Contractor's Documents and other design documents made by or on behalf of the Contractor.*

The Employer shall then give notice that the Contractor's Equipment and Temporary Works will be released to the Contractor at or near the Site. The Contractor shall promptly arrange their removal, at the risk and cost of the Contractor. However, if by this time the Contractor has failed to make a payment due to the Employer, these items may be sold by the Employer in order to recover this payment. Any balance of the proceeds shall then be paid to the Contractor.

OVERVIEW OF KEY FEATURES

15–011 • The Employer may terminate in the following circumstances, namely if the Contractor:

    (a) – Fails to obtain a Performance Security in accordance with sub-cl.4.2.

          – Fails to remedy defects as set out in a sub-clause 15.1 notice from the Engineer.

    (b) – Abandons or refuses to continue the Works.

    (c) – Fails to proceed with the Works as required by clause 8.

          – Fails to comply with a notice requiring the rectification of Works rejected by the Engineer under sub-cl.7.6.

          – Fails to comply with a notice requiring the Contractor to carry out remedial work under sub-cl.7.6.

    (d) – Sub-contracts or otherwise assigns the Works to another.

    (e) – Becomes insolvent.

    (f) – Offers or gives bribes.

• The Employer may only terminate immediately in respect of the conduct described in sub-paras (e) and (f).

• In respect of sub-paras (a)–(d) the Employer must give 14 days notice prior to termination.

• If termination is carried out in accordance with the requirements of the sub-clause, none of the Employer's rights under the Contract will be adversely affected.

• Upon termination:

    – The Contractor must leave site;

    – The Employer may complete the works itself or arrange for others to do

so;
– The Employer may use the Contractor's Documents and Equipment to effect completion.

COMMENTARY

Termination is a very serious step and it is not one to be taken lightly. Parties would be well-advised to obtain proper legal advice whenever termination is contemplated. **15–012**

Termination is likely to delay the completion and increase the costs of the project. Sub-clause 15.2 deals with the mechanism of termination and sets out the circumstances in which the Employer may terminate. In addition, the Employer may use sub-cl.15.5 to terminate at its "convenience".

It is important that determination provisions are precisely followed. If a dispute arises, those procedures particularly in relation to notices or time limits will usually be carefully considered and strictly applied. The case of *Brown & Docherty v Whangarei Country*,[4] provides a typical example of the principles that apply when considering termination clauses. Here Smellie J. held that:

(i)   Determination clauses must be interpreted strictly.

(ii)   For a determination to be valid under the contract, the correct procedure must be complied with. **15–013**

(iii) A professional consultant (such as an engineer or architect) must act fairly and impartially in the exercise of any discretion to issue a contractual certificate or notice that may be relied upon by the Employer as grounds for determination.

(iv)   The contractor must be given fair warning that continuation of his conduct may result in determination and should not be lulled into assuming that he would be permitted to continue with the work.

(v)   A certificate or notice issued by the architect or engineer in reliance upon incorrect or irrelevant information or grounds (such as claims for additional payment and requests for further extensions of time) will be invalid.

The Employer must therefore take care to follow any notice requirements in the Contract. Who should the notice be served upon? Where should the notice be served? How should the notice be delivered? The purpose of this contractual provision is to ensure that one party knows what the other wishes

---

[4] [1988] 1 N.Z.L.R 33.

to communicate. There is conflicting case law on the effect of serving the notice in the wrong way.

In the case of *Central Provident Fund Board v Ho Bock Kee*,[5] the Court of Appeal in Singapore found that a notice was invalid on the grounds that it failed to comply with the contract being both sent by the wrong body (the Superintending Officer and not the Chairman of the Board) and being delivered by hand and not by registered post. However, in England, in the case of *J.M. Hill v London Borough of Camden*[6] the Court of Appeal took a different line when a notice was served by recorded delivery rather than registered post.[7] Ormrod L.J. said:

> *Nothing is more distasteful to me than to construe a business contract in this formalistic sense. . .Everybody concerned knew perfectly well what was happening. No-one was in the very slightest degree prejudiced by it. . .So that it is the most purely formal point.*

15–014 Nevertheless it is always better to follow the requirements of the contract. Different countries may have different legal approaches and the costs and time involved in asking the court to resolve any dispute that arises as a result of uncertainty might be considerable. With time-limits, the 14-day period must be followed. If the notice is served a day early it will not count. The consequence of a wrongful termination will be significant, typically amounting to a repudiatory breach of contract, which will leave the Employer vulnerable to a significant claim by the aggrieved Contractor.

This contractual right to terminate exists in addition to the common law right to repudiation. In England, the Court Appeal in the case of *Lockland Builders v Rickwood* accepted that contractual and common law rights can sit side by side provided the defaulting party demonstrates an intention not to be bound by the Contract.[8] Hirst L.J. adopted the general principle stated in *Chitty on Contracts*[9] that:

> *The fact that one party is contractually entitled to terminate the agreement in the event of a breach by the other party does not preclude that party from treating the agreement as discharged by reason of the other's repudiation or breach of condition, unless the agreement itself expressly or impliedly provides that it can only be terminated by exercise of the contractual right.*

---

[5] (1981) 17 BLR 21.
[6] (1980) 18 BLR 35.
[7] HHJ Gilliland Q.C. followed this approach in the recent case of *Construction Partnership UK Ltd v Leek Developments Ltd* [2006] CILL 2357, where the notice was served by fax not special or recorded delivery.
[8] 77 BLR 38.
[9] Paragraph 22–044 – 27th edn.

Nevertheless where, as here, the Contract contains an express termination clause, the Party seeking termination would be best advised to rely on that clause. It is likely to be the more practical and straightforward approach. Indeed it was in recognition of this, that the Court of Appeal noted in the *Lockland* case that Mr Rickwood could have saved himself considerable time and expense by adopting the simple procedure available to him under condition 2 of the Contract.

Under sub-cl.15.2, the majority of reasons which justify termination are reasonably clear: **15–015**

(a)   Under sub-cl.4.2, the Contractor must provide the Employer with a Performance Security within 28 days of receiving the Letter of Acceptance. Whilst, as noted above, the key to the notice referred to in sub-cl.15.1 is not the nature of the defect, but the failure to remedy it.

(b)   Likewise whether Works have been abandoned should be a relatively straightforward question to answer. Equally, whilst there is more room for argument about what constitutes plainly demonstrating a failure to proceed with the Works, if the Contractor is refusing to carry-out a particular part of the Work then that would clearly fall within this sub-paragraph. The wording here is similar to the common law right to repudiation which talks of the need to demonstrate an intention not to be bound by the Contract.

(c)   Under sub-cl.8.1, the Contractor is required to proceed *"with due expedition and without delay"*.[10] The Contractor has to produce a programme by sub-cl.8.3, and this must be regularly updated in the monthly progress reports required by sub-cl.4.21. These will give a good summary of the rate of progress. If the Employer wants to give the Contractor an opportunity to demonstrate how it can recover delays to progress, then the Engineer could request a revised programme under sub-cl.8.6, whereby the Contractor has to show how progress can be expedited to achieve completion on time.

With the notices, it is important that the Employer is content that the requirements of sub-cl.7.6 have been complied with. Indeed following the *Brown & Docherty* case, the Employer perhaps ought to consider whether the information in the notice is correct.

This sub-paragraph only applies where the Contractor does not have *"reasonable excuse"*. Although no definition of this is provided, the most obvious reasonable excuse would be one of the grounds set out at sub-cl.8.4 which justify an entitlement to an extension of time.

(d)   Again the fact of an assignment or other form of transfer should be **15–016** straightforward to ascertain. To qualify here, the assignment must for

---

[10] For a full discussion of what this means see sub-cl.8.1 below.

the whole of the Works. Indeed sub-cl.1.7 expressly prevents the assignment of the whole or part of the Works on the part of either the Employer or the Contractor, without agreement.

(e) The Employer is likely to be alert to the possibility of insolvency and so the fact of a Contractor failing in this way, should not come out of the blue. The sub-clause is widely drafted and covers a variety of different types of insolvency. The reason the Employer does not need to give 14 days' notice here will be to protect the Employer from making any payment to the Contractor which it may not be able to recover subsequently.

(f) This sub-paragraph deals with the giving (or even the offering) of bribes. It is also widely drawn and makes the Contractor responsible for the actions of not only its own Personnel but also its Sub-Contractors as well. The Contractor will be judged by the wording of the sub-paragraph and not whether its conduct is illegal in accordance with the laws of the project. As the sub-paragraph makes reference to "gifts and things of value", in theory this sub-clause could be implemented over items of minor value. Thus, the Contractor might want to take care to implement a policy of complete transparency.

15–017 It is important that the Employer remembers that it may only terminate immediately in respect of the insolvency or bribery as described in sub-paras (e) and (f). In respect of the other sub-paragraphs (a)–(d) the Employer must give 14 days' notice prior to termination. This thereby gives both parties a short period to try and resolve their differences. The Contractor should be aware that there is nothing in sub-cl.15.2 which says that if the default specified in the notice is remedied then the Employer cannot still go ahead and terminate.

If termination is carried out in accordance with the requirements of the sub-clause, none of the Employer's rights under the Contract will be adversely affected. Upon termination, the Contractor must leave site. The sub-clause does not say how quickly the Contractor must leave, but it is to be presumed from the use of the word "*expel*" that the Employer will expect the Contractor to depart promptly. However, the Contractor may be required to take certain steps to assign its sub-contracts and also to secure and make safe its works before it departs.

Finally, upon termination under sub-cl.15.2, the Employer may complete the works itself or arrange for others to do so. It may also make use of the Contractor's Documents and Goods to effect completion. Under sub-cl.1.1.5.2, Goods will include equipment, plant, materials and temporary works. Provided the Contractor does not owe the Employer any money, these must be returned to the Contractor, once the Employer has finished using them, although the Contractor must collect them at its own cost. Strictly there is no provision here for what might happen if the Contractor refuses to comply with any such instruction to collect its equipment. However if the

Contractor does owe money, then the Employer is entitled to sell them, accounting to the Contractor for any excess balance.

MDB HARMONISED EDITION

There is no change. However reference should be made to sub-cl.15.6 which **15–018** is a feature of the MDB version only. This gives the Employer the right to terminate where the Contractor engages in corrupt or fraudulent practice. There is one key difference, under sub-cl.15.6, 14 days' notice must be given; under the similar sub-cl.15.1(f) a termination notice may be served immediately.

15.3 VALUATION AT DATE OF TERMINATION

As soon as practicable after a notice of termination under Sub-Clause 15.2 **15–019** *[Termination by Employer] has taken effect, the Engineer shall proceed in accordance with Sub-Clause 3.5 [Determinations] to agree or determine the value of the Works, Goods and Contractor's Documents, and any other sums due to the Contractor for work executed in accordance with the Contract.*

OVERVIEW OF KEY FEATURES

After termination, the Engineer must agree or determine the value of any **15–020** sums due to the Contractor.

COMMENTARY

This straightforward sub-clause requires the Engineer to, if necessary, deter- **15–021** mine the value of the Contractor's works upon termination. The sub-clause requires that the valuation take place as soon as it is practicable and there are clear advantages to all parties in this happening, not least everyone knowing where they stand.

Under sub-cl.14.2, upon termination, any advance payment becomes immediately due and payable to the Employer. This must be taken account of by the Engineer in his determination.

MBD HARMONISED EDITION

15–022   There is no change.

15.4 PAYMENT AFTER TERMINATION

15–023   *After a notice of termination under Sub-Clause 15.2 [Termination by Employer] has taken effect, the Employer may:*

(a)   *proceed in accordance with Sub-Clause 2.5 [Employer's Claims],*
(b)   *withhold further payments to the Contractor until the costs of execution, completion and remedying of any defects, damages for delay in completion (if any), and all other costs incurred by the Employer, have been established, and/or*
(c)   *recover from the Contractor any losses and damages incurred by the Employer and any extra costs of completing the Works, after allowing for any sum due to the Contractor under Sub-Clause 15.3 [Valuation at Date of Termination]. After recovering any such losses, damages and extra costs, the Employer shall pay any balance to the Contractor.*

OVERVIEW OF KEY FEATURES

15–024   • The Employer does not have to make any immediate payment to the Contractor upon termination.
   • No payment needs to be made until the Works are complete and account has been taken of the Employer's entitlement (if any) to claims.

COMMENTARY

15–025   This is an important sub-clause for the Employer. It provides that it does not need to make any payment to the Contractor, upon termination, until the project has been completed and a balancing exercise has been carried out taking into account the costs of completion and the value of any claims the Employer may have. Indeed it is almost inevitable that the Employer will act in this way. In addition, upon termination most Employers will look, if they are able, to call upon the Performance Guarantee. Thus Contractors, if they suspect termination is likely, will want to take steps (if any are available to them) to prepare to resist any such call. The consequence of all this, is that the Contractor will have to wait for some time before any sums to which they may be entitled fall due for payment.

MDB HARMONISED EDITION

There is no change. **15–026**

15.5 EMPLOYER'S ENTITLEMENT TO TERMINATION

The Employer shall be entitled to terminate the Contract, at any time for the **15–027**
Employer's convenience, by giving notice of such termination to the Contractor.
The termination shall take effect 28 days after the later of the dates on which
the Contractor receives this notice or the Employer returns the Performance
Security. The Employer shall not terminate the Contract under this Sub-Clause
in order to execute the Works himself or to arrange for the Works to be
executed by another contractor.
  After this termination, the Contractor shall proceed in accordance with Sub-
Clause 16.3 [Cessation of Work and Removal of Contractor's Equipment] and
shall be paid in accordance with Sub-Clause 19.6 [Optional Termination] and
shall be paid in accordance with Sub-Clause 19.6 [Optional Termination,
Payment and Release].

OVERVIEW OF KEY FEATURES

- The Employer may terminate at any time, absent Contractor default, **15–028**
  upon 28 days' notice.
- If the Employer terminates on this basis, the Works cannot be completed
  by another.
- The Contractor is entitled to be paid in accordance with sub-cll 16.3 and
  19.6.

COMMENTARY

Although this is a new provision, termination at will clauses, such as sub- **15–029**
cl.15.5 here, are fairly common.[11] The idea is that the Employer is able to

---

[11] For an example of this, see Hadley *Design Associates v The Lord Mayor and Citizens of the City of Westminster* [2003] E.W.H.C. 1617.

bring the contract to an end without there having been any default on the part of the Contractor. The most likely reason the Employer will chose to operate this sub-clause will be an inability to fund and thereby finish the project.

However, there is a deliberate restraint on the Employer's ability to implement this sub-clause, as it will be unable to finish the project either by itself or by engaging a new Contractor. The key words here are "in order to." Thus the operation of this sub-clause cannot be carried out if the Employer's intention is to remove the Contractor. Presumably, this would not prevent an Employer from resuming the project at a later date if it had a genuine alternative reason at the time of the operation of this sub-clause.

The Contractor is also entitled to be paid in accordance with the force majeure provisions of sub-cl.19.6. In other words the Engineer must determine the value of the works done and issue a payment certificate. This valuation will favour the Contractor, including for example, the cost of materials already ordered and of the repatriation of personnel. However the sub-clause does not refer to loss of profits.

MDB HARMONISED EDITION

15–030 The words "or to avoid a termination of the Contract by the Contractor under Clause 16.2 [Termination by Contractor]" have been added to the first paragraph. The reason for the introduction of a new restriction on the Employer's ability to terminate would appear to be to prevent the Employer from stepping in to prevent the Contractor from operating its own right to terminate.

In addition the Contractor is now entitled to be paid under sub-cl.16.4, which deals with payment on termination rather than 19.6. However although this change might initially seem to be of little consequence, as sub-cl.16.4 cross-refers to sub-cl.19.6, sub-para.(c) in fact gives the Contractor the right to "loss of profit or other loss or damages sustained" as a consequence of the termination.

15.6 CORRUPT OR FRAUDULENT PRACTICES

15–031 The following sub-clause appears in the MDB Harmonised Version only:

If the Employer determines that the Contractor has engaged in corrupt, fraudulent, collusive or coercive practices, in competing for or in executing the Contract, then the Employer may, after given 14 days notice to the Contractor, terminate the Contractor's employment under the Contract and expel him from the Site, and the provisions of Clause 15 shall apply as if such expulsion had been made under Sub-Clause 15.2 [Termination by Employer].

Should any employee of the Contractor be determined to have engaged in corrupt, fraudulent or coercive practice during the execution of the work then that employee shall be removed in accordance with Sub-Clause 6.9 [Contractor's Personnel].

For the purposes of this Sub-Clause:

See Notes for definitions of corrupt, fraudulent, collusive or coercive practices for each Participating Bank.

The sub-clause will be slightly different for each Participating Bank as each **15–032** has its own definition of corrupt, fraudulent, collusive or coercive practice. The sub-clause has some similarities with sub-para.(f) of sub-cl.15.6; however unlike sub-cl.15.1(f), here 14 days' notice must be given. This new sub-clause is also more widely drawn, for example, making it clear that the tendering process must be fair as it refers to both "competing for" and "executing" the Works. In the case of *Cameroon Airlines v Trasnet Ltd*[12], an arbitration tribunal ruled that Trasnet had to repay commission monies it had added to its tender sum – the commission monies being money paid as bribes to officials.

This extension to cl.15 is entirely in keeping with the global trend in seeking to clamp down on this type of behaviour. For example, in the UK, the Anti-Terrorism, Crime and Security Act 2001[13] provides that a UK citizen can be guilty of an offence in the UK if he is involved in offering or receiving bribes abroad, provided that what he has done would amount to an offence in the UK.

---

[12] [2004] E.W.H.C. 1829.
[13] Thereby adopting the 1999 Organisation for Economic Co-operation and Development Convention on Combating Bribery of Foreign Public Officials in International Business Transactions.

CLAUSE 16 – SUSPENSION AND TERMINATION BY CONTRACTOR

16.1 CONTRACTOR'S ENTITLEMENT TO SUSPEND WORK

If the Engineer fails to certify in accordance with Sub-Clause 14.6 [Issue of   **16–001**
*Interim Payment Certificates] or the Employer fails to comply with Sub-Clause
2.4 [Employer's Financial Arrangements] or Sub-Clause 14.7 [Payment], the
Contractor may, after giving not less than 21 days' notice to the Employer,
suspend work (or reduce the rate of work) unless and until the Contractor has
received the Payment Certificate, reasonable evidence of payment, as the case
may be and as described in the notice.*

*The Contractor's action shall not prejudice his entitlements to financing
charges under Sub-Clause 14.8 [Delayed Payment] and to termination under
Sub-Clause 16.2 [Termination by Contractor].*

*If the Contractor subsequently receives such Payment Certificate, evidence or
payment (as described in the relevant Sub-Clause and in the above notice)
before giving a notice of termination, the Contractor shall resume normal
working as soon as is reasonably practicable.*

If the Contractor suffers delay and/or incurs Cost as a result of suspending   **16–002**
*work (or reducing the rate of work) in accordance with this Sub-Clause, the
Contractor shall give notice to the Engineer and shall be entitled subject to
Sub-Clause 20.1 [Contractor's Claims] to:*

(a)   *an extension of time for any such delay, if completion is or will be
delayed, under Sub-Clause 8.4 [Extension of Time for Completion], and*
(b)   *payment of any such Cost plus reasonable profit, which shall be included
in the Contract Price.*

*After receiving this notice, the Engineer shall proceed in accordance with
Sub-Clause 3.5 [Determinations] to agree or determine these matters.*

OVERVIEW OF KEY FEATURES

- The Contractor, may only suspend work in the following circumstances: **16–003**

    - if the Engineer fails to issue a timely interim certificate; or
    - if the Employer fails to provide financial information; or

    – if the Employer fails to pay any sums due.

- Before the Contractor can suspend, it must give 21 days' notice.
- If the default is remedied, the Contractor must resume work as soon as possible.
- Where the suspension is valid, the Contractor is entitled to claim for delay or cost plus reasonable profit in accordance with cl.20.

COMMENTARY

16–004   Sub-clause 16.1 which gives the right to suspend work in the event of non-payment by the Employer is similar to the clauses found in many of the standard form contracts. For contracts which are performed under UK law, this clause is compliant with s.112 of the Housing Grants, Construction and Regeneration Act 1996 (HGCRA).

    Sub-clause 16.1 requires the Contractor to give 21 days' notice before it may suspend the work, as opposed to the 7 days stipulated in s.112 of the HGCRA. It is anticipated that this may cause cash-flow difficulties for some Contractors who will have to continue working despite not being in receipt of payment.

    The notice must be in writing (see sub-clause 1.3) and must set out the grounds for the suspension. This is not specifically stated in the wording of the sub-clause but it is practical and the final words of the first paragraph use the words "as described in the notice".

16–005   If following notice or actual suspension, payment is forthcoming from the Employer, and provided that the Contractor has not terminated under cl.16.2, then the Contractor must resume work. Where the work has been suspended, the Contractor is entitled to make a claim against the Employer for an extension of time and loss incurred as a result of the suspension. It is not specifically stated in the sub-clause, but it is suggested that this would include re-mobilization costs as these can only have been incurred as a direct result of the suspension.

    One of the difficulties with s.112 of the HGCRA was that the compensation to which the suspending party was entitled, under the legislation in the event of a legitimate suspension, was not generous. Subsection 112(4) simply confirmed that the suspending party was entitled to an extension of time for completion of the works covering the period during which performance was suspended. Any extension did not necessarily extend to the 7-day notice period prior to the right to suspend becoming operative, and did not apply to the time that it might take to remobilise following the suspension. This was important since the right to suspend ceased on payment of the amount "due" in full. Here there is an entitlement to claim for an extension of time and cost plus reasonable profit.

At the time of writing, it has been recommended[1] that the statutory right to suspend performance in the UK should be supplemented with a right to reclaim the reasonable costs of suspension and remobilisation, provided nothing compromised the ability of a payer to reject a claim for such costs where a suspension was unjustified. Slightly more controversially, it has been suggested that insofar as the statutory Scheme for Construction Contracts was concerned, the reasonable costs of suspension and re-mobilisation should not be linked to the actual costs incurred but instead it was proposed they should be calculated by formula and should not exceed 5 per cent of the value of the payments in default. It was also suggested that the time taken to re-mobilize ought not to exceed 7 days. This would not apply to the FIDIC form here.

Such claims must be compliant with sub-cl.20.1 which sets strict time-limits within which particulars must be placed before the Engineer. Where these time-limits are not complied with, the Contractor will not be able to dispute this before the Dispute Adjudication Board. Since reference to the DAB is a condition precedent to arbitration, where the time-limits are not complied with or where proper particulars are not placed before the Engineer, the Contractor could be without a remedy. It is thus questionable whether this is compliant with s.112(4) of the HGCRA which states that periods of suspension must be disregarded during the computation of time limits. **16–006**

MDB HARMONISED EDITION

The following new paragraph has been added to follow the opening one: **16–007**

Notwithstanding the above, if the Bank has suspended disbursements under the loan or credit from which payments to the Contractor are being made, in whole or in part, for the execution of the Works, and no alternative funds are available as provided for in Sub-Clause [Employer's Financial Arrangements], the Contractor may by notice suspend work or reduce the rate of work at any time, but not less than 7 days after the Borrower having received the suspension notification from the Bank.

This has presumably been introduced to provide additional protection to the Contractor.

[1] Government Consultation Paper, March 22, 2005.

16.2 TERMINATION BY CONTRACTOR

16–008 *The Contractor shall be entitled to terminate the Contract if:*

>   *(a) the Contractor does not receive the reasonable evidence within 42 days after giving notice under Sub-Clause 16.1 [Contractor's Entitlement to Suspend Work] in respect of a failure to comply with Sub-Clause 2.4 [Employer's Financial Arrangements],*

16–009 *(b) the Engineer fails, within 56 days after receiving a Statement and supporting documents, to issue the relevant Payment Certificate,*

>   *(c) the Contractor does not receive the amount due under an Interim Payment Certificate within 42 days after the expiry of the time stated in Sub-Clause 14.7 [Payment] within which payment is to be made (except for deductions in accordance with Sub-Clause 2.5 [Employer's Claims]),*

16–010 *(d) the Employer substantially fails to perform his obligations under the Contract,*

>   *(e) the Employer fails to comply with Sub-Clause 1.6 [Contract Agreement] or Sub-Clause 1.7 [Assignment],*

>   *(f) a prolonged suspension affects the whole of the Works as described in Sub-Clause 8.11 [Prolonged Suspension], or*

>   *(g) the Employer becomes bankrupt or insolvent, goes into liquidation, has a receiving or administration order made against him, compounds with his creditors, or carries on business under a receiver, trustee or manager for the benefit of his creditors, or if any act is done or event occurs which (under applicable Laws) has a similar effect to any of these acts or events.*

In any of these events or circumstances, the Contractor may, upon giving 14 days' notice to the Employer, terminate the Contract. However, in the case of sub-paragraph (f) or (g), the Contractor may by notice terminate the Contract immediately.

16–011 *The Contractor's election to terminate the Contract shall not prejudice any other rights of the Contractor, under the Contract or otherwise.*

OVERVIEW OF KEY FEATURES

*   The grounds under which the Contractor may terminate are as follows:

    (a) the Employer fails to provide reasonable evidence of its financial arrangements;

    (b) the Engineer fails to issue a Payment Certificate on time;

    (c) the Employer fails to make due payment on time;

(d)   the Employer fails to perform its contractual obligations;

(e)   the Employer fails to enter into the Contract Agreement or assigns its entire interest in the project;

(f)   there is a prolonged suspension in accordance with sub-cl.8.11;

(g)   the Employer goes into insolvency

- In the case of sub-paras (f) and (g) the Contractor may give immediate notice to termination.
- In the case of sub-paras (a) to (e) the Contractor must give 14 days' notice.   **16–012**

COMMENTARY

The comments made in cl.15 about the seriousness of termination apply equally here. Just as for the Employer, it is not a step to be taken lightly. Of course, the reason why a Contractor will most likely want to bring the Contract to an end, namely that it has under-priced the Contract, is not dealt with here. That will not stop Contractors in such circumstances, looking for ways to exploit this clause.

The provisions of sub-cl.16.2 are fairly self-explanatory. Where the circumstances exist as set out in the sub-clause, the Contractor is entitled to terminate after giving notice of 14 days other than for the circumstances set out in (f) and (g) where no such notice is required. That said, with a number of the potential termination grounds, a substantial period of default must often pass before a termination notice can be given. The grounds under which the Contractor may terminate are as follows:

(a)   Under sub-cl.2.4, the Employer is required to provide reasonable   **16–013** evidence that adequate financial arrangements have been made to finance the project. As noted above, it will take a period of 105 days from the date of the initial request for reasonable evidence before the Contractor could determine the Contract. This is a lengthy period if there is concern about the Employer's ability to pay.

(b)   Sub-clause 14.6 deals with the issue of interim payment certificates. The Engineer has 28 days to issue such a certificate from the time it receives the Contractor's payment application. Sub-paragraph (b) gives the Engineer a further 28 days before the Contractor is entitled to issue a suspension notice.

(c)   Under sub-cl.14.7, the Employer must pay the Contractor within 56 days from the date the payment application is made. Under sub-cl.14.8 the Contractor has a right to interest as soon as payment is late. Here the Employer has a further 42 days to make payment before the Contractor can issue a notice of termination.

(d)   The ability to terminate if the Employer "substantially fails to perform" its obligations under the Contract is likely to be the most contentious of the sub-paras here. The Contractor will need to be especially sure of its grounds. The wording used is different to its counterpart sub-cl.15.2(b) which requires a plain demonstration of an intention not to continue performing. There is no definition of what "substantially fails" means. Some limited guidance can be found from the objective test for "substantial performance" applied by L.J. Hirst in *RJ Young v Thomas Properties.*[2] Thus, the Contractor should (and almost certainly the DAB and arbitral tribunal thereafter will) compare what has been done against what the Employer has contracted to do. Of the Employer's contractual obligations, the most important one as far as the Contractor is concerned is likely to be payment, and there may be other conditions of the contract which the Contractor may be able to use with more certainty in the event of any failure to pay.

16–014   (e)   Under sub-cl.1.6, the Contract Agreement should, unless agreed otherwise, be entered into within 28 days of the Contractor receiving the Letter of Acceptance. Under sub-cl.1.7, neither party can assign its interest in a part or whole of the contract to another, without agreement.

(f)   Under sub-cl.8.11, when the Engineer has suspended the Work in accordance with sub-cl.8.8 and that suspension has lasted for 84 days or 12 weeks, the Contractor may request permission to proceed. If that permission is not received within 28 days, then the Contractor may give notice of termination;

(g)   The definition of insolvency is the same broad definition to be found at sub-cl.15.2(e).

16–015   The sub-clause states that the Contractor's election to terminate does not prejudice its rights under the Contract. It is assumed that this will extend to the right under the Contract to refer disputes to the DAB. However, it must be questioned whether any party would be content with the decision of a DAB once, effectively, the Contract is at an end. It is envisaged that the parties at this stage would prefer to proceed directly to arbitration. That said, assuming that the dispute resolution provisions still operate even after termination, a DAB decision will be necessary before arbitration can be commenced.

---

[2] Court of Appeal January 21, 1999 – although substantial performance is more often a term used when considering whether a party has actually carried out and/or completed and performed its obligations under a contract.

MDB HARMONISED EDITION

Sub-paragraph (d) has been extended and now reads: **16–016**

> *the Employer substantially fails to perform his obligations under the <u>Contract</u>*
> <u>*in such manner as to materially and adversely affect the economic balance of*</u>
> <u>*the Contract and/or the ability of the Contractor to perform the Contract,*</u>

The question here is whether this adds any clarity to the meaning of sub-paragraph (d). If anything, the new words increase the scope for argument. Certainly the new words serve to restrict the Contractor's ability to terminate as the Employer's breach must now have one of two specified consequences.
   In addition two new sub-paragraphs have been added:

> *(h)   In the event the Bank suspends the loan or credit from which part or whole of the payments to the Contractor are being made, if the Contractor has not received the sums due to him upon expiration of the 14 days referred to in Sub-Clause 14.7 [Payment] for payments under Interim Payment certificates, the Contractor may, without prejudice to the Contractor's entitlement to financing charges under Sub-Clause 14.8 [Delayed Payment], take on of the following actions, namely (i) suspend work or reduce the rate of work, or (ii) terminate his employment under the Contract by giving notice to the Employer, with a copy to the Engineer, such termination to take effect 14 days after the giving of the notice.*

> *(i) The Contractor does not receive the Engineer's instruction recording the agreement of both Parties on the fulfilment of the conditions for the Commencement of Works under Sub-Clause 8.1 [Commencement Works].*

The Contractor must give 14 days' notice of termination under both of these   **16–017**
new sub-paragraphs.
   Sub-paragraph (h) is an extension from the new sub-paragraph added in the MDB version to sub-cl.16.1. Sub-paragraph (i) deals with the change to sub-cl.8.1, which is itself one of the more significant changes to be found in the MBD Version. As discussed above, it sets out a number of pre-conditions, stated to be conditions precedent, which must be fulfilled before the project can commence. If the Engineer's instruction to commence is not received within 108 days from the receipt of the Letter of Acceptance, then the Contractor is entitled to serve a notice of termination.

16.3 CESSATION OF WORK AND REMOVAL OF CONTRACTOR'S EQUIPMENT

16–018   *After a notice of termination under Sub-Clause 15.5 [Employer's Entitlement to Termination], Sub-Clause 12 [Termination by Contractor] or Sub-Clause 19.6 [Optional Termination, Payment and Release] has taken effect, the Contractor shall promptly:*

    (a)   *cease all further work, except for such work as may have been instructed by the Engineer for the protection of life or property or for the safety of the Works,*

    (b)   *hand over Contractor's Documents, Plant, Materials and other work, for which the Contractor has received payment, and*

    (c)   *remove all other Goods from the Site, except as necessary for safety, and leave the Site.*

OVERVIEW OF KEY FEATURES

16–019   Upon termination the Contractor must cease work, hand over all goods for which it has received payment and remove any other goods from the site.

COMMENTARY

16–020   Sub-clause 16.3 seems to adopt a pragmatic position. Once the Contract has been terminated, there seems to be little point for the Contractor to continue working or for the Contractor to keep its documents, plant, materials and other work for which it has been paid. Thus, it should take prompt steps to leave the site. Further, as recognized by this sub-clause, there is no need for the Contractor's presence on site other than in the interests of safety.

MDB HARMONISED EDITION

16–021   There is no change.

16.4 PAYMENT ON TERMINATION

After a notice of termination under Sub-Clause 16.2 [Termination by   **16–022**
Contractor] has taken effect, the Employer shall promptly:

(a)   *return the Performance Security to the Contractor,*
(b)   *pay the Contractor in accordance with Sub-Clause 19.6 [Optional*
       Termination, Payment and Release], and
(c)   *pay to the Contractor the amount of any loss of profit or other loss or*
       damage sustained by the Contractor as a result of this termination.

OVERVIEW OF KEY FEATURES

Upon termination, the Employer must promptly return the Performance   **16–023**
Security and pay the Contractor, including in the respect of the costs of
termination.

COMMENTARY

Where the Contract has been terminated under sub-cl.16.2 (i.e. on account of   **16–024**
a failing by the Employer), it is for the Employer to pay the Contractor immediately for any loss as described in sub-cl.16.4. The costs of termination
might include the cost of materials already on order, the costs of removing
any temporary works and the costs of repatriating personnel. These are set
out in more detail in sub-cl.19.6. This differs from sub-cl.16.1 for suspension
of the works where the Contractor must make an application for payment in
respect of its losses within a rigid timeframe. These payments must be made
promptly. Presumably, disagreement as to what loss has been suffered would
qualify as a dispute arising out of the Contract and as such would fall to be
decided by a DAB prior any reference to arbitration being possible.

MDB HARMONISED EDITION

The words "of profit or other losses" have been deleted from sub-para.(c),   **16–025**
thereby deleting the qualifications on the type of loss the Contractor can seek
to recover.

CLAUSE 17 – RISK AND RESPONSIBILITY

17.1 INDEMNITIES

The Contractor shall indemnify and hold harmless the Employer, the   **17–001**
Employer's Personnel, and their respective agents, against and from all claims,
damages, losses and expenses (including legal fees and expenses) in respect of:

(a)   *bodily injury, sickness, disease or death, of any person whatsoever arising*
      out of or in the course of or by reason of the Contractor's design (if any),
      the execution and completion of the Works and the remedying of any
      defects, unless attributable to any negligence, wilful act or breach of the
      Contract by the Employer, the Employer's Personnel, or any of their
      respective agents, and

(b)   *damage to or loss of any property, real or personal (other than the*
      Works), to the extent that such damage or loss:

     (i)   *arises out of or in the course of or by reason of the Contractor's*   **17–002**
         design (if any), the execution and completion of the Works and
         the remedying of any defects, and

     (ii)   *is attributable to any negligence, wilful act or breach of the*
         Contract by the Contractor, the Contractor's Personnel, their
         respective agents, or anyone directly or indirectly employed by any
         of them.

The Employer shall indemnify and hold harmless the Contractor, the
Contractor's Personnel, and their respective agents, against and from all claims,
damages, losses and expenses (including legal fees and expenses) in respect to
of (1) bodily injury, sickness disease or death, which is attributable to any negli-
gence, wilful act or breach of the Contract by the Employer, the Employer's
Personnel, or any of their respective agents, and (2) the matters for which
liability may be excluded from insurance cover, as described in sub-paragraphs
(d)(i), (ii) and (iii) of Sub-Clause 18.3 [Insurance Against Injury to Persons
and Damage to Property].

OVERVIEW OF KEY FEATURES

17–003
- The risks for which the Contractor will be held liable for are:
  - personal injury arising out of the Contractor's design and/or execution of the works unless attributable to the negligence, wilful act or breach of contract by the Employer;
  - damage to property arising out of the Contractor's design and/or execution of the works which is attributable to the negligence, wilful act or breach of contract of the Contractor or its Personnel.

- The risks for which the Employer will be held liable for are:
  - personal injury arising out of its negligence, wilful act or breach of contract;
  - losses arising as a consequence of the Employer's right to have Work executed on and/or to occupy any land as per sub-clause 18.3(d)(i);
  - damage which is the unavoidable consequence of the Contractor fulfilling its contractual obligations, as per sub-cl.18.3(d)(ii);
  - losses arising as a consequence of items listed as Employer risks in sub-cl.17.3, as per sub-cl.18.3(d)(iii).

COMMENTARY

17–004 Clauses 17 and 18 are closely related. Their aim is to protect both parties, through indemnities and insurance from claims arising out of the project. Clause 17 as a whole covers a wide-range of risks and responsibilities. Whilst, sub-cl.17.1 defines those risks primarily in relation to loss and damage which the Contractor and Employer will be held responsible for, it would be fair to say that the Contractor is responsible for the majority of the risks. In practice, the type of risks set out here are those for which the Contractor and Employer are likely to obtain insurance.

Claims for personal injury will unsurprisingly be born by the responsible party. There is however one significant difference in how this sub-clause treats the Contractor and Employer. The Employer's obligation to indemnify in respect of personal injury is subject to default on the part of the Employer; the Contractor's obligation is all-embracing regardless of fault. The nature of the Contractor's obligation can be shown by the contrast with the way in which damage to property caused by the Contractor is treated. Here the Contractor is only liable to the Employer for damage which is attributable to its negligence, wilful act or breach of contract.

To appreciate fully those items for which the Employer will be responsible, reference will need to be made to sub-cl.18.3(d). The underlying feature is that the Contractor will not be liable for claims for damage which is the "unavoidable result of the Contractor's obligations to execute the Works." In practice this might be a difficult sub-clause for the Contractor to take advantage of. Typically, the Contractor has a free choice as to how it carries out its works and there may be difficulties in establishing that damage could have been avoided had the Contractor adopted a different approach.

MDB HARMONISED EDITION

Sub-paragraph (b) has been changed so that it now reads: **17–005**

> *damage to or loss of any property, real or personal (other than the Works), to the extent that such damage or loss arises out of or in the course of or by reason of the Contractor's design (if any), the execution and completion of the Works and the remedying of any defects, <u>unless and to the extent that any such damage or loss</u> is attributable to any negligence, wilful act or breach of the Contract by the <u>Employer</u>, the <u>Employer's</u> Personnel, their respective agents, or anyone directly or indirectly employed by any of them.*

The effect of this amendment is to bring the Contractor's obligation to indemnify in relation to damage to property in line with its obligations in respect of personal injury. Under the FIDIC form, as noted above, the indemnity only applied where it was attributable to the Contractor's negligence, wilful act or breach of contract. This obligation has been extended and now applies to all damage and loss, howsoever caused, unless that damage is caused by the negligence, wilful act of breach of contract to the Employer. According to one commentator,[1] this amendment (presumably because of the way in which the Contractor's indemnity obligations have been increased), is "not liked" by FIDIC.

17.2 CONTRACTOR'S CARE OF THE WORKS

The Contractor shall take full responsibility for the care of the Works and **17–006**
Goods from the Commencement Date until the Taking-Over Certificate is issued (or is deemed to be issued under Sub-Clause 10.1 [Taking Over of the Works and Sections]) for the Works, when responsibility for the care of the

---

[1] Christopher Wade – Presentation Notes on The FIDC Contract Forms and the new MDB Contract, ICC-FIDIC Conference, Paris October 17/18, 2005.

Works shall pass to the Employer. If a Taking-Over Certificate is issued (or is so deemed to be issued) for any Section or part of the Works, responsibility for the care of the Section or part shall then pass to the Employer.

After responsibility has accordingly passed to the Employer, the Contractor shall take responsibility for the care of any work which is outstanding on the date stated in a Taking-Over Certificate, until this outstanding work has been completed.

If any loss or damage happens to the Works, Goods or Contractor's Documents during the period when the Contractor is responsible for their care, from any cause not listed in Sub-Clause 17.3 [Employer's Risks], the Contractor shall rectify the loss or damage at the Contractor's risk and cost, so that the Works, Goods and Contractor's Documents conform with the Contract.

17–007   *The Contractor shall be liable for any loss or damage caused by any actions performed by the Contractor after a Taking-Over Certificate has been issued. The Contractor shall also be liable for any loss or damage which occurs after a Taking-Over Certificate has been issued and which arose from a previous event for which the Contractor was liable.*

OVERVIEW OF KEY FEATURES

17–008
- The Contractor is responsible for the Works and Goods until the Taking-Over Certificate is issued.
- The Employer is responsible for the Works and Goods once the Taking-Over Certificate has been issued.

COMMENTARY

17–009   Sub-clause 17.2 has the effect that, while it is performing actions on site, the Contractor will be responsible for those actions, unless the damage is caused by one of the Employer risks set out at sub-cl.17.3. Once the Taking-Over Certificate has been issued, responsibility switches to the Employer. The practical importance of this is that insurance responsibilities may well switch as well. It is therefore important that enough time has been allowed to ensure that the project is not left uninsured for even a short period of time.

This sub-clause makes it clear that the Contractor is responsible for remedying, at its own risk and cost, any damage for which it is responsible. Those costs might ultimately translate into delay damages if progress is delayed as a consequence. Although the Contractor will not be held responsible for damage caused by an Employer Risk, this and the subsequent sub-cl.17.4, make it clear that provided the proper notice it given, the Contractor must rectify that damage.

The Contractor should also appreciate that its liability to indemnify does not necessarily end at the issue of the Taking-Over Certificate. The final paragraph states that the Contractor will remain liable for any loss or damage that arises from an event, prior to the issue of the certificates, for which the Contractor was responsible. In addition, understandably, the Contractor will also be liable for any work it undertakes consequently, notwithstanding the issue of the certificate.

As described above in relation to cl.10, when only a section of the works is **17–010** taken over, it is important that the section is clearly defined. Here, if it is not adequately defined, a part of the works might remain uninsured.

The parties should also note that there is a difference between the way the Contractor's Goods, Documents and Works are treated. It is responsibility for the Contractor's Goods (i.e. as per sub-cl.1.1.5.1, equipment, materials, plant and temporary works) and Works *only* which passes at the issue of the Taking-Over Certificate. The omission from the first paragraph of the Contractors' Documents was deliberate and the only reference to Documents comes in the third paragraph which refers to the need for the Contractor to make good any damage or loss which occurs whilst it is responsible for the Works. The FIDIC Guide suggests that this is so because under sub-cl.1.8, the Contractor is responsible for the Contractors' Documents "unless and until" they are taken over by the Employer. Thus they may be taken over at a different time to the issue of the Taking-Over Certificate.

MDB HARMONISED EDITION

There is no change.   **17–011**

17.3 EMPLOYER'S RISKS

The risks referred to in Sub-Clause 17.4 below are: **17–012**

(a)   war, hostilities (whether war be declared or not), invasion, act of foreign enemies,

(b)   rebellion, terrorism, revolution, insurrection, military or usurped power, or civil war, within the Country,

(c)   riot, commotion or disorder within the Country by persons other than the Contractor's Personnel and other employees of the Contractor and Subcontractors,

(d)   munitions of war, explosive materials, ionising radiation or contamination   **17–013**
by radio-activity, within the Country, except as may be attributable to the Contractor's use of such munitions, explosives, radiation or radio-activity,

(e) *pressure waves caused by aircraft or other aerial devices travelling at sonic or supersonic speeds,*

(f) *use or occupation by the Employer of any part of the Permanent Works, except as may be specified in the Contract,*

(g) *design of any part of the Works by the Employer's Personnel or by others for whom the Employer is responsible, and*

(h) *any operation of the forces of nature which is Unforeseeable or against which an experienced Contractor could not reasonably have been expected to have taken adequate preventative precautions.*

OVERVIEW OF KEY FEATURES

17–014   The risks for which the Employer is responsible, to the extent defined in sub-cl.17.4, are listed at sub-paras (a)–(h).

COMMENTARY

17–015   There is a significant overlap between the Employer's risks set out in sub-cl.17.3 and the definition of force majeure given in cl.19 below. These are events which the Contractor, although in control of the site, would be power-less to protect itself against. The risk must be one which directly affects the execution of the project works. In reality these are risks which are beyond the control of both Contractor and Employer. The list of risks can be split into two. The first five (a)–(e) are typical risks for which no sensible or commer-cially priced insurance can be obtained.

Under sub-cl.19.4, the Contractor can claim additional payment for the costs incurred as a result of the events listed in sub-cl.19.1, sub-paras (i) to (iv). These four categories of events are exactly the same as those listed under sub-cl.17.3, sub-paras (a) to (d) for which the Contractor can also claim addi-tional payment under sub-cl.17.4. There is therefore a complete overlap between these four provisions (and the Contractor will need to follow the procedure under sub-cl.20.1 in both cases). The only distinction is that sub-cl.17.3 lists one further category at sub-para.(e), not listed as a type of force majeure event under cl.19.1, which deals with pressure waves caused by aircraft or other aerial devices travelling at sonic or supersonic speeds.

The remaining three sub-paragraphs here have an element of Party default, subparas (f)–(g) are ones which relate to Employer default, whilst with (h) although it refers to forces of nature, there is arguably room for Contractor default, which would mean that it could not make any claim in respect of this risk item. These last three risks (f)–(h) do not appear in the Silver Book.

There have been some changes to the Employer risks listed in the Old Red **17–016** Book FIDIC 4th edn, which in part are a consequence of the changes in global society over the passage of time. "Terrorism" has been added to sub-para.(b). The words "within the Country" have also been added to that sub-paragraph. The same restriction has been added to sub-paras (c) and (d). It is perfectly possible that acts of terrorism or riots in Countries other than where the project is carried out will affect the Contractor's progress. This is particularly the case if equipment, which is being transported or manufactured in other countries, is destroyed or seized as a consequence of civil war. It should be noted that this "country-based" restriction does not appear in the definition of force majeure to be found at sub-para.19.1.

With sub-para.(c), the exclusion whereby disorder on the part of the Contractor or its personnel is understandably not an Employer risk remains. However, under the Old Red Book, FIDC 4th edn, disorder or rioting on the part of the Contractor which arose from the conduct of the Works was excluded as an Employer risk. This qualification has been removed.

The words "munitions of war" and "explosive materials" have been added to sub-para.(d). This must primarily be a reference to landmines, which can take many many years to be removed. The sub-paragraph also makes it clear that the Contractor cannot assert that damage which may be caused by its use of explosives or radioactivity is an Employer risk.

Sub-paragraphs (f) and (g) understandably mark the occupation of the **17–017** Works and any design carried out by the Employer, as Employer risks. It is envisaged that (f) above will be the most likely cause of disputes. It is not immediately apparent what will constitute "use or occupation by the Employer" within the meaning of that clause. The Employer should take special care when it has more than one contractor on site, as this is typically when such difficulties are likely to arise. Both parties should also bear in mind that sub-para.(g) will include any design carried out by the Engineer who by sub-cl.1.1.2.6 is included within the definition of Employer's Personnel.

Subparagraph (h) refers to the forces of nature which are either unforeseeable or against which the experienced contractor could not have been expected to take reasonable preventive measures. It will be recalled that sub-cl.1.1.6.8 defines "unforeseeable" as meaning not reasonably foreseeable by an experienced contractor by the date of the submission of the tender. It is possible that there could be Employer default if inadequate tender information is provided. However, the real onus here lies with the Contractor. For example seismic reports establish the probability of earthquakes and will give the Contractor an idea of the likely frequency of seismic activity. Therefore the experienced Contractor will know that if minor earthquakes occur every three years and the project is scheduled to last for four years, it will be expected to make approximate provision.

The consequences of Employer's Risk are dealt with in sub-cl.17.4.

MDB HARMONISED EDITION

17–018   The words "below, insofar as they directly affect the execution of the Works in the Country" have been added to the first paragraph. The effect of this is to narrow further the nature of the Employer risks.

In addition, the words "sabotage by persons other than the Contractor's Personnel" have been added to sub-para.(b).

17.4 CONSEQUENCES OF EMPLOYER'S RISK

17–019   *If and to the extent that any of the risks listed in Sub-Clause 17.3 above results in loss or damage to the Works, Goods or Contractor's Documents, the Contractor shall promptly give notice to the Engineer and shall rectify this loss or damage to the extent required by the Engineer.*

If the Contractor suffers delay and/or incurs Cost from rectifying this loss or damage, the Contractor shall give a further notice to the Engineer and shall be entitled subject to Sub-Clause 20.1 [Contractor's Claims] to:

(a)   an extension of time for any such delay, if completion is or will be delayed, under Sub-Clause 8.4 [Extension of Time for Completion], and

17–020   *(b)   payment of any such Cost, which shall be included in the Contract Price. In the case of sub-paragraphs (f) and (g) of Sub-Clause 17.3 [Employer's Risks], reasonable profit on the Cost shall also be included.*

After receiving this further notice, the Engineer shall proceed in accordance with Sub-Clause 3.5 [Determinations] to agree or determine these matters.

OVERVIEW OF KEY FEATURES

17–021
- If the Contractor suffers loss or damage as a consequence of any item of Employer risk, then he must give prompt notice to the Engineer.
- The Engineer may require the Contractor to remedy that loss or damage.
- If the Contractor suffers delay or increased cost, it may be entitled to an extension of time and to recover its cost, or its cost plus reasonable profit in respect of use of the works or design by the Employer.

COMMENTARY

This sub-clause deals with how the Contractor must proceed if an Employer **17–022** Risk causes loss or damage. The sub-clause provides for a two-part notice procedure. First, the Contract must give notice to the Engineer of the damage caused by the sub-cl.17.3 risk item. The notice must be in writing and ought to define the risk and damage caused by that risk.

Second, if the Contractor suffers loss as a consequence of rectifying the damage caused by a sub-cl.17.3 event, notice must be given to the Engineer adopting the procedure contained in sub-cl.20.1. However, although the Contractor can recover cost, he is only entitled to reasonable profit in respect of sub-paras 17.3.1 (f) and (g). These are the items which refer to Employer responsibility and/or fault. Where the time- limits and requirement for providing the Engineer with particulars have not been complied with, it is suggested that no remedy will be available to the Contractor in respect of those losses.

MDB HARMONISED EDITION

The word "reasonable" has been deleted before "profit" in sub-para.(b).   **17–023**

17.5 INTELLECTUAL AND INDUSTRIAL PROPERTY RIGHTS

In this Sub-Clause, "infringement" means an infringement (or alleged infringe- **17–024** *ment) of any patent, registered design, copyright, trade mark, trade name, trade secret or other intellectual or industrial property right relating to the Works; and "claim" means a claim (or proceedings pursuing a claim) alleging an infringement.*

Whenever a Party does not give notice to the other Party of any claim within 28 days of receiving the claim, the first Party shall be deemed to have waived any right to indemnity under this Sub-Clause.

The Employer shall indemnify and hold the Contractor harmless against and from any claim alleging an infringement which is or was:

(a)   an unavoidable result of the Contractor's compliance with the Contract, **17–025** *or*
(b)   a result of any Works being used by the Employer.

17–026

> *(ii)   for a purpose other than that indicated by, or reasonably to be inferred from, the Contract, or*
>
> *(iii) in conjunction with anything not supplied by the Contractor, unless such use was disclosed to the Contractor prior to the Base Date or is stated in the Contract.*

The Contractor shall indemnify and hold the Employer harmless against and from any other claim which arises out or in relation to (i) the manufacture, use, sale or import of any Goods, or (ii) any design for which the Contractor is responsible.

If a Party is entitled to be indemnified under this Sub-Clause, the indemnifying Party may (at its cost) conduct negotiations for the settlement of the claim, and any litigation or arbitration which may arise from it. The other Party shall, at the request and cost of the indemnifying Party, assist in contesting the claim. This other Party (and its Personnel) shall not make any admission which might be prejudicial to the indemnifying Party, unless the indemnifying Party failed to take over the conduct of any negotiations, litigation or arbitration upon being requested to do so by such other Party.

OVERVIEW OF KEY FEATURES

17–027
- Where a claim is made alleging an infringement of copyright, the party receiving the claim must, within 28 days, notify the other of that claim. If no notice is given, then the right to any indemnity is lost.
- The Employer must indemnify the Contractor in respect of any such claim which was the unavoidable result of the Contractor executing the Works or as a result of the Employer using the Works for purposes not related to the Contract.
- The Contractor must indemnify the Employer in respect of any such claim arising out of its design or works.
- The parties must co-operate together to deal with any claims that may arise.

COMMENTARY

17–028   Sub-clause 17.5 sets out the respective rights and responsibilities of the parties regarding intellectual property. Again these rights are often the subject of insurance. It is an expansion of sub-cl.28.1 of the Old Red Book FIDIC, 4th edn. The sub-clause works by requiring the party who receives a claim to notify the other within 28 days. The consequences of failing to make the required notification are serious and would mean the party in default losing the right to any indemnity. Thereafter, the last paragraph of the sub-clause stipulates that the parties are to co-operate in resisting claims against one of them. Such co-operation would benefit the project as a whole. This consideration of the well-being of the project rather than the individual is a clear feature running through the FIDIC form.

MDB HARMONISED EDITION

There is no change.   **17–029**

17.6 LIMITATION OF LIABILITY

Neither Party shall be liable to the other Party for loss of use of any Works, loss   **17–030**
of profit, loss of any contract or for any indirect or consequential loss or damage which may be suffered by the other Party in connection with the Contract, other than under Sub-Clause 16.4 [Payment on Termination] and Sub-Clause 17.1 [Indemnities].

The total liability of the Contractor to the Employer, under or in connection with the Contract other than under Sub-Clause 4.19 [Electricity, Water and Gas], Sub-Clause 4.20 [Employer's Equipment and Free-Issue Material], Sub-Clause 17.1 [Indemnities] and Sub-Clause 17.5 [Intellectual and Industrial Property Rights], shall not exceed the sum stated in the Particular Conditions or (if a sum is not so stated) the Accepted Contract Amount.

This Sub-Clause shall not limit liability in any case of fraud, deliberate default or reckless misconduct by the defaulting Party.

OVERVIEW OF KEY FEATURES

- Claims for indirect or consequential loss or damage, loss of use of the   **17–031**
  Works, loss of any contract, or loss of profit can only be made in respect of sub-cll 16.4 and 17.1.
- The Contractor's total liability is capped and shall not exceed either the sum stated in the Contract or the Accepted Contract Amount.

- There are no limitations on liability in cases of fraud, or deliberate default or reckless misconduct.

COMMENTARY

032   Whilst the introduction of restrictions or limitations on the Contractor's liability is new to the FIDIC Red Book scheme, it is not new to FIDIC. Some limitations on the Contractor's liability, including a cap (which in default of any other provision was the Contract Sum) were included in the Old Yellow Book 3rd edn, which was produced for mechanical and electrical works.

The intention of the first paragraph is to limit liability in respect of the four listed items, save as provided for in sub-paras 16.4, (which deals with termination by the Contractor) and 17.1 (which deals with general indemnification obligations). The first paragraph deals with a variety of direct and indirect losses. The direct losses are self-explanatory. In particular, the first paragraph provides that neither Party shall be liable for any indirect or consequential loss or damage. Consequential damages fall within the second limb of the classic 1854 case of *Hadley v Baxendale*[2] which provides that:

> *Where two parties have made a contract which one of them has broken, the damages which the other party ought to receive in respect of such breach of contract should be such as may fairly and reasonably be considered either:*
>
> *(1)   arising naturally, ie according to the usual course of things from such breach of contract itself; or*
> *(2)   such as may reasonably be supposed to have been in the contemplation of both parties at the time they made the contract as the probable result of the breach of it.*

033   In reality there is little, if any, difference in meaning between the words "*indirect*" and "*consequential*" in exclusion of liability clauses. In the case of *Hotel Services Limited v Hilton International Hotels (UK) Ltd*[3] Sedley L.J. commented that where an exclusion clause bundled together the words "*any indirect or consequential loss, damages or liability*" then "*consequential*" was clearly used as a synonym for "*indirect*".

One cause of claims in relation to exclusions such as these relates to loss of profit. One reason for its inclusion here in the FIDIC form is probably in recognition of the difficulty potential tenderers might have in providing for the costs of this risk into the tender price. Another reason for the specific reference is because there have been many disputes which have turned on

---

[2] See also sub-cl.14.8 below.
[3] [2000] BLR 235.

whether loss of profit can be said to be an indirect or consequential loss. For example in *British Sugar Plc v NEI Power Projects Ltd*,[4] Waller L.J. rejected the submission that consequential loss would in the view of the typical average businessman include loss of profit. In part he felt bound by authority.[5] However the FIDIC form here expressly removes that possibility including loss of profit as one of the four specified exclusions. In the case of *Deepak Fertilisers v Davy McKee*,[6] the Court of Appeal had to consider an exclusion clause which expressly excluded liability for both loss of profits and for consequential loss. Stuart Smith L.J. came to the logical conclusion that:

> *The direct and natural result of the destruction of the plant was that Deepak was left without a methanol plant, the reconstruction of which would cost money and take time, losing for Deepak any methanol production in the meantime. Wasted overheads incurred during the reconstruction of the plant, as well as profits lost during that period, are no more remote as losses than the cost of reconstruction. Lost profits cannot be recovered because they are excluded in terms, not because they are too remote.*

The second paragraph places a cap on the liability of the Contractor to the Employer. No limit is placed on the cap. This may be stated in the Particular Conditions or, if not, will be the Accepted Contract Amount. The Accepted Contract Amount, per sub-cl.1.1.4.1, is the original contract sum as agreed in the Letter of Acceptance. It is not the ultimate contract price, which may well have increased or decreased depending on the nature of any variations or omissions. Liability caps are frequently encountered in process engineering contracts. One reason for this is the difference between the potential losses from such plants and not only the contract sum but also the level of profit made by many contractors. Another consideration might be the level of any professional indemnity insurance cover. Although this would primarily apply to design and build contracts, it would be odd if any indemnity cap was for a sum in excess of the level of the insurance cover.

17–034

The cap is new to the Red Book. One reason for it, aside from the obvious reassurance to the Contractor, is that it should encourage amicable settlement. Both parties would know that there is a limit to the Employer's ultimate level of recovery. The cap only applies to losses which arise "under or in connection with the Contract". There are certain exclusions to the cap which fall into two categories. The potential costs of power, water and any equipment provided under sub-cll 4.20 and 4.21 are likely to be minor, whilst the indemnities under sub-cl.17.1 are likely to be covered by insurance. Although the costs of copyright infringement can and are likely to be covered

---

[4] 87 BLR 42.
[5] *Croudace Construction v Cawoods Concrete Products Ltd* 8 BLR 20.
[6] [1999] 1 Lloyd's Rep 387.

by insurance, it stands in a slightly different category, outside the mainstream of everyday project liabilities.

As discussed above[6a] in the case of *Decoma UK Limited v Haden Drysys International Limited*,[6b] Haden's liability was capped at 5 per cent of the contract price. Haden's potential liability was far greater than this sum. Decoma attempted to argue that if Haden was allowed to rely on the cap to limit its liability, it would be taking advantage of its own wrong. The Court rejected this. Both parties had clearly agreed the level of the cap. Thus the extent of any damages in excess of that sum was irrelevant.

As with any contract, care needs to be taken in agreeing the contract terms, especially when they relate to a contractual liability cap. Take the Court of Appeal decision in *Ove Arup & Partners International Limited v Mirant Asia-Pacific Construction (Hong Kong) Limited*.[7] The central issue between the parties was whether the agreements between them (in relation to engineering design and ground investigations) incorporated the terms of the 1991 FIDIC Client/Consultant Model Services Agreement. This was significant because the contract included a five-year limitation of liability period and a limit of compensation of £4m. Arup started engineering design work on the basis of a letter of intent, but kept pressing for agreement of the services agreement. Several months later, although no agreement had been reached, the employer decided to engage Arup for ground investigation work. Arup sent the employer its proposals, which provided that it would be governed by the model services agreement. A manager wrote "okay" on the proposal, initialled it and sent it back to Arup. L.J. May here found that a signature accompanied by the word "okay" did constitute an acceptance of the proposal of those FIDIC terms which were part of the proposal before that individual at the time. In other words, "okay" meant the parties intended the ground investigation works (and those works only) to be covered by the FIDIC terms, *including* the cap and limitation on liability.

17–035   As the FIDIC Guide notes, the limitations within this sub-clause may well be affected by the chosen applicable law, which might in particular impose a higher or lower liability cap or may limit the duration of any such liability. In fact sub-cl.11.10 confirms that there is no time-limit in the FIDIC form here on the Contractor's liability. This provides that, both the Employer and Contractor shall, after issue of the Performance Certificate, remain liable for any obligation that remains unperformed. For the Contractor, this will include any hidden or latent defect. As a consequence of this, the FIDIC Guide notes the following four points:

[6a] See sub-cl.8.7, above.
[6b] [2005] EWHC 2429.
[7] (2004) BLR 75.

(i)   Under some common law jurisdictions, the period of liability might not commence until the Employer ought reasonably to have been aware of the defective work.

(ii) Under some civil law jurisdictions, decennial liability[8] will apply. This provides that the Contractor will be liable absolutely for any hidden defects for a period of ten years from completion.

(iii) Unless the Works include major items of Plant it may be inappropriate to limit the length of the duration of the Contractor's liability.

(iv) Conversely, if the Works do include major items of Plant then it would be appropriate to limit the length of the Contractor's liability for a fixed period.

The reason given for the difference between (iii) and (iv) is that after a few years of operation it becomes increasingly difficult to establish whether any defects are attributable to design, manufacture or operation, or a combination of all three. However, it is difficult to see why there should be a difference in policy only if there is a major item of plant. The way in which a completed project is operated and maintained will have a significant impact across the board, if that operation is of a poor standard.

Finally, the sub-clause ends by confirming that the cap and any limitation **17–036** of liability will not apply in cases of fraud, deliberate default or reckless misconduct on the part of the defaulting party. These are equally likely to be reasons which would lead to insurance cover being declined.

MDB HARMONISED EDITION

The following has been added to the first paragraph.   **17–037**

> . . .*other than as specifically provided in Sub-Clause 8.7 [Delay Damages]; Sub-Clause 11.2 [Cost of Remedying Defects]; Sub-Clause 15.4 [Payment after Termination]; Sub-Clause 17.1 [Indemnities]; Sub-Clause 17.4(b) [Consequences of Employer's Risks] and Sub-Clause 17.5 [Intellectual Employer's Risks] and Sub-Clause 17.5 [Intellectual and Industrial Property Rights].*

The purpose of this addition is to make it absolutely clear that certain items, for example delay damages, are not covered by the limitation of liability set out in the first paragraph. In addition the following has been added to the second paragraph:

---

[8] For example, art.1792 of the French Civil Code.

17–038   *. . .resulting from the application of a multiplier (less or greater than one) to the Accepted Contract Amount, as stated in the Contract Data, or (if such multiplier or other sum is not so stated), the Accepted Contract Amount.*

In order to avoid any possibility of dispute, it might be thought simpler if a fixed limit, rather than a multiplier or any other form of calculation as suggested here, was adopted.

17.7 USE OF THE EMPLOYER'S ACCOMMODATION FACILITIES

17–039   The following self-explanatory sub-clause appears in the MDB Harmonised Edition only. It is another example of a sub-clause being adopted from the Particular Conditions.

The Contractor shall take full responsibility for the care of the Employer provided accommodation and facilities, if any, as detailed in the Specification, from the respective dates of hand-over to the Contractor until cessation of occupation (where hand-over or cessation of occupation may take place after the date stated in the Taking-Over Certificate for the Works).

If any loss of damage happens to any of the above items while the Contractor is responsible for their care arising from any cause whatsoever other than those for which the Employer is liable, the Contractor shall, at his own cost, rectify the loss or damage to the satisfaction of the Engineer.

CLAUSE 18 – INSURANCE

18.1 GENERAL REQUIREMENTS FOR INSURANCES

In this Clause "insuring Party" means, for each type of insurance, the Party **18–001**
responsible for effecting and maintaining the insurance specified in the relevant
Sub-Clause.

Wherever the Contractor is the insuring Party, each insurance shall be
effected with insurers and in terms approved by the Employer. These terms shall
be consistent with any terms agreed by both Parties before the date of the Letter
of Acceptance. This agreement of terms shall take precedence over the
provisions of this Clause.

Wherever the Employer is the insuring Party, each insurance shall be effected
with insurers and in terms consistent with the details annexed to the Particular
Conditions.

If a policy is required to indemnify joint insured, the cover shall apply sepa- **18–002**
rately to each insured as though a separate policy had been issued for each of
the joint insured. If a policy indemnifies additional joint insured, namely in
addition to the insured specified in this Clause, (i) the Contractor shall act
under the policy on behalf of these additional joint insured except that the
Employer shall act for Employer's Personnel, (ii) additional joint insured shall
not be entitled to receive payments directly from the insurer or to have any
other direct dealings with the insurer, and (iii) the insuring Party shall require
all additional joint insured to comply with the conditions stipulated in the
policy.

Each policy insuring against loss or damage shall provide for payments to be
made in the currencies required to rectify the loss or damage. Payments received
from insurers shall be used for the rectification of the loss or damage.

The relevant insuring Party shall, within the respective periods stated in the
Appendix to Tender (calculated from the Commencement Date), submit to the
other Party:

(a) *evidence that the insurances described in this Clause have been effected,* **18–003**
and
(b) *copies of the policies for the insurances described in Sub-Clause 18.2*
[Insurance for Works and Contractor's Equipment] and Sub-Clause 18.3
[Insurance against Injury to Persons and Damage to Property].

When each premium is paid, the insuring Party shall submit evidence of
payment to the other Party. Whenever evidence or policies are submitted, the
insuring Party shall also give notice to the Engineer.

18–004   *Each Party shall comply with the conditions stipulated in each of the insurance policies. The insuring Party shall keep the insurers informed of any relevant changes to the execution of the Works and ensure that insurance is maintained in accordance with this Clause.*

Neither Party shall make any material alteration to the terms of any insurance without the prior approval of the other Party. If an insurer makes (or attempts to make) any alteration, the Party first notified by the insurer shall promptly give notice to the other Party.

If the insuring Party fails to effect and keep in force any of the insurances it is required to effect and maintain under the Contract, or fails to provide satisfactory evidence and copies of policies in accordance with this Sub-Clause, the other Party may (at its option and without prejudice to any other right or remedy) effect insurance for the relevant coverage and pay the premiums due. The insuring Party shall pay the amount of these premiums to the other Party, and the Contract Price shall be adjusted accordingly.

18–005   *Nothing in this Clause limits the obligations, liabilities or responsibilities of the Contractor or the Employer, under the other terms of the Contract or otherwise. Any amounts not insured or not recovered from the insurers shall be borne by the Contractor and/or the Employer in accordance with these obligations, liabilities or responsibilities. However, if the insuring Party fails to effect and keep in force an insurance which is available and which it is required to effect and maintain under the Contract, and the other Party neither approves the omission nor effects insurance for the coverage relevant to this default, any moneys which should have been recoverable under this insurance shall be paid by the insuring Party.*

Payments by one Party to the other Party shall be subject to Sub-Clause 2.5 [Employer's Claims] or Sub-Clause 20.1 [Contractor's Claims], as applicable.

OVERVIEW OF KEY FEATURES

18–006   • Any terms in relation to insurance agreed by the parties before the Contract will over-ride this sub-clause.
  • Any insurance policy taken out by the Contractor must be in a form agreed with the Employer.
  • Any insurance policy taken out by the Employer must be in a form set out in the Particular Conditions.
  • Any insurance payment made for loss or damage must be utilised in respect of that loss or damage.
  • The insuring party must supply proof to the other, including evidence of payment of premiums that the policy has been taken out.
  • If either party fails to do this, the other may take out an appropriate policy and seek reimbursement for this.
  • Each party should comply with any insurance conditions.

- Any dispute shall be decided in accordance with sub-clauses 2.5, (although the Contractor will in the main be responsible for obtaining insurance), and 20.1.

COMMENTARY

In keeping with what appears to be the general ethos of the FIDIC form, **18–007** although the Contractor will in the main be responsible for obtaining cover, insurance is to be taken out for the benefit of the project rather than to protect a party against the defaults of the other. Where insurance is taken out in joint names, there is little motivation for disputes to arise between the parties as to matters covered by the policy. The same policy would pay in either outcome. The basic intention of the clause here is that the Contractor is to arrange cover, as the "insuring party", unless stated otherwise in the Particular Conditions. However, some concern has rightly been expressed that there is scope for confusion with the drafting of this clause because the "insuring party" is not necessarily the same for every policy.[1] Accordingly, care must be taken by both parties to ensure that there are no policy gaps, which might lead to parts of the project being uninsured or overlaps which might lead to over-payment of premiums and increased costs.

The parties must agree when the various insurance polices must be in place. Indeed the sub-clause envisages that there will be discussion between the parties before the Letter of Acceptance is issued. The sub-clause also provides that the Employer must approve any policy the Contractor takes out. So there is no room for doubt, the Appendix to Tender should set out when evidence must be provided that the insurance polices have been taken out. This will be expressed in a period of days from the Commencement Date. That said, it would obviously be preferable if the insurance is taken out before the project commences.

The parties should be aware that the penultimate paragraph expressly states that the obtaining of insurance cover will not limit the obligations, liabilities or responsibilities of either the Contractor or Employer.

MDB HARMONISED EDITION

The words "Contract Data" have been inserted in place of "Appendix to **18–008** Tender". Under the FIDIC Form, whilst the policies taken out by the Contractor must be approved by the Employer, the opposite was not true of

[1] Professor Nael Bunni – cll 17–18 of the FIDIC new contracts – June 2001.

Employer policies. That imbalance has been corrected here and the end of the third paragraph has been altered so that the paragraph reads as follows:

> *Wherever the Employer is the insuring Party, each insurance shall be effected with the insurers and in terms acceptable to the Contractor. These terms shall be consistent with any terms agreed by both Parties before the date of the Letter of Acceptance. This agreement of terms shall take precedence over the provisions of this Clause.*

In addition, the following additional paragraph has been added to the end of the sub-clause

18–009   *The Contractor shall be entitled to place all insurance relating to the Contract (including, but not limited to the insurance referred to Clause 18) with insurers from any eligible source country.*

As discussed above at sub-cl.4.1, the World Bank maintains a list of countries from which bidders, goods and services, are not eligible to participate in procurement which is financed by the Bank.

18.2 INSURANCE FOR WORKS AND CONTRACTOR'S EQUIPMENT

18–010   *The insuring Party shall insure the Works, Plant, Materials and Contractor's Documents for not less than the full reinstatement cost including the costs of demolition, removal of debris and professional fees and profit. This insurance shall be effective from the date by which the evidence is to be submitted under sub-paragraph (a) of Sub-Clause 18.1 [General Requirements for Insurances], until the date of issue of the Taking-Over Certificate for the Works.*

     The insuring Party shall maintain this insurance to provide cover until the date of issue of the Performance Certificate, for loss or damage for which the Contractor' is liable arising from a cause occurring prior to the issue of the Taking-Over Certificate, and for loss or damage caused by the Contractor in the course of any other operations (including those under Clause 11 [Defects Liability]).

     The insuring Party shall insure the Contractor's Equipment for not less than the full replacement value, including delivery to Site. For each item of Contractor's Equipment, the insurance shall be effective while it is being transported to the Site and until it is no longer required as Contractor's Equipment.

18–011   *Unless otherwise stated in the Particular Conditions, insurances under this Sub-Clause:*

(a) *shall be effected and maintained by the Contractor as insuring Party,*

(b) *shall be in the joint names of the Parties, who shall be jointly entitled to receive payments from the insurers, payments being held or allocated between the Parties for the sole purpose of rectifying the loss or damage,*

(c) *shall cover all loss and damage from any cause not listed in Sub-Clause 17.3 [Employer's Risks],*　　**18–012**

(d) *shall also cover loss or damage to a part of the Works which is attributable to the use or occupation by the Employer of another part of the Works, and loss or damage from the risks listed in sub-paragraphs (c), (g) and (h) of Sub-Clause 17.3 [Employer's Risks], excluding (in each case) risks which are not insurable at commercially reasonable terms, with deductibles per occurrence of not more than the amount stated in the Appendix to Tender (if an amount is not so stated, this sub-paragraph (d) shall not apply), and*

(e) *may however exclude loss of, damage to, and reinstatement of:*

    (i) *a part of the Works which is in a defective condition due to a defect in its design, materials or workmanship (but cover shall include any other parts which are lost or damaged as a direct result of this defective condition and not as described in sub-paragraph (ii) below),*　　**18–013**

    (ii) *a part of the Works which is lost or damaged in order to reinstate any other part of the Works if this other part is in a defective condition due to a defect in its design, materials or workmanship,*

    (iii) *a part of the Works which has been taken over by the Employer, except to the extent that the Contractor is liable for the loss or damage, and*

    (iv) *good while they are not in the Country, subject to a Sub-Clause 14.5 [Plant and Materials intended for the Works].*　　**18–014**

If, more than one year after the Base Date, the cover described in sub-paragraph (d) above ceases to be available at commercially reasonable terms, the Contractor shall (as insuring Party) give notice to the Employer, with supporting particulars. The Employer shall then (i) be entitled subject to Sub-Clause 2.5 [Employer's Claims] to payment of an amount equivalent to such commercially reasonable terms as the Contractor should have expected to have paid for such cover, and (ii) be deemed, unless he obtains the cover at commercially reasonable terms, to have approved the omission under Sub-Clause 18.1 [General Requirements for Insurances].

OVERVIEW OF KEY FEATURES

18–015
- The insuring Party in respect of the Works and Contractor's Equipment will be the Contractor, although the insurance must be in the joint names of both parties.
- The Works must be insured for a sum not less than the full reinstatement value.
- This insurance must be maintained until the date the Performance Certificate is issued.
- The insuring party shall put in place full replacement value insurance for equipment to cover the entire period from while it is transported to the site until it is no longer required.
- The insurance must cover all loss and damage to the Works save for certain items listed as an Employer Risk.
- The insurance need not include loss or damage as a consequence of a defect in the Employer's design, materials and workmanship, or parts of the works taken over by the Employer.
- If certain insurance cover is not available at commercially reasonable terms, the Contractor *may* be entitled to dispense with the requirement to insure.

COMMENTARY

18–016 The insurance of the Contractor's Works, Goods and Documents is to be in place from the time specified in the Appendix to Tender, typically from the Commencement Date, until the Taking-Over Certificate is issued. It is important that the insurance is in place from the time work proceeds. The scheme behind the sub-clause is that the various goods are insured until they are no longer required on the project. The Contractor is to take out the cover and again the insurance is required to be in joint names. In the event that payments are made out under the policy, their sole purpose is to be for rectifying any loss or damage that may have been caused.

The Works, Plant, Materials and Contractor's Documents are defined at sub-cll 1.1.5.8, 1.1.5.5, 1.1.5.3 and 1.1.6.1 respectively. This cover must be for the full reinstatement cost which includes fees and recoverable profit. Some care is required with sub-cl.1.1.5.1. Contractor's Equipment means all apparatus, machinery, vehicles and anything else required to execute the Works. It does not include any Temporary Works, Plant and Materials which are intended to form part of the Permanent Works.

Sub-paragraph (d) requires that the cover under this sub-clause shall be for all loss and damage attributable to the Employer's use or occupation of the Works including for three only of the Employer Risks listed in sub-cl.17.3. These are riot or other disorder by persons other than the Contractor (sub-para.c), design by the Employer's Personnel (sub-para.g) and unforeseeable operations of the forces of nature (sub-para.h). Sub-paragraph (d) here gives some protection to the Contractor if it is unable to find insurance "at commercially reasonable terms". If this is the case, then the Contractor must pay to the Employer a sum equivalent to the commercially reasonable price for such cover. This sum is to be calculated according to sub-clause 2.5. In these circumstances, the Employer will be deemed to have approved the omission, unless he obtains suitable cover himself. However, that cover must be at commercially reasonable terms and the sub-clause make no reference to what might happen if the Employer chooses to obtain cover at a higher price.

It is the intention of this sub-clause that the Contractor take out the cover. For example, the Employer may have some difficulties in effecting cover for the Contractor's Equipment. However, the Particular Conditions do suggest an alternative final paragraph if the Employer is to take out insurance here: **18–017**

> However the insurances described in the first two paragraphs of Sub-Clause 18.2 shall be effected and maintained by the Employer as insuring Party, and not by the Contractor.

MDB HARMONISED EDITION

The words "to the Party actually bearing the costs" replace "between the Parties for the sole purpose" in sub-para.(b). **18–018**

In sub-para.(d), the qualification "to the extent specifically required in the bidding documents of the Contract" has been added. The words "Contract Data" again replace "Appendix to Tender".

18.3 INSURANCE AGAINST INJURY TO PERSONS AND DAMAGE TO PROPERTY

The Insuring Party shall insure against each Party's liability for any loss, damage, death or bodily injury which may occur to any physical property (except things insured under Sub-Clause 18.2 [Insurance for Works and Contractor's Equipment] or to any person (except persons insured under Sub-Clause 18.4 [Insurance for Contractor's Personnel], which may arise out of the Contractor's performance of the Contract and occurring before the issue of the Performance Certificate. **18–019**

This insurance shall be for a limit per occurrence of not less than the amount stated in the Appendix to Tender, with no limit on the number of occurrences. If an amount is not stated in the Appendix to Tender, this Sub-Clause shall not apply.

Unless otherwise stated in the Particular Conditions, the insurances specified in this Sub-Clause:

18–020   (a)   *shall be effected and maintained by the Contractor as insuring Party,*

(b)   *shall be in the joint names of the Parties,*

(c)   *shall be extended to cover liability for all loss and damage to the Employer's property (except things insured under Sub-Clause 18.2) arising out of the Contractor's performance of the Contract, and*

(d)   *may however exclude liability to the extent that it arises from:*

18–021   (i)   *the Employer's right to have the Permanent Works executed on, over, under, in or through any land, and to occupy this land for the Permanent Works,*

(ii)   *damage which is an unavoidable result of the Contractor's obligations to execute the Works and remedy any defects, and*

(iii)   *a cause listed in Sub-Clause 17.3 [Employer's Risks], except to the extent that cover is available at commercially reasonable terms.*

OVERVIEW OF KEY FEATURES

18–022   • Insurance against injury to persons and damage to property shall be in joint names but effected by the Contractor.

• The insurance amount shall be limited as set out in the Appendix to Tender.

• The insurance amount shall include cover for loss and damage to the Employer's property.

COMMENTARY

18–023   This is another sub-clause which will only apply if the amount of the insurance required is stated in the Appendix to Tender.

The insurance here includes damage to property not covered by sub-cl.18.2, in addition to third-party liability. As with sub-cl.17.1, the Employer will indemnify the Contractor in respect of damage which is the "unavoidable result" of the Contractor's obligations.

MDB HARMONISED EDITION

As usual, the words "Contract Data" replace "Appendix to Tender".   **18–024**

18.4 INSURANCE FOR CONTRACTOR'S PERSONNEL

The Contractor shall effect and maintain insurance against liability for claims,   **18–025**
damage, losses and expenses (including legal fees and expenses) arising from
injury, sickness, disease or death of any person employed by the Contractor or
any other of the Contractor's Personnel.

The Employer and the Engineer shall also be indemnified under the policy of
insurance, except that this insurance may exclude losses and claims to the extent
that they arise from any act or neglect of the Employer or of the Employer's
Personnel.

The insurance shall be maintained in full force and effect during the whole
time that these personnel are assisting in the execution of the Works. For a
Subcontractor's employees, the insurance may be effected by the Subcontractor,
but the Contractor shall be responsible for compliance with this Clause.

OVERVIEW OF KEY FEATURES

- The Contractor must take out insurance for personal injury, sickness and   **18–026**
  death for its Personnel during the entire time they are working on the
  project.
- The Contractor must indemnify the Employer and Engineer in respect of
  any such claims that may arise, unless they arise as a consequence of any
  act or neglect on the part of said Employer or Engineer.

COMMENTARY

The insurance here is a typical legal requirement in many countries. It must   **18–027**
remain in place throughout the currency of the Works. Like the other sub-
clauses, here it is the Contractor who has to take out the insurance. The cover
must include the Contractor, the Contractor's Personnel and provide an
indemnity for the Employer and Engineer. In accordance with the requirements of clause 5, the Contractor will be responsible for any Sub-Contractor
and its employees, even if it is the Sub-Contractor who takes out the relevant

CLAUSE 18 – INSURANCE

insurance under this sub-clause. However, the cover need not include claims arising out of any act or neglect of the Employer and its personnel.

MDB HARMONISED EDITION

18–028   In order that there is no doubt that the intention of the sub-clause is to provide the Employer and Engineer with an indemnity for personal-injury type claims (provided there was no act or neglect on their part) the second paragraph has been replaced with the following:

> *The insurance shall cover the Employer and the Engineer against liability for claims, damages, losses and expenses (including legal fees and expenses) arising from injury, sickness, disease or death of any person employed by the Contractor or any other of the Contractor's Personnel, except that this insurance may exclude losses and claims to the extent that they arise from any act or neglect of the Employer or of the Employer's Personnel.*

CLAUSE 19 – FORCE MAJEURE

19.1 DEFINITION OF FORCE MAJEURE

In this Clause, "Force Majeure" means an exceptional event or circumstance: **19–001**

(a)   which is beyond a Party's control,
(b)   which such Party could not reasonably have provided against before entering into the Contract,
(c)   which, having arisen, such Party could not reasonably have avoided or overcome, and
(d)   which is not substantially attributable to the other Party.

Force Majeure may include, but is not limited to, exceptional events or   **19–002**
circumstances of the kind listed below, so long as conditions (a) to (d) above
are satisfied:

(i)   war, hostilities (whether war be declared or not), invasion, act of foreign enemies,
(ii)   rebellion, terrorism, revolution, insurrection, military or usurped power, or civil war,
(iii) riot, commotion, disorder, strike or lockout by persons other than the   **19–003**
Contractor's Personnel and other employees of the Contractor and Sub-Contractors,
(iv)   munitions of war, explosive materials, ionising radiation or contamination by radio-activity, except as may be attributable to the Contractor's use of such munitions, explosives, radiation or radio-activity, and
(v)   natural catastrophes such as earthquake, hurricane, typhoon or volcanic activity.

OVERVIEW OF KEY FEATURES

- The term "force majeure" is widely defined under the new FIDIC Form   **19–004**
  meaning an exceptional event which satisfies the criteria set out in
  sub-paras (a)–(d).
- The term Force Majeure includes the list of specific events as set out at
  items (i)–(v).
- This list is non-exhaustive and the events specified are examples only.

COMMENTARY

19–005 There is a difference in the way that force majeure is treated in common and civil law jurisdictions. Whilst most civil codes make provisions for force majeure events, at common lawforce majeure is not a term of art and its meaning is far from clear. No force majeure provision will be implied in the absence of specific contractual provisions and the extent to which the parties deal with unforeseen events will be defined in the contract between them. Thus without cl.19 there would not necessarily be relief for force majeure events.

There are many definitions of force majeure. For example, in the case of *Atlantic Paper Stock Ltd v St Anne-Nackawic Pulp and Paper Co*,[1] Dickson J. in the Supreme Court of Canada said that:

> *An act of God or force majeure clause, . . . generally operates to discharge a contracting party when a supervening, sometimes supernatural, event, beyond the control of either party, makes performance impossible. The common thread is that of the unexpected, something beyond reasonable human foresight and skill.*

19–006 A force majeure clause serves to exempt a party from performance on the occurrence of a force majeure event. That reflects the broad definition of force majeure to be found in the FIDIC form.

As noted above, the contractual definition of force majeure provided in sub-cl.19.1 has a significant overlap with the Employer's risks specified in sub-cl.17.3, although elements which can be attributed to the fault of another party, for example, design, have been omitted. Force majeure does not arise if the event is largely attributable to the other party to the contract. There is also a risk of potential overlap and/or contradiction between sub-cl.19.1 and the definition of force majeure, which one can find in the civil codes of most, if not all, civil law jurisdictions. For example, the definition of force majeure under the Quebec civil code is much narrower in scope. Article 1470 simply provides that:

> *A superior force [in the French version, force majeure] is an unforeseeable and irresistible event, including external causes with the same characteristics.*

19–007 This has lead one commentator to express caution that,

> *incorporating a clause such as Clause 19 into a contract not only duplicates what is usually provided for in the civil code of a civil law jurisdiction, but also*

---

[1] [1976] 1 SCR 580.

enlarges the scope of the meaning and application of force majeure. This could result in the Parties getting into a muddle and a contradictory situation.[2]

In any event, the Particular Conditions note that the Employer should verify, before inviting tenders, that the wording of Clause 19 is compatible with the law governing the Contract.

There was no specific force majeure clause in the Old Red Book FIDIC 4th edn. However, the Contractor was afforded some protection by Clause 65 which dealt with special risks including the outbreak of war and Clause 66 which dealt with payment when the Contractor was released from performance of its contractual obligations. The scheme of the FIDIC form is that the party affected, which is usually the Contractor but could here also be the Employer, is entitled to an extension of time and (with exceptions) additional cost where a "force majeure" occurs. To constitute an act of force majeure, the following needs to have occurred:

(i)   there must be an exceptional event;
(ii)   which is beyond the control of the party affected;
(iii)   about which the party affected could neither have foreseen or provided against before entering into the contract nor avoided once it had arisen;
(iv)   which was not the fault of the other party; and
(v)   about which, proper notice in accordance with sub-paragraph 19.2 has been provided.

Although examples of force majeure events are provided at items (i)–(iv) of sub-cl.19.1, they are specifically stated to be examples only. The key is that the above conditions are satisfied.   **19–008**

The definitions at (i)–(ii) which relate to war, rebellion and terrorism, are in keeping with cl.19 as a whole, broadly defined. For example, there is debate as to whether or not the 2003 invasion of Iraq can be defined as a war or not. However, that debate would be of no concern to the parties of the FIDIC form here. The definition of hostilities at sub-para.(i) includes the words in brackets "whether war be declared or not". Whilst items (i), (iii) and (iv) are the same as the equivalent items of Employer Risk to be found at sub-cl.17.3, namely items (a), (c) and (d), there is one difference between sub-cl.17.3(b) and 19.1(ii). The words "within the Country" do not appear in sub-cl.19.1(ii). Thus, this sub-clause recognises that certain acts of terrorism or civil war outside of the country where the project is proceeding may still have a considerable effect on progress.

It must be an exceptional event to qualify as an act of force majeure. It must also be an event against which a party could not provide. These are

---

[2] Professor Nael Bunni – FIDIC's New Suite of Contracts – Cll 17 to 19 – available on the FIDIC website.

requirements which will particularly affect consideration of item (v) which deals with natural catastrophes such as earthquakes or typhoons. The Tsunami which devastated parts of South East Asia at the end of 2004 clearly meets both criteria. However, a hurricane may well not fall within that definition if it is a typical (i.e. not exceptional) occurrence in the part of the world where the work is being done. It might also be argued that it was an event which the experienced Contractor ought to have foreseen and taken precautions against. Ultimately of course, what constitutes a force majeure event will depend on the circumstances and thus the notice which the party affected by the event is required to give by virtue of sub-clause 19.2 is of particular importance.

MDB HARMONISED EDITION

19–009 The words "sabotage by persons other than the Contractor's Personnel" have been added to sub-para.(d)(ii) and the words "and other employees of the Contractor and Subcontractors" deleted from sub-para.(d)(iii). The first change mirrors the addition made to sub-cl.17.3(b). The second change has come about because the words are unnecessary as a consequence of the definition of Contractor's Personnel to be found at sub-cl.1.1.2.7.

19.2 NOTICE OF FORCE MAJEURE

19–010 *If a Party is or will be prevented from performing any of its obligations under the Contract by Force Majeure, then it shall give notice to the other Party of the event or circumstances constituting the Force Majeure and shall specify the obligations, the performance of which is or will be prevented. The notice shall be given within 14 days after the Party became aware, or should have become aware, of the relevant event or circumstance constituting Force Majeure.*

The Party shall, having given notice, be excused performance of such obligations for so long as such Force Majeure prevents it from performing them.

Notwithstanding any other provision of this Clause, Force Majeure shall not apply to obligations of either Party to make payments to the other Party under the Contract.

OVERVIEW OF KEY FEATURES

19–011 • If a force majeure event prevents a party from performing any of its obligations, this party must give notice within 14 days of becoming aware, or of the date when it should have become aware, of this Force Majeure event.

- With the exception of payment obligations, the delivery of a Force Majeure notice will excuse the affected party from performing its obligations for the duration of the Force Majeure event.

COMMENTARY

For cl.19 to apply, the force majeure event must prevent a Party from performing any of its obligations under the Contract. The now classic example of this is the refusal of the English and American courts to grant relief as a consequence of the Suez crisis during the 1950s. Those who had entered into contracts to ship goods were not prevented from carrying their contractual obligations as they could go via the Cape of Good Hope even though the closure of the Suez Canal made the performance of that contract far more onerous. As noted above, the affected party must give a notice to the Engineer, which must be in writing as per sub-cl.1.3, of the force majeure event. That notice needs to set out both details of the event and also of what the effect of the event is, in other words, which obligations can no longer be performed. That notice must be given within 14 days of the affected party becoming aware of the event, or within 14 days after the party affected should have become aware of the said event.

The drafting of the notice is important as, for the duration of the force majeure event, the affected party will be excused from performing the contractual obligations specified in the notice which the force majeure event has rendered impossible. For example, thought will need to be given to obligations which depend on the prior performance of the immediate obligation which has been affected by the force majeure event. There is no provision for the giving of a further notice. The starting point is therefore to consider that which must be performed, and then to look at the extent to which that performance might be excused by the force majeure clause.

Finally, it should be noted that a party cannot use a force majeure event as an excuse not to make any payment which may have fallen due. In other words neither party can use financial hardship, even if it caused by an exceptional event, to excuse non-payment.

MDB HARMONISED EDITION

The opening part of the first sentence has been changed so that instead of a party being prevented from performing "any of its obligations" as a consequence of the force majeure event, the party must be prevented from performing "its *substantial* obligations" by such an event. Under the FIDIC form, it did not matter what part of a party's obligations were affected by the force majeure event. This has been tightened up here by the restriction introduced by the MDB edition. Now the impact must have a substantial effect on that party's obligations.

19.3 DUTY TO MINIMISE DELAY

19–014 *Each Party shall at all times use all reasonable endeavours to minimise any delay in the performance of the Contract as a result of Force Majeure.*

A Party shall give notice to the other Party when it ceases to be affected by the Force Majeure.

OVERVIEW OF KEY FEATURES

19–015
- Each party must minimise the delay caused by the Force Majeure event.
- Notice must be given when the Force Majeure event is no longer of any effect.

COMMENTARY

19–016 Although this sub-clause does not specifically say so, it clearly envisages that the parties will work together to minimise any delay caused by the force majeure event. In other words, the sub-clause sets out the requirement, already existing under the law of most jurisdictions that a party should, where possible, take steps to mitigate the potential loss. Working separately would not seem to equate with using reasonable endeavours to try and solve the problems caused by the force majeure event. Under sub-cl.1.1.2.1, "Party" is defined as Employer and Contractor. In addition, practically, using reasonable endeavours must also include working with the Engineer.

The party affected by the force majeure event must remember to give notice notifying that the effect of the force majeure event has ceased. As it might not be in the interests of the affected party to serve such a notice, the other party would be well advised to keep matters under review and if thought appropriate, signal when it considers the effect of the force majeure event to be over.

MDB HARMONISED EDITION

There is no change. **19–017**

19.4 CONSEQUENCES OF FORCE MAJEURE

If the Contractor is prevented from performing any of his obligations under the **19–018**
Contract by Force Majeure of which notice has been given under Sub-Clause
19.2 [Notice of Force Majeure], and suffers delay and/or incurs Cost by reason
of such Force Majeure, the Contractor shall be entitled subject to Sub-Clause
20.1 [Contractor's Claims] to:

(a)   an extension of time for any such delay, if completion is or will be
delayed, under Sub-Clause 8.4 [Extension of Time for Completion], and
(b)   if the event or circumstance is of the kind described in sub-paragraphs (i)
to (iv) of Sub-Clause 19.1 [Definition of Force Majeure] and, in the case
of sub-paragraphs (ii) to (iv), occurs in the Country, payment of any
such Cost.

After receiving this notice, the Engineer shall proceed in accordance with
Sub-Clause 3.5 [Determinations] to agree or determine these matters.

OVERVIEW OF KEY FEATURES

If the Contractor is the party affected by the force majeure event, the **19–019**
Contractor may claim an extension of time, and, in certain circumstances,
additional monies.

COMMENTARY

Under sub-cl.19.4, the Contractor may be allocated both time and money, **19–020**
depending on the circumstances. In order to be able to claim money, in addition to an extension of time, the force majeure event must *not* be an Act of
God (i.e. an event outside human control). In other words, this exclusion
covers the natural catastrophes referred to in sub-cl.19.1(v). Instead, to be
able to claim money, the event must fall within sub-cll 19.1(i)–(iv) and be
categorised as either a war (even if it is not happening in the country where
the site is located) or a lesser type of event such as a rebellion or act of

terrorism, industrial action, or land-contamination. For this lesser type of event (i.e. sub-paras 19.1(ii)–(iv)) to warrant any award for additional monies, however, the event must occur in the country where the site is located. As discussed above at sub-cl.19.1, the definition of the category of war events includes "hostilities (whether war be declared or not), invasion, [and] act of foreign enemies". Thus, to take another example, the conflict that opposed Israel and the Hezbollah in Southern Lebanon in the summer of 2006 would be classified as a war regardless of the fact Israel and Lebanon never officially declared war on one another and the official Lebanese army never engaged in combat against the Tsahal.

MDB HARMONISED EDITION

19–021   The same amendment to the opening paragraph has been made here as to sub-para.19.2. Accordingly, instead of a party being prevented from performing "any of its obligations" as a consequence of the force majeure event, the party must be prevented from performing "its substantial obligations" by such an event.

In addition, the following has been added to sub-para.(b):

> ... *including the costs of rectifying or replacing the Works and/or Goods damaged or destructed by Force Majeure, to the extent they are not indemnified through the insurance policy referred to in Sub-Clause 18.2 [Insurance for Works and Contractor's Equipment].*

19.5 FORCE MAJEURE AFFECTING SUBCONTRACTOR

19–022   *If any Subcontractor is entitled under any contract or agreement relating to the Works to relief from force majeure on terms additional to or broader than those specified in this Clause, such additional or broader force majeure events or circumstances shall not excuse the Contractor's non-performance or entitle him to relief under this Clause.*

OVERVIEW OF KEY FEATURES

19–023   The Contractor cannot rely on a broader force majeure clause in a sub-contract to claim force majeure relief against the Employer.

COMMENTARY

Sub-clause 19.5 is an extension of sub-cl.4.4, which provides that the **19–024**
Contractor shall be responsible for the acts or defaults of any sub-contractor.
It also provides an example of why the Contractor should ensure that its
contracts with its sub-contractors mirror the obligations of the contract with
the Employer. If the terms of the sub-contract have a wider definition of
force majeure then the Contractor is not able to rely on that in order to
obtain relief under the FIDIC Form here.

MDB HARMONISED EDITION

There is no change. **19–025**

19.6 OPTIONAL TERMINATION, PAYMENT AND RELEASE

If the execution of substantially all the Works in progress is prevented for a **19–026**
continuous period of 84 days by reason of Force Majeure of which notice has
been given under Sub-Clause 19.2 [Notice of Force Majeure], or for multiple
periods which total more than 140 days due to the same notified Force Majeure,
then either Party may give to the other Party a notice of termination of the
Contract. In this event, the termination shall take effect 7 days after the notice
is given, and the Contractor shall proceed in accordance with Sub-Clause 16.3
[Cessation of Work and Removal of Contractor's Equipment].

Upon such termination, the Engineer shall determine the value of the work
done and issue a Payment Certificate which shall include:

(a) the amounts payable for any work carried out for which a price is stated
in the Contract;
(b) the Cost of Plant and Materials ordered for the Works which have been **19–027**
delivered to the Contractor, or of which the Contractor is liable to accept
delivery: this Plant and Materials shall become the property of (and be
at the risk of) the Employer when paid for by the Employer, and the
Contractor shall place the same at the Employer's disposal;
(c) any other Cost or liability which in the circumstances was reasonably
incurred by the Contractor in the expectation of completing the Works;
(d) the Cost of removal of Temporary Works and Contractor's Equipment
from the Site and the return of these items to the Contractor's works in
his country (or to any other destination at no greater cost); and

19–028   *(e)   the Cost of repatriation of the Contractor's staff and labour employed wholly in connection with the Works at the date of termination.*

OVERVIEW OF KEY FEATURES

19–029   • Any party may serve a termination notice if the Force Majeure event prevents the execution of substantially all the works for: (i) a continuous period of 84 days; or (ii) multiple periods of more than 140 days.
   • If such notice is served, the Contractor will be entitled to specific payment.

COMMENTARY

19–030   If the force majeure event prevents the affected party from progressing substantially with its contractual obligations for a period of more than 84 days in one stretch, or for a combined period of 140 days if the delay caused is intermittent, then either party may give seven day notice of termination. The FIDIC Guide confirms that the pre-condition of the first sentence relates to an actual impediment to progress and not to the extent to which the Works are substantially complete.

Where termination occurs in these circumstances, the second half of this sub-clause sets out the way in which the Engineer is to determine how much, if anything, the Contractor is entitled to be paid. The Engineer must make a similar determination if the Employer exercises its right to terminate at will in accordance with sub-cl.15.5. Essentially, the Contractor is entitled to payment for any work carried out "for which a price is stated in the Contract." The Contractor is also entitled to payment in respect of any goods ordered, but not yet delivered. Once the Employer has paid for these, they will belong to the Employer. The sub-clause does not make provision for where the Contractor might be able to negotiate a reduced fee. On the one hand, the Employer would have to pay less, on the other, the Employer would not receive the benefit of the Goods in question. This might well be a matter for negotiation.

The three other sub-paras (c)–(e) deal with the consequential costs of the force majeure event. It may be that the Contractor has already incurred certain advance costs in the contemplation and expectation of completing the Works. Equally, if the Contractor has to leave the site, there will be removal costs both in respect of materials and personnel.

19–031   In accordance with sub-cl.16.3, in the event of termination under this sub-clause, the Contractor shall cease all further work, (except for any work instructed by the Engineer for health and safety reasons), hand over the

Contractor's Documents, Plant, Materials and other work, (provided Contractor has received the appropriate payment), and remove all other Goods from the Site, and finally leave the Site. This must all be done promptly.

MDB HARMONISED EDITION

The words "and necessarily" have been added to sub-para.(c). Thus, the costs incurred by the Contractor in the expectation of completing the Works must not only have been reasonably incurred, they must also have been necessarily incurred. The presumed intention of this addition is to limit the Contractor's right to payment. However, it is not thought that this will lead to any significant change. An expense which has been incurred necessarily will obviously be one which is reasonable. Put another way, what reasonable cost will also be one which was unnecessary? **19–032**

19.7 RELEASE FROM PERFORMANCE UNDER THE LAW

Notwithstanding any other provision of this Clause, if any event or circumstance outside the control of the Parties (including, but not limited to, Force Majeure) arises which makes it impossible or unlawful for either or both Parties to fulfil its or their contractual obligations or which, under the law governing the Contract, entitles the Parties to be released from further performance of the Contract, then upon notice by either Party to the other Party of such event or circumstance: **19–033**

(a) the Parties shall be discharged from further performance, without prejudice to the rights of either Party in respect of any previous breach of the Contract, and

(b) the sum payable by the Employer to the Contractor shall be the same as would have been payable under Sub-Clause 19.6 [Optional Termination, Payment and Release] if the Contract had been terminated under Sub-Clause 19.6.

OVERVIEW OF KEY FEATURES

The parties will also be released from performance (and the Contractor entitled to specific payment) if (i) any irresistible event (not limited to force **19–034**

371

majeure) makes it impossible or unlawful for the parties to fulfil their contractual obligations, or (ii) the governing law so provides.

COMMENTARY

19–035   Sub-clause 19.7 is particularly interesting in that it acts as a fall-back provision for extreme events (i.e., events rendering contractual performance illegal or impossible) which do not fit within the strict definition of force majeure laid out under sub-cl.19.1. It also grants the party seeking exoneration the right to rely on any alternative relief-mechanism contained in the law governing the contract. Because of this last provision, local legal advice will be indispensable to ascertain the precise boundaries of sub-cl.19.7. In the event of this sub-clause applying, the Contractor will be entitled to payment in the same terms as sub-cl.19.6 above.

If English law applies, following the landmark case of *Davis Contractors v Fareham UDC*, the affected party will be able to rely on the common law concept of frustration, which,

> *occurs whenever the law recognises that without the default of either party a contractual obligation has become incapable of being performed because the circumstances in which performance is called for would render it a thing radically different from that which was undertaken by the contract.*[3]

Here, the contract was to build 78 houses for a fixed price in eight months. Because of labour shortages and bad weather, it took the contractor 22 months to build the houses. It was held by the House of Lords that the contract had not been frustrated. To claim frustration, therefore, it will not be enough for a contractor to establish that new circumstances have rendered its contractual performance more onerous or even dangerously uneconomic.

For frustration, what is required is a radical turn of events completing changing the nature of the contractual obligations. It is a difficult test to fulfil, but not as difficult as that of sub-cl.19.4 or the first limb of sub-cl.19.7 which both refer to the concept of impossibility (or illegality). To take the example put forward by A. Puelinckx of a wine connoisseur signing a contract for the construction under his house of a very sophisticated wine cellar,[4] if the house is burned down before execution of the contract, leaving the basement part in perfect condition, this will certainly be considered frustration under English law. However, no claim could be put forward under a

---

[3] *Davis Contractors Ltd v Fareham UDC 2* All E.R. 145 at 160, HL, per Lord Radcliffe.
[4] Frustration, Hardship, Force Majeure, Imprévision, Wegfall der Geschä-ftsgrundlage, Unmöglichkeit, Changed Circumstances: A Comparative study in English, French, German and Japanese Law Journal of International Arbitration, Vol.3 No.2 (1986), pp.47–66.

strict interpretation of sub-cll 19.4 or 19.7 as the house could in theory be rebuilt and the contractual obligation to build the cellar performed. French law would apply the same reasoning as sub-cll 19.4 or 19.7 and because performance is still possible, would hold the above-described events as a mere *imprévision*, which will not afford any financial relief to the affected party.[5]

What is common to both the notion of frustration and that of force **19–036** majeure as interpreted under English law though, is that no relief will be granted in case of economic unbalance. A recent illustration concerning the interpretation of a force majeure clause under English law can be found in the case of *Thames Valley Power Limited v Total Gas & Power Limited.*[6] Here there was a 15-year exclusive gas supply contract between Thames Valley Power Limited (buyer) and Total Gas & Power Limited (supplier) for the operation of a combined heat and power-plant at Heathrow Airport. Clause 15 of the supply contract provided in part as follows:

> *if either party is by reason of force majeure rendered unable wholly or in part to carry out any of its obligations under this agreement then upon notice in writing [. . .] the party affected shall be released from its obligations and suspended from the exercise of its rights herender to the extent that they are affected by the circumstances of force majeure and for the period that those circumstances exist [. . .].*

The supplier sought to rely on cl.15 to stop supplying gas at the contract price **19–037** as the market price for gas had increased significantly and rendered it "uneconomic" for the supplier to supply gas. Christopher Clarke J. however, found that:

> *The force majeure event has to have caused Total to be unable to carry out its obligations under the [agreement]. [. . .] Total is unable to carry out that obligation if some event has occurred as a result of which it cannot do that. The fact that it is much more expensive, even greatly more expensive for it to do so, does not mean that it cannot do so.*

Clause 19 would certainly be interpreted in much the same way by English courts. In large projects where the performance of the parties' contractual obligations is spread over several years, the parties might thus consider whether or not to add a hardship clause to the contract which will stipulate when and how the parties will rearrange the contractual terms in the event the contract loses its economic balance.

---

[5] Cour de Cassation, 6 mars 1876 (Canal de Craponne). This decision is to be contrasted with the position taken by the French Administration Tribunals regarding contracts with a French Public Body, where the affected party may be able to claim an additional sum of money to compensate for the effect of the *"imprévision"* [Conseil d'Etat, Gaz de Bordeaux, 30 mars 1916].

[6] (2006) 1 Lloyd's Rep 441.

The FIDIC form has, however, made some potential provision for this by way of sub-cll 13.7 and 13.8. If included as part of the Contract, these sub-clauses entitle a Contractor to time and money as a consequence of any impact of changes in legislation and make provision for increased cost due to inflation.

MDB HARMONISED EDITION

19–038   There is no change.

CLAUSE 20 – CLAIMS, DISPUTES AND ARBITRATION

20.1 CONTRACTOR'S CLAIMS

If the Contractor considers himself to be entitled to any extension of the Time for Completion and/or any additional payment, under any Clause of these Conditions or otherwise in connection with the Contract, the Contractor shall give notice to the Engineer, describing the event or circumstance giving rise to the claim. The notice shall be given as soon as practicable, and not later than 28 days after the Contractor became aware, or should have become aware, of the event or circumstance. **20–001**

If the Contractor fails to give notice of a claim within such period of 28 days, the Time for Completion shall not be extended, the Contractor shall not be entitled to additional payment, and the Employer shall be discharged from all liability in connection with the claim. Otherwise, the following provisions of this Sub-Clause shall apply.

The Contractor shall also submit any other notices which are required by the Contract, and supporting particulars of the claim, all as relevant to such event or circumstance.

The Contractor shall keep such contemporary records as may be necessary to substantiate any claim, either on the Site or at another location acceptable to the Engineer. Without admitting the Employer's liability, the Engineer may, after receiving any notice under this Sub-Clause, monitor the record-keeping and/or instruct the Contractor to keep further contemporary records. The Contractor shall permit the Engineer to inspect all these records, and shall (if instructed) submit copies to the Engineer. **20–002**

Within 42 days after the Contractor became aware (or should have become aware) or the event or circumstance giving rise to the claim, or within such other period as may be proposed by the Contractor and approved by the Engineer, the Contractor shall send to the Engineer a fully detailed claim which includes full supporting particulars of the basis of the claim and of the extension of time and/or additional payment claimed. If the event or circumstance giving rise to the claim has a continuing effect:

(a) this fully detailed claim shall be considered as interim; **20–003**
(b) the Contractor shall send further interim claims at monthly intervals, giving the accumulated delay and/or amount claimed, and such further particulars as the Engineer may reasonably require; and
(c) the Contractor shall send a final claim within 28 days after the end of the effects resulting from the event or circumstance, or within such other

period as may be proposed by the Contractor and approved by the Engineer.

Within 42 days after receiving a claim or any further particulars supporting a previous claim, or within such other period as may be proposed by the Engineer and approved by the Contractor, the Engineer shall respond with approval, or with disapproval and detailed comments. He may also request any necessary further particulars, but shall nevertheless give his response on the principles of the claim within such time.

20–004 *Each Payment Certificate shall include such amounts for any claim as have been reasonably substantiated as due under the relevant provision of the Contract. Unless and until the particulars supplied are sufficient to substantiate the whole of the claim, the Contractor shall only be entitled to payment for such part of the claim as he has been able to substantiate.*

The Engineer shall proceed in accordance with Sub-Clause 3.5 [Determination] to agree or determine (i) the extension (if any) of the Time for Completion (before or after its expiry) in accordance with Sub-Clause 8.4 [Extension of Time for Completion], and/or (ii) the additional payment (if any) to which the Contractor is entitled under the Contract.

The requirements of this Sub-Clause are in addition to those of any other Sub-Clause which may apply to a claim. If the Contractor fails to comply with this or another Sub-Clause in relation to any claim, any extension of time and/or additional payment shall take account of the extent (if any) to which the failure has prevented or prejudiced proper investigation of the claim, unless the claim is excluded under the second paragraph of this Sub-Clause.

OVERVIEW OF KEY FEATURES

20–005
- The Contractor must give notice to the Engineer of time or money claims, as soon as practicable and not later than 28 days after the date on which the Contractor became aware, or should have become aware, of the relevant event or circumstance.
- Any claim to time or money will be lost if there is no notice within the specified time limit.
- Supporting particulars should be served by the Contractor and the Contractor should also maintain such contemporary records as may be needed to substantiate claims.
- The Contractor should submit a fully particularised *claim* after 42 days.
- The Engineer is to respond, in principle at least, within 42 days.
- The claim shall be an interim claim. Further interim updated claims are to be submitted monthly. A final claim is to be submitted, unless agreed otherwise, within 28 days of the end of the claim event.

- Payment Certificates should reflect any sums acknowledged in respect of substantiated claims.

COMMENTARY

The use of the term "dispute board" or Dispute Adjudication Board **20–006** ("DAB") whether or not it includes "adjudication" is relatively recent. It describes a dispute resolution procedure which is normally established at the commencement of a project and remains in place throughout the project's duration. The intention behind this is so that the members of the DAB can become acquainted with the contract and project and, if appropriate, provide informal assistance, recommendations about how disputes should be resolved, and ultimately binding (if only on a temporary basis) decisions.

The term DAB should not be confused with Dispute Review Boards, which originated in the mid 1970s in the USA. The difference is that the function of the Dispute Review Board is to make a recommendation which the parties are able to accept or reject. The DAB can issue written decisions of a (temporarily) binding nature, which must be implemented immediately during the course of the project.

It is only recently that FIDIC introduced the DAB concept. Before that the Engineer, for example under cl.67 of the Old Red Book FIDIC 4th edn, was given the responsibility of resolving disputes, prior to formal arbitration. FIDIC first introduced a DAB in its Orange Book in 1995. The DAB was then introduced as an option in the Red Book in 1996. This lead to the dispute resolution system under the FIDIC form here which retains the Engineer[1] in accordance with the provisions of sub-clause 3.5 but also made the DAB mandatory.

The DAB procedures under the FIDIC form consist of the following: **20–007**

(i)   Clause 20 – the Dispute Adjudication Board.
(ii)   Appendix – General Conditions of Dispute Adjudication Agreement.
(iii)   Annex 1 – Procedural Rules.
(iv)   The Dispute Adjudication Agreement.

Of all the provision to be found in the FIDIC form, the provisions of cl.20 have attracted by far the most comment. This is clear from the list of further reading to be found at the end of this work. That in itself is unsurprising, in that if disputes do arise, they can quite quickly become very costly. Of course, the better an understanding both parties have of how the entire contract works, then the less likelihood there is for disputes to arise. However, when

---

[1] Although as discussed below, the Particular Conditions include an option for the Engineer retaining his traditional role.

disputes do arise, it is of crucial importance that both parties follow the provisions of cl.20 with some care. A failure to do so could quite possibly prevent an aggrieved party from bringing a claim.

Sub-clause 20.1 requires that the Contractor, if it considers it has a claim for an extension of time and/or any additional payment, must give notice to the Engineer "as soon as practicable, and not later than 28 days after the event or circumstance giving rise to the claim". This makes it clear that the Contractor must submit its claims during the course of the project. The initial notice at first instance does not need to indicate, (for the very good reason that usually it cannot) the total extension or payment sought. The scheme of the FIDIC form is thus that where possible disputes should be resolved during the course of the works.

20–008 Compliance with the notice provisions is intended to be a condition precedent to recovery of time and/or money and, without notices, the Employer has no liability to the Contractor.[2] Certainly parties should treat the sub-clause in this way. It is possible that an argument could be brought under UK law that the Contractor will not lose its right to bring a claim if the claim is not brought within the stipulated timescale. Generally, in the UK the courts will take the view that timescales in construction contracts are directory rather than mandatory,[3] unless the contract clause in question clearly states that the party with a claim will lose the right to bring that claim if it fails to comply with the required timescale.[4] Although sub-cl.20.1 does not include wording in that precise form, the meaning of the second paragraph, it is submitted, is perfectly clear. Thus, it is thought that such an argument will not succeed and the prudent Contractor should take care to comply with the timescales set out in this sub-clause and submit the required notice during the course of the works and within the prescribed period of 28 days.

The Contractor also needs to remember that where the effects of a particular event are on-going then, rather unusually, the Contractor is specifically required to continue submitting notices at monthly intervals. Thereafter, the Contractor has a further period in which to submit a fully particularised claim. There is no condition precedent attached to this part of sub-cl.20.1. Although the Engineer can request additional particulars, the Contractor should not rely on any such request to flag up potential areas of weakness in its claim. The Contractor's claim will stand and fall by the quality of the evidence and the time within which it is produced.

The Contractor is required to keep contemporary records to substantiate its claim.[5] In the case of *Attorney General for the Falkland Islands v Gordon*

---

[2] See comments by Christopher Seppala – Contractor's Claims Under the FIDIC Contracts for Major Works, paper given at the International Construction Contracts and Dispute Resolution Conference in Cairo April 2005. Paper available on FIDIC website.
[3] *Temloc v Errill Properties* (1987) 39 BLR 30, CA per Croom Johnson L.J.
[4] In England *Bremer Handelsgesellschaft mbH v Vanden Avenue-Izegem PVBA* (1978) 2 Lloyd's Rep 109, and in Scotland *City Inn Limited v Shepherd Construction Limited* (2002) S.L.T. 781.
[5] See also discussion above at sub-cll 6.10 and 8.4.

Forbes Construction (Falklands) Ltd,[6] Acting Judge Sanders ruled that you could not attempt to get around this requirement by producing simple witness statements after the event. Such statements would not be the equal of either statements taken at the relevant time. He defined "contemporary records" thus:

> *original or primary documents, or copies thereof, produced or prepared at or about the time giving rise to a claim, whether by or for the contractor or the employer.*

20–009

The sub-clause ends by noting that the success of the Contractor's claim *"shall take account of the extent (if any) to which the failure"* to provide for example contemporary evidence, *"has prevented or prejudiced proper investigation of the claim"*. Thus, in the case of *London Borough of Merton v Stanley Hugh Leach Ltd*,[7] whilst the giving of a notice was not a condition precedent to the architect considering whether an extension of time should be granted under the relevant contractual clause, the failure to give such a notice was a breach of contract. Thus, if such a breach had caused a delay which would otherwise have been avoidable, then the defaulting party would not be entitled to recover for that avoidable delay. This what the final paragraph of sub-clause 20.1 envisages would happen under the FIDIC form here.

There is one final cut-off point which this sub-clause fails to mention. Under sub-cl.14.10(c), the Contractor's Statement at Completion, which must be provided within 84 days of receiving the Taking-Over Certificate, must include an estimate of any future sum the Contractor considers due. This will include an estimate in respect of claims. Likewise sub-cl.14.11(b) requires that the Contractor's application for a Final Payment Certificate, which must be submitted within 56 days of receiving the Performance Certificate, includes a similar estimate. The importance of this is that sub-cl.14.14 says that the Employer shall not be liable to the Contractor for anything under the Contract unless the Contractor shall have made provision for this in the Statement at Completion and Final Statement.

Under sub-cl.20.1, the Engineer has a number of responsibilities. **20–010** Although the burden of proof lies with the Contractor to prove its claim, the Engineer can require that the Contractor submit additional particulars. The Engineer can also inspect the Contractor's records and instruct the Contractor to keep further records. The Engineer must also respond within 42 days both to the Contractor's claim and to a further interim claim. At the very least the Engineer must provide his comments on the principles of the claim. The Engineer's determination must be made in accordance with the principle of the sub-cl.3.5. Thus it must be fair, in accordance with the contract and the take due regard of all relevant circumstances.

---

[6] (2003) 6 BLR 280. See also sub-cl.6.10, above.
[7] (1985) 32 BLR 51.

MDB HARMONISED EDITION

20–011   The MDB Harmonised edition throughout refers to the Dispute Board, the word adjudication having been deleted.

The words "such time" in the paragraph which follows sub paragraph (c) have been replaced with "the above defined time period." A new paragraph has been introduced which follows directly on from that paragraph:

> *Within the above defined period of 42 days, the Engineer shall proceed in accordance with Sub-Clause 3.5 [Determinations] to agree or determine (i) the extension (if any) of the Time for Completion (before or after its expiry) in accordance with Sub-Clause 8.4 [Extension of Time for Completion], and/or (ii) the additional payment (if any) to which the Contractor is entitled under the Contract.*

20–012   The penultimate paragraph has been replaced with the following which deals with what might happen if the Engineer fails to respond to a claim:

> *If the Engineer does not respond within the timeframe defined in this Clause, either Party may consider that the claim is rejected by the Engineer and any of the Parties may refer to the Dispute Board in accordance with Sub-Clause 20.4 [Obtaining Dispute Board's Decision].*

20.2 APPOINTMENT OF THE DISPUTE ADJUDICATION BOARD

20–013   *Disputes shall be adjudicated by a DAB in accordance with Sub-Clause 20.4 [Obtaining Dispute Adjudication Board's Decision]. The Parties shall jointly appoint a DAB by the date stated in the Appendix to Tender.*

The DAB shall comprise, as stated in the Appendix to Tender, either one or three suitably qualified persons ("the members"). If the number is not so stated and the Parties do not agree otherwise, the DAB shall comprise three persons.

If the DAB is to comprise three persons, each Party shall nominate one member for the approval of the other Party. The Parties shall consult both these members and shall agree upon the third member, who shall be appointed to act as chairman.

20–014   *However, if a list of potential members is included in the Contract, the members shall be selected from those on the list, other than anyone who is unable or unwilling to accept appointment to the DAB.*

The agreement between the Parties and either the sole member ("adjudicator") or each of the three members shall incorporate by reference the General Conditions of Dispute Adjudication Agreement contained in the

Appendix to these General Conditions, with such amendments as are agreed between them.

The terms of the remuneration of either the sole member or each of the three members, including the remuneration of any expert whom the DAB consults, shall be mutually agreed upon by the Parties when agreeing the terms of appointment. Each Party shall be responsible for paying one-half of this remuneration.

If at any time the Parties so agree, they may appoint a suitably qualified person or persons to replace (or to be available to replace) any one or more members of the DAB. Unless the Parties agree otherwise, the appointment will come into effect if a member declines to act or is unable to act as a result of death, disability, resignation or termination of appointment.

20–015

If any of these circumstances occurs and no such replacement is available, a replacement shall be appointed in the same manner as the replaced person was required to have been nominated or agreed upon, as described in this Sub-Clause.

The appointment of any member may be terminated by mutual agreement of both Parties, but not by the Employer or the Contractor acting alone. Unless otherwise agreed by both Parties, the appointment of the DAB (including each member) shall expire when the discharge referred to in Sub-Clause 14.12 [Discharge] shall have become effective.

OVERVIEW OF KEY FEATURES

- Sub-clause 20.2 establishes the Dispute Adjudication Board or DAB.   **20–016**
- The DAB shall consist of one or three people who must be suitably qualified.
- The composition of the DAB shall be by nomination and then joint selection.
- DAB members are to be remunerated jointly by the parties with each paying half of any fees.
- DAB members can only be replaced by mutual agreement.

COMMENTARY

Sub-clause 20.2 deals with the establishment of the DAB. The DAB will   **20–017** either be named in the Contract or must be constituted by a date set out in the Appendix to Tender. The DAB procedure is to be considered as the primary method for dispute resolution under the contract and is a development of real significance in the area of Dispute Resolution Procedures, as noted above, replacing the process of decision-making by the Engineer.

Referral to the DAB must occur prior to any reference to arbitration. It is hoped that referral to the DAB will occur at a practical "job" level with the members of the DAB being able to see the disputes referred to it on a practical level rather than on the more abstract level encountered in, say, arbitration. It is also hoped that parties will refer matters to the DAB at an earlier stage so that any dispute can be nipped in the bud before it develops into something more time consuming and costly.

As the DAB is appointed by a date specified in the Appendix to Tender it is highly likely that the DAB will be appointed before work has begun. This also results in consistency throughout a project as all disputes should be referred to the same DAB. The early appointment of the DAB will bring the benefit that the DAB will over time become familiar with and better understand any complexities of the project. For example, under the first Procedural Rule, the DAB is required to visit the site at intervals of not less than 140 days.

For further discussion on the selection of the DAB and the responsibilities of the DAB members see the discussion below on the General Conditions of the Dispute Adjudication Agreement and the Procedural Rules.

20–018   The parties do not have to agree to the full-time appointment of a DAB. These can be costly, particularly in the early part of a project when there is little construction activity going on. The Particular Conditions suggest that if it is intended that the Parties want to defer the appointment of the DAB until a dispute actually arises then they should use the wording to be found in sub-cl.20.2 of the FIDIC Form for Plant and Design-Build. This would be achieved primarily by adding the following sentence to the first paragraph of this sub-clause:

> *The Parties shall jointly appoint a DAB by the date of 28 days after a Party gives notice to the other Party of its intention to refer a dispute to a DAB in accordance with sub-clause 20.4.*

It can be seen that one potential difficulty with this ad-hoc procedure will be the ability to achieve the swift composition of the DAB.

20–019   Depending on the agreement between the parties, the DAB consists of either one (then referred to as an adjudicator) or three members. On smaller projects, in the interests of cost, a one-person DAB might well be thought sufficient. The FIDIC Guide says that a three-person DAB would be appropriate for projects involving an average monthly Payment Certificate exceeding US $2 million. It does not say for how long this spend should last. In the authors' view, parties to projects with a value of more than £10 million would be advised to consider a three-person DAB. The default, where the number has not been agreed, is for three members. An odd numbered DAB means that it is unlikely that a position of stalemate will be reached from which there is no impasse. Where the DAB consists of three members, each party chooses one of the members with the approval of the other party. The third member of the DAB is then chosen by agreement of the parties, after consultation with the two party appointed members. This will, presumably, allay any fear by either party that the DAB will favour one party's interests over the others.

The Contract contemplates that the parties produce a list of potential members for the DAB. Where they have done so, the DAB must then be selected from that list. Where three members have been specified, members will be chosen from the list as set out above. As the DAB is to be appointed by the date set in the Tender documents and not merely when a dispute arises, it is not certain what effect this provision of the clause will have. Presumably, the parties must agree to the list from which DAB members are to be chosen. Circumstances can be contemplated in which the time interval between the agreeing the list from which DAB members are to be chosen and appointing the DAB could be short. Compilation of the list of potential members will, however, be unlikely to be subject to the same scrutiny by the parties and further does not require the agreement of those suggested DAB members.

The FIDIC Guide makes the point that it would be contrary to the (spirit of) the adjudication provisions for a member of the DAB to act as an advocate for one party. Paragraph 9(6) of the procedural rules stresses that the members of the DAB should endeavour to reach an unanimous decision. The FIDIC Guide says that each Party should aim to appoint *"a truly independent expert with the ability and freedom to act impartially, develop a spirit of team work within the DAB, and make fair unanimous decisions."* In reality these may only be some of the qualities a party looks for.

The sub-clause states that the agreement between the each member of the **20–020** DAB, or adjudicator, as the case may be, should incorporate provisions contained in the Appendix to the Contract. This must, however, be subject to agreement by the DAB member or adjudicator. Thus, it will remain to be seen what value the provisions contained in the Appendix will be.

The terms for the remuneration of the DAB members and experts who the DAB uses to help it, must be decided before the appointment and are to be paid in equal proportions by each party. This seems to be in keeping with the general ethos of the DAB that the DAB is decided on and ready to be used prior to any dispute arising. It also reinforces the idea that the DAB is to be considered as a pragmatic rather that litigious dispute resolution procedure in which the cost of the DAB is to be carried by both parties rather than by the unsuccessful party. It is common for members of the DAB to be paid on a monthly retainer.

Although the sub-clause does provide for replacement of a member of the DAB, this seems to be discouraged. Firstly, the agreement of both parties is required. A party acting alone cannot terminate the DAB once it is constituted. Secondly, the clause only anticipates the replacement of the DAB for death, disability, resignation or termination of appointment of the DAB. Presumably, this can occur only rarely and will invariably be in circumstances beyond the control of the parties.

20–021   Where no replacement is available, the same method is used to appoint the replacement as such member of the DAB would have been originally appointed. This carries with it the implication that it will rarely be in a party's interests to lose a DAB member as the replacement is unlikely to be more advantageous to that party's cause than the DAB member being replaced.

Termination of the appointment of a DAB is by mutual agreement of the parties. This is not surprising given that the DAB appears to have been contemplated to serve the good of the project rather than the individual parties.

The decision of the DAB may be regarded as being of interim binding effect since it *must* be complied with until the further steps in cl.20 have been taken. Under Procedural Rule 8(a), the DAB can also decide upon provisional relief such as interim or conservatory measures. In some other international forms of contract, the equivalent of the DAB only has a power to make a non-binding recommendation. The decision of a DAB, in contrast, has greater impact and provides greater commercial certainty and thus has much to recommend it.

20–022   Most importantly, the DAB represents a significant improvement on the Engineer's Decision. As a result of the increasing perception of the Engineer as hardly independent, the Engineer's Decision had almost become a mere formality on the route of the dispute resolution procedure, offering little possibility of a permanent solution to disputes. The DAB offers the real possibility of early dispute *resolution* and, in doing so, would seem to justify the additional costs to the project which DABs will represent.

That said, the Particular Conditions do provide for an alternative wording whereby the Engineer can be substituted for the DAB in respect of the pre-arbitral decision-making. If the Engineer is empowered in this way, the Particular Conditions make it clear that the Engineer must act impartially, notwithstanding that the Engineer generally acts for the Employer. The alternative wording is this:

> *Delete Sub-Clauses 20.2 and 20.3*
>
> *Delete the second paragraph of Sub-Clause 20.4 and substitute:*
>
> *The Engineer shall act as the DAB in accordance with this sub-clause 20.4 acting fairly, impartially and at the cost of the Employer. In the event that the Employer intends to replace the Engineer, the Employer's notice under sub-clause 3.4 shall include detailed proposals for the appointment of a replacement DAB.*

MDB HARMONISED EDITION

20–023   There have been a number of changes. The first three paragraphs now read as follows:

Disputes shall be referred to a DB for decision in accordance with Sub-Clause 20.4 [Obtaining Dispute Board's Decision]. The Parties shall appoint a DB by the date stated in the Contract Data.

The DB shall comprise, as stated in the Contract Data, either one or three suitably qualified persons ("the members"), each of whom shall be fluent in the language for communication defined in the Contract and shall be a professional experienced in the type of constructions involved in the Works and with the interpretation of contractual documents. If the number is not so stated and the Parties do not agree otherwise, the DB shall comprise three persons.

If the Parties have not jointly appointed the DB 21 days before the date started in the Contract Data and the DB is to comprise three persons, each Party shall nominate one member for the approval of the other Party. The first two members shall recommend and the Parties shall agree upon the third member, who shall act as chairman.

The additions to the second paragraph reflect the requirements of para.3 **20–024** (Warranties) of the General Conditions of Dispute Adjudication Agreement.

The words "has been agreed by the Parties" has been added to the fourth paragraph. It might be thought that this was an unnecessary amendment as under the FIDIC form, the list of potential members was required to appear in the Contract, which suggests a degree of agreement. Finally, the two penultimate paragraphs have been deleted and been replaced with the following one paragraph:

If a member declines to act or is unable to act as a result of death, disability, resignation or termination of appointment, a replacement shall be appointed in the same manner as the replaced person was required to have been nominated or agreed upon, as described in this Sub-Clause.

20.3 FAILURE TO AGREE DISPUTE ADJUDICATION BOARD

If any of the following conditions apply, namely: **20–025**

(a) *the Parties fail to agree upon the appointment of the sole member of the DAB by the date stated in the first paragraph of Sub-Clause 20.2,*

(b) *either Party fails to nominate a member (for approval by the other Party) of a DAB of three persons by such date,*

(c) *the Parties fail to agree upon the appointment of the third member (to act as chairman) of the DAB by such date, or*

20–026   *(d)*   *the Parties fail to agree upon the appointment of a replacement person within 42 days after the date on which the sole member or one of the three members declines to act or is unable to act as a result of death, disability, resignation or termination of appointment,*

then the appointing entity or official named in the Appendix to Tender shall, upon the request of either or both of the Parties and after due consultation with both Parties, appoint this member of the DAB. This appointment shall be final and conclusive. Each Party shall be responsible for paying one-half of the remuneration of the appointing entity or official.

OVERVIEW OF KEY FEATURES

20–027   Where the parties fail or are otherwise unable to agree upon the appointment, nomination or replacement of any member of the DAB, then the appointing official so named in the Appendix to Tender shall make the appointment.

COMMENTARY

20–028   Sub-clause 20.3 is concerned with any problems with the establishment of the DAB. It should be remembered that the DAB process is mandatory. The DAB appears to be established for the collective good of the project and to ensure its smooth running rather than to satisfy the litigious desires of the individual members. However, that clause does not consider the possibility that the Parties are unable to agree on a DAB or single adjudicator. The mechanism to appoint a DAB or adjudicator in such circumstances is addressed here in sub-cl.20.3.

    This sub-clause again appears to place emphasis on the good of the project rather than the individual parties. The parties must be consulted by the appointing body with regard the appointment. However, thereafter, the appointment made by that appointing body is final and conclusive. Presumably, this appointment may be altered under the provisions of sub-cl.20.2. However, where this is to occur, the mutual consent of both parties is required. Where this is available, it seems unlikely that the parties would have resorted to the mechanism of sub-cl.20.3 rather than that of 20.2.

    In common with sub-cl.20.2, sub-cl.20.3 envisages that the cost of the DAB be shared between the parties. As stated above, the DAB is created to allow the project to run smoothly and efficiently.

20–029   The Particular Conditions note that for the adjudication procedure to succeed, a key factor is the confidence of the parties in the individuals who will serve on the DAB. Thus, where there is a failure to agree, it is important that the selection process for any replacement is impartial. This is why the Appendix to Tender suggests the FIDIC President (or someone appointed by the President) as the default nominating body. FIDIC keeps a list of dispute adjudicators for this purpose.

MDB HARMONISED EDITION

The words "[Appointment of the Dispute Board]" have been added to sub-paragraph (a). This just adds the title for sub-cl.20.2 for clarity. A further qualification has been added to sub-para.(b), namely "or fails to approve a member nominated by the other Party." Finally, the words "Appendix to Tender" are replaced by "Contract Data." **20–030**

20.4 OBTAINING DISPUTE ADJUDICATION BOARD'S DECISION

If a dispute (of any kind whatsoever) arises between the Parties in connection **20–031** *with or arising out of, the Contract or the execution of the Works, including any dispute as to any certificate, determination, instruction, opinion or valuation of the Engineer, either Party may refer the dispute in writing to the DAB for its decision, with copies to the other Party and Engineer. Such reference shall state that it is given under this Sub-Clause.*

For a DAB of three persons, the DAB shall be deemed to have received such reference on the date when it is received by the chairman of the DAB.

Both Parties shall promptly make available to the DAB all such additional information further access to the Site, and appropriate facilities, as the DAB may require for the purposes of making a decision on such dispute. The DAB shall be deemed to be not acting as arbitrator(s).

Within 84 days after receiving such reference, or with such other period as **20–032** *may be proposed by the DAB and approved by both Parties, the DAB shall give its decision, which shall be reasoned and shall state that it is given under this Sub-Clause. The decision shall be binding on both Parties, who shall promptly give effect to it unless and until it shall be revised in an amicable settlement or an arbitral award as described below. Unless the Contract has already been abandoned, repudiated or terminated, the Contractor shall continue to proceed with the Works in accordance with the Contract.*

If either Party is dissatisfied with the DAB's decision, then either Party may, within 28 days after receiving the decision, give notice to the other Party of its dissatisfaction. If the DAB fails to give its decision within the period of 84 days (or as otherwise approved) after receiving such reference, then either Party may, within 28 days after this period has expired, give notice to the other Party of its dissatisfaction.

In either event, this notice of dissatisfaction shall state that it is given under this Sub-Clause, and shall set out the matter in dispute and the reason(s) for dissatisfaction Except as stated in Sub-Clause 20.7 [Failure to Comply with Dispute Adjudication Board's Decision] and Sub-Clause 20.8 [Expiry of Dispute Adjudication Board's Appointment], neither Party shall be entitled to commence arbitration of a dispute unless a notice of dissatisfaction has been given in accordance with this Sub-Clause.

20–033   *If the DAB has given its decision as to a matter in dispute to both Parties, and no notice of dissatisfaction has been given by either Party within 28 days after it received the DAB's decision, then the decision shall become final and binding upon both Parties.*

OVERVIEW OF KEY FEATURES

20–034
- If any dispute arises between the parties then either party may refer that dispute to the DAB.
- The reference must be in writing and copies must be provided to the other party *and* the Engineer.
- The DAB shall be entitled to whatever access it requires, including access to information and the Site.
- The DAB will not act as an arbitral panel.
- Unless otherwise agreed, the DAB shall reach its reasoned decision within 84 days.
- That decision shall be binding unless it is overturned by agreement or by the decision of an arbitral panel.
- If a party disagrees with the decision of the DAB, it should serve a Notice of Dissatisfaction in accordance with sub-cl.20.7 and 20.8.
- If no such notice is served, then the decision of the DAB shall become final and binding.

COMMENTARY

20–035   Sub-clause 20.4 provides the mechanism whereby disputes are referred to the DAB. The DAB must follow the Procedural Rules which are discussed below. In many respects, the DAB takes the dispute resolution capacity of the engineer in standard forms such as the ICE 5th edn. In those contracts, it was the function of the engineer specified at the outset of the project to provide an initial dispute resolution service.

The sub-clause provides that either party might refer a dispute of any kind whatsoever arising between the Parties to the DAB. The definition of what can be referred is very broad indeed and is to include the review of any determination made by the Engineer. As such, the role of the Engineer becomes secondary to the DAB in which there is an appeal from decisions of the Engineer.

No definition is provided as to what will constitute a "dispute" within the meaning of clause sub-cl.20.4. That there is a dispute is, it is submitted, essential in the event that there is a later challenge to the jurisdiction of the DAB if its decision is to be enforced. Where the contract is entered into under English law, recent decisions of the English courts on the meaning of "dispute" found in the case law relating to the meaning of that term in the context of arbitration and adjudication should prove useful.[8]

A reference to a DAB of three members is deemed to have been received   **20–036**
on the date that the chairman receives it. Since the chairman of a three member DAB is chosen with the consent of both of the parties, this provision will help to reduce the fear that one party is obtaining an advantage over the other. The parties should direct their correspondence to the Chairman, but with copies to the other members, as well as providing a copy to the other party and Engineer.

The parties are to co-operate with the DAB in its decision making process. This is another manifestation of how the DAB is designed to be integral in and to the smooth running of the project.

A deadline is set for the DAB to produce its decision. That decision must be reasoned. The 84 days specified is longer than often provided for by adjudication clauses in English construction contracts. The decision of the DAB then becomes binding on the parties until settlement or arbitration. Throughout the DAB process, unless the contract has been ended, the parties must continue with the operation of the contract. Again the emphasis is on the smooth running of the contract as a whole.

Once the decision of the DAB has been produced, the parties have 28 days   **20–037**
to register their dissatisfaction. It would appear that decisions made by the DAB after the 84 days given to produce such decisions has expired will still be valid. In the UK there has been a divergence of opinion in relation to the validity of adjudication decisions delivered late. In Scotland,[9] the Court of Session held that an adjudicator's jurisdiction expired at the end of a 28-day period allowed to reach a decision. In contrast the English Courts[10] have held

---

[8] For example see the seven principles outlined by Mr Justice Jackson in *AMEC Civil Engineering Ltd v The Secretary of State for Transport* [2005] CILL 2189, endorsed and slightly expanded by the Court of Appeal in the same case [2005] CILL 2228 and the case of *Collins (Contractors) Ltd v Baltic Quay Management* (1994) Ltd [2005] CILL 2213.

[9] *Ritchie Brothers (PWC) Ltd v David Phillip (Commercials) Ltd* (2005) BLR 384.

[10] *Barnes & Elliott Ltd v Taylor Woodrow* [2004] CILL 2057.

that the giving of such a decision after the expiry of the time period will not be fatal. However, the 28 days to register dissatisfaction will run from the expiration of that 84 days. If the DAB has not produced a decision by the expiration of 28 days from the end of the 84 days to produce its decision, it seems that this would in itself be a reason for dissatisfaction.

A notice of dissatisfaction with the decision of the DAB is a condition precedent to commencing arbitration proceedings. A referral to arbitration will therefore not be valid where the DAB procedure has not been attempted first. Similarly, court proceedings (in England) will not be possible since the presence of an arbitration clause will entitle the defendant to a stay of proceedings under the Arbitration Act 1996, s.9. It is not known at this stage to what extent parties will be encouraged to refer to the DAB in situation where it is well known that the DAB will not produce the desired result simply as a way to arbitration. It can be well imagined that bogus referrals to the DAB will be made so that the party can proceed directly to arbitration once the time limits have expired.

It is important for parties to be aware of the 28 day limit for registering a notice of dissatisfaction in the prescribed form as failure to do so will cause the DAB's decision to become final and binding.

MDB HARMONISED EDITION

20–038   The words "and intention to commence arbitration" have been added to the fifth paragraph. In other words, the Notice of Dissatisfaction must include confirmation that the disaffected party intends to arbitrate should an amicable decision not be reached in accordance with sub-cl.20.5.

20.5 AMICABLE SETTLEMENT

20–039   *Where notice of dissatisfaction has been given under Sub-Clause 20.4 above, both Parties shall attempt to settle the dispute amicably before the commencement of arbitration. However, unless both Parties agree otherwise, arbitration may be commenced on or after the fifty-sixth day after the day on which notice of dissatisfaction was given, even if no attempt at amicable settlement has been made*

OVERVIEW OF KEY FEATURES

- In the event that a Notice of Dissatisfaction is served both parties must try and resolve that dispute amicably.   **20–040**
- An arbitration may not be commenced until 56 days after the Notice of Dissatisfaction has been served.

COMMENTARY

An attempt to obtain an amiable settlement for a prescribed time of 56 days   **20–041**
is also a condition precedent to a referral to arbitration. This is a further instance of where the FIDIC contract places emphasis on the smooth running of the project in which disputes are resolved on a local level. It is anticipated that, where a party has not waited and then complied with the 56 day "cooling off" period, then any reference to arbitration would be invalid. That said, it is conceded by the final sentence that a party does not need to make an attempt to achieve an amicable settlement.

The parties would be well advised to try and achieve such a settlement. There is not the space to discuss the variety of forms of alternative dispute resolution (otherwise known as ADR) available to the parties. These include mediation, conciliation and even simple negotiation. There may be benefits in obtaining specialist advice on what would be the best format to adopt. Whatever format is chosen it should be confidential. The advantages of achieving such settlement are plain – certainty and the saving of legal costs (some of which may be recoverable), and management time and costs (which are probably irrecoverable).

MDB HARMONISED EDITION

As with sub-cl.20.4, the words "and intention to commence arbitration" have   **20–042**
been added within the final sentence.

20.6 ARBITRATION

Unless settled amicably, any dispute in respect of which the DAB's decision (if   **20–043**
any) has not become final and binding shall be finally settled by international arbitration. Unless otherwise agreed by both Parties:

(a)   the dispute shall be finally settled under the Rules of Arbitration of the International Chamber of Commerce,

(b)   the dispute shall be settled by three arbitrators appointed in accordance with these Rules, and

(c)   the arbitration shall be conducted in the language for communications defined in Sub-Clause 1.4 [Law and Language].

20–044   *The arbitrator(s) shall have full power to open up, review and revise any certificate, determination, instruction, opinion or valuation of the Engineer, and any decision of the DAB, relevant to the dispute. Nothing shall disqualify the Engineer from being called as a witness and giving evidence before the arbitrator(s) on any matter whatsoever relevant to the dispute.*

Neither Party shall be limited in the proceedings before the arbitrator(s) to the evidence or arguments previously put before the DAB to obtain its decision, or to the reasons for dissatisfaction given in its notice of dissatisfaction. Any decision of the DAB shall be admissible in evidence in the arbitration.

Arbitration may be commenced prior to or after completion of the Works. The obligations of the Parties, the Engineer and the DAB shall not be altered by reason of any arbitration being conducted during the progress of the Works.

OVERVIEW OF KEY FEATURES

20–045   • If a dispute remains following the decision of the DAB and any attempt at amicable settlement, then that dispute is to be settled by international arbitration under the rules of the International Chamber of Commerce.
• Any arbitral decision is to be final and binding.
• The arbitral tribunal will have full powers to open up and revise any decision of both the Engineer and the DAB.
• The Engineer may be called as a witness.
• Provided the necessary time limits have been complied with, an arbitration may take place during the currency of the project.

COMMENTARY

20–046   Similar clauses to 20.6 can be found in many of the standard form contracts. A decision of the DAB which has not become final and binding shall be settled by international arbitration. The forum chosen by FIDIC is the International Chamber of Commerce or ICC. FIDIC does not nominate arbitrators. Before a dispute can be subject to arbitration it must first have been referred to the DAB and the appropriate Notice of Dissatisfaction must have been served. This would include in respect of any potential counter-claim. Given the length of time the DAB process will take, a party should be alive to the need to commence the appropriate procedures to get the counterclaim underway.

The clause specifically states that the arbitrator will have the power to open up and review any decision of the DAB which is relevant to the dispute. The arbitral panel will not be bound by any previous decision. Circumstances may therefore arise in which the arbitrator will open up decisions of the DAB other than the one for which the notice of dissatisfaction has been served insofar as it is related to the dispute. It is conceivable that the arbitrator is faced with the quandary of whether to open up and review parts of the DAB decision which, where it not for a notice of dissatisfaction having been served in a related dispute, would be final and binding on the parties.

Where arbitration proceedings are undertaken, the parties are not to be restricted to the evidence presented to the DAB. This is to be expected in that arbitration is not strictly an appeal from the DAB. The DAB and arbitration serve distinct functions. That said the DAB decision is admissible in evidence. It will not, of course, be binding.   **20–047**

Where arbitration is commenced prior to completion of the works, this should be done in such a way so as not to affect the progress of the works. This Contract appears to make every endeavour to facilitate the smooth running of the project.

The Particular Conditions recommend that the place of arbitration should be situated in a country other than that of the Employer or Contractor. This is presumably for reasons of impartiality as well as (in)convenience, if both parties have to travel. It is important that the parties are confident that any arbitral award could be ratified and enforced in their homeland. Thus the parties should check the applicability of the 1958 New York Convention.

Sub-clause 20.6 leaves it to the parties to decide the seat of the Arbitral Tribunal, which at least under English law,[11] in the absence of a clear choice from the parties will also decide the procedural law governing the arbitration. Note that under art.14.1 of the ICC rules, if the parties fail to choose a seat, the place of arbitration will be fixed by the ICC Court. Under article 16.1 of the LCIA rules, in the same situation, London will be the seat.

Finally, it must be pointed out that as sub-cl.20.6 is an arbitration clause, it is separable from the rest of the contract. For that reason, it will be valid even if the rest of the contract is declared invalid. Also, although there is a general assumption that it will,[12] the law governing the contract will not necessarily apply to it and the parties may wish to specify under which law clause 20.6 should be construed, or provide at the beginning of the contract that the governing law will also apply to cl.20.6.   **20–048**

---

[11] *Channel Tunnel Group Ltd v Balfour Beatty Construction Ltd* [1993] A.C. 334.
[12] Law and Practice of International Commercial Arbitration, A Redfern, M Hunter, N Blackaby & C Partasides 4th edn, para 2.86.

MDB HARMONISED EDITION

20–049 There have been a number of changes to the first two paragraphs which now read as follows:

> *Unless indicated otherwise in the Particular Conditions, any dispute not settled amicably and in respect of which the DB's decision (if any) has not become final and binding shall be finally settled by arbitration. Unless otherwise agreed by both Parties:*
>
> *(a) for contracts with foreign contractors, international arbitration with proceedings administered by the institution appointed in the Contract Data conduced in accordance with the rules of arbitration of the appointed institution, if any, or in accordance with UNCITRAL arbitration rules, at the choice of the appointed institution,*
>
> *(b) the place of arbitration shall be the city where the headquarters of the appointed arbitration institution is located,*
>
> *(c) the arbitration shall be conducted in the language for communications defined in Sub-Clause 1.4 [Law Language], and*
>
> *(d) for contracts with domestic contractors, arbitration with proceedings conducted in accordance with the laws of the Employer's country.*

20–050 The third paragraph, now states that there is nothing to stop the Engineer or "representatives of the Parties" from giving evidence before the tribunal.

The Banks would thus prefer that any arbitration be carried out by an institution with which they have some familiarly or failing that UNCITRAL. In addition the change makes certain that any dispute will be resolved in accordance with the law of the Employer's country.

20.7 FAILURE TO COMPLY WITH DISPUTE ADJUDICATION BOARD'S DECISION

20–051 *In the event that:*

> *(a) neither Party has given notice of dissatisfaction within the period stated in Sub-Clause 20.4 [Obtaining Dispute Adjudication Board's Decision],*
>
> *(b) the DAB's related decision (if any) has become final and binding, and*
>
> *(c) a Party fails to comply with this decision,*

then the other Party may, without prejudice to any other rights it may have, refer the failure itself to arbitration under Sub-Clause 20.6 [Arbitration]. Sub-Clause 20.5 [Amicable Settlement] shall not apply to this reference.

OVERVIEW OF KEY FEATURES

If a Party fails to comply with the decision of the DAB, then the other party may refer this failure to arbitration in accordance with sub-cl.20.6.　　**20–052**

COMMENTARY

Where a decision of the DAB has not been complied with and no notice of **20–053** dissatisfaction served, the other party is entitled to arbitrate and need not attempt an amicable settlement. This provision gives force to the DAB. It is envisaged that arbitration proceedings commenced for the enforcement of a DAB decision will be relatively quick. The arbitrator will be asked, in effect, to give summary judgment to enforce that DAB decision.

In situations, however, where a notice of dissatisfaction has been served by a party and that party also refuses to comply with the DAB's decision (contrary to the clear provisions of cl.20.4), then there appears to be a gap in cl.20.7, as first identified by Professor Nael Bunni.[13]

Sub-clause 20.7 only deals with the situation where both parties are satisfied with the DAB decision. If not (i.e. if a Notice of Dissatisfaction has been served) then there is no immediate recourse for the aggrieved party to ensure the DAB decision can be enforced.

Note that the arbitral proceedings envisaged by sub-cl.20.7 are quite **20–054** distinct from those under sub-cl.20.6. Whereas sub-cl.20.6 provides for a fresh procedure to decide the merits of the dispute between the parties, the only purpose of sub-cl.20.7 is to enable the enforcement of a DAB decision, without considering the substantial dispute between the parties.

MDB HARMONISED EDITION

Sub-paragraphs (a), (b) and (c) have been deleted so that the sub-clause now **20–055** reads as follows:

> *In the event that a Party fails to comply with a final and binding DB decision, then the other Party may, without prejudice to any other rights it may have, refer the failure itself to arbitration under Sub-Clause 20.6 [Arbitration]. Sub-Clause 20.4 [Obtaining Dispute Board's Decision] and Sub-Clause 20.5 [Amicable Settlement] shall not apply to this reference.*

---

[13] Nael Bunni – The Gap in Sub-Clause 20.7 of the 1999 FIDIC Contract for Major Works – ICLR [2000] p.272.

20.8 EXPIRY OF DISPUTE ADJUDICATION BOARD'S APPOINTMENT

20–056   *If a dispute arises between the Parties in connection with, or arising out of, the Contract or the execution of the Works and there is no DAB in place, whether by reason of the expiry of the DAB's appointment or otherwise:*

   (a)   Sub-Clause 20.4 [Obtaining Dispute Adjudication Board's Decision] and Sub-Clause 20.5 [Amicable Settlement] shall not apply, and
   (b)   the dispute may be referred to arbitration under Sub-Clause 20.6 [Arbitration].

OVERVIEW OF KEY FEATURES

20–057   Where no DAB exists then parties may proceed directly to arbitration.

COMMENTARY

20–058   The reason for this sub-clause is simply that, were this provision not in place, then the parties would have no other way of settling disputes other than by amicable settlement. It is suggested that this would be detrimental to the smooth running of the project.

MDB HARMONISED EDITION

20–059   There has been no change save for the usual deletion of the word "Adjudication".

GENERAL CONDITIONS OF DISPUTE ADJUDICATION AGREEMENT

1 DEFINITIONS

Each "Dispute Adjudication Agreement" is a tripartite agreement by and **20–060**
between:

(a)   the "Employer";
(b)   the "Contractor"; and
(c)   the "Member" who is defined in the Dispute Adjudication Agreement as being:

> (i)   *the sole member of the "DAB" (or "adjudicator") and, where this is the case, all references to the "Other Members" do not apply, or*
> (ii)   *one of the three persons who are jointly called the "DAB" (or "dispute adjudication board") and, where this is the case, the other two persons are called the "Other Members".*

The Employer and the Contractor have entered (or intend to enter) into a contract, which is called the "Contract" and is defined in the dispute Adjudication Agreement, which incorporates this Appendix. In the Dispute Adjudication Agreement, words and expressions which are not otherwise defined shall have the meanings assigned to them in the Contract.

COMMENTARY

Sub-clause 20.2 provides that the agreement between the Employer, **20–061**
Contractor and DAB shall incorporate by reference the General Conditions of Dispute Agreement. Thus, any term to be found in the Agreement which is defined in the main FIDIC form will have that meaning, unless specifically stated elsewhere. The General Conditions of Dispute Agreement is usually to be found in the Appendices to the FIDIC form General Conditions. Annexed to the General Conditions of Dispute Agreement will be a series of procedural rules. The following commentary deals with both the Agreement and those Rules, although the Rules are set out at the end of this section.

The DAB is not constituted until the Agreement is signed. As stated below, the Agreement will take place on the latest of:

(i)   the Commencement Date as per the Contract; or

(ii)   when all parties have signed the dispute agreement whether it is in the FIDIC form or some other format.

MDB HARMONISED EDITION

20–062   The word "Board" replaces "Adjudication". This is common throughout the entire agreement.

2 GENERAL PROVISIONS

20–063   *Unless otherwise stated in the Dispute Adjudication Agreement, it shall take effect on the latest of the following dates:*

(a)   the Commencement Date defined in the Contract,

(b)   when the Employer, the Contractor and the Member have each signed the Dispute Adjudication Agreement, or

(c)   when the Employer, the Contractor and each of the Other Members (if any) have respectively each signed a dispute adjudication agreement.

When the Dispute Adjudication Agreement has taken effect, the Employer and the Contractor shall each give notice to the Member accordingly. If the Member does not receive either notice within six months after entering into the Dispute Adjudication Agreement, it shall be void and ineffective.

20–064   *This employment of the Member is a personal appointment. At any time, the Member may give not less than 70 days' notice of resignation to the Employer and to the Contractor, and the Dispute Adjudication Agreement shall terminate upon the expiry of this period.*

No assignment or subcontracting of the Dispute Adjudication Agreement is permitted without the prior written agreement of all the parties to it and of the Other Members (if any).

COMMENTARY

20–065   When the Dispute Agreement has taken effect, an appropriate notice must be sent to the Members. This is important, as it can be seen that if no notice is received within six months then the Agreement is void. The appointment of a DAB member is a personal appointment. The Agreement will continue until it is terminated, in accordance with cl.7 below or until the Contractor issues a written discharge in accordance with sub-cl.14.12. If a Member wishes to resign, he must give 70 days' notice. This lengthy notice period is presumably designed to give time to find a replacement.

MDB HARMONISED EDITION

The second and fourth paragraphs have been deleted. The second paragraph **20–066** might be deemed to be an unnecessary technicality. However, it is unclear why the fourth paragraph has been deleted. It is not thought to be in the Parties' interests for the DAB Agreement to be assigned without agreement.

3 WARRANTIES

The Member warrants and agrees that he/she is and shall be impartial and inde- **20–067** *pendent of the Employer, the Contractor and the Engineer. The Member shall promptly disclose, to each of them and to the Other Members (if any), any fact or circumstance which might appear inconsistent with his/her warranty and agreement of impartiality and independence.*

When appointing the Member, the Employer and the Contractor relied upon the Member's representations that he/she is:

(a) experienced in the work which the Contractor is to carry out under the Contract,
(b) experienced in the interpretation of contract documentation, and
(c) fluent in the language for communications defined in the Contract.

COMMENTARY

This is an important clause. By cl.3, the members of the DAB warrant that **20–068** they are impartial, independent and will disclose anything which might happen to cast doubt on that. This disclosure must be prompt. This point is discussed above at sub-cl.20.2 and below at cl.4 which sets out, at length, the obligations of impartiality of the DAB members.

MDB HARMONISED EDITION

20–069   There is no change. However the requirement in sub-paras (a)–(c) can be found in the MDB version of sub-cl.20.2.

4 GENERAL OBLIGATIONS OF THE MEMBER

20–070   *The Member shall:*

(a)   *have no interest financial or otherwise in the Employer, the Contractor or the Engineer, nor any financial interest in the Contract except for payment under the Dispute Adjudication Agreement;*

(b)   *not previously have been employed as a consultant or otherwise by the Employer, the Contractor or the Engineer, except in such circumstances as were disclosed in writing to the Employer and the Contractor before they signed the Dispute Adjudication Agreement;*

(c)   *have disclosed in writing to the Employer, the Contractor and the Other to his/her best knowledge and recollection, any professional or personal relationships with any director, officer or employee of the Employer, the Contractor or the Engineer, and any previous involvement in the overall project of which the Contract forms part;*

20–071   (d)   *not, for the duration of the Dispute Adjudication Agreement, be employed as a consultant or otherwise by the Employer, the Contractor or the Engineer, except as may be agreed in writing by the Employer, the Contractor and the Other Members (if any);*

(e)   *comply with the annexed procedural rules and with Sub-Clause 20.4 of the Conditions of Contract;*

(f)   *not give advice to the Employer, the Contractor, the Employer's Personnel or in accordance with the annexed procedural rules;*

20–072   (g)   *not while a Member enter into discussions or make any agreement with the Employer, the Contractor or the Engineer regarding employment by any of them, whether as a consultant or otherwise, after ceasing to act under the Dispute Adjudication Agreement;*

(h)   *ensure his/her availability for all site visits and hearings as are necessary;*

(i)   *become conversant with the Contract and with the progress of the Works (and of any other parts of the project of which the Contract forms part) by studying all documents received which shall be maintained in a current working file;*

20–073   (j)   *treat the details of the Contract and all the DAB's activities and hearings as private and confidential, and not publish or disclose them without the prior written consent of the Employer, the Contractor and the Other Members (if any); and be available to give advice and opinions, on any* *matter relevant to the Contract when requested by both the Employer and the Contractor, subject to the agreement of the Other Members (if any).*

(k)   Be available to give advice and opinions, on any matter relevant to the Contract when requested by both the Employer and the Contractor, subject to the agreement of the Other Members (if any).

COMMENTARY

As noted above, the first seven sub-paragraphs focus on the members' impar-   **20–074**
tiality and neutrality. This is one reason why all communications between the
DAB and the Contractor must be copied to the other Party. This impartiality
is the key to the success of the DAB procedure. These obligations run
throughout the project. Procedural Rule 5 reinforces this by requiring that the
members of the DAB shall *"act, fairly and impartially as between the
Employer and the Contractor, giving each of them a reasonable opportunity of
putting his case and responding to the other's case."* These requirements accord
with the common law rules of natural justice set out in by the English Court
of Appeal in the case of *AMEC Capital Projects Limited v Whitefriars City
Estates Ltd.*[14]

Case-law tends to draw a distinction between "actual bias", which will
depend on the facts and "apparent bias" for which the test was stated by Lord
Hope in the case of *Porter v. Magill*[15] to be:

> *The question is whether the fair minded and informed observer, having consid-
> ered the facts, would conclude that there was a real possibility that the
> tribunal was biased.*

For example, the House of Lords[16] ruled that Lord Hoffman ought to have   **20–075**
declared his links to Amnesty International before sitting in judgment on the
question of the extradition of General Pinochet. By failing to do so, although
there was no question that he actually was biased, he could not be seen to be
impartial.

The need for impartiality might well be why Procedural Rule 9 states that
the DAB shall not express any opinions on merits during any hearing. It
should be noted that Procedural Rule 7 does place a potential limit on the
right to make representations. Here the DAB is empowered with the discretion to proceed with a hearing in the absence of a Party, if it is satisfied that
the Party in question did receive proper notice of the hearing. Thus, a Party

---

[14] (2005) BLR 1.
[15] [2002] 2 A.C. 375.
[16] (1999) 2 W.L.R. 272.

cannot subsequently complain if it chooses not to turn up to a hearing convened by the DAB.

The DAB members do not fulfil the same functions as arbitrators. Nevertheless, the obligations set out in the sub-clause, conform with the much more detailed IBA guidelines on Conflict of Interest in International Arbitration, first published in 2005.

20–076 The final four sub-paragraphs note that the DAB Member must be prepared to devote sufficient time to the project to become conversant with it and be prepared to provide opinions where required. The acceptance of a retainer, in accordance with Clause 6 below, is said to signify an agreement to do this. Potential members must therefore take time to consider whether they will be able to make themselves available and denote the appropriate time to the project. Some projects will last for a number of years. Some projects will involve considerable travel.

Some idea of the likely time commitment can be found from the Procedural Rules, which suggest site visits at intervals of not more than 70 days but not less than 140 days. The DAB must then prepare a report on what it has done during the visit. DAB members should also be provided with copies of the Contract documents, progress reports and any other documents pertinent to the performance of the Contract. There is the suggestion at sub-para.(k) that the DAB members might also be asked to give ongoing advice about problems which arise outside of the formal DAB procedures. However, in accordance with the principles of natural justice, as sub-para.(f) notes, this must be done openly and by agreement in response to a request from both parties.

Finally, all DAB hearings are to be treated as confidential, although as per sub-cl.20.6, DAB decisions are admissible during any arbitration proceedings.

MDB HARMONISED EDITION

20–077 There is no change. However, there is a small addition to cl.2 of the Procedural Rules which confirm that one of the purposes of the site visits is to, as far as reasonable, *"endeavour to prevent potential problems or claims from becoming disputes"*

5 GENERAL OBLIGATIONS OF THE EMPLOYER AND THE CONTRACTOR

20–078 *The Employer, the Contractor, the Employer's Personnel and the Contractor's Personnel shall not request advice from or consultation with the Member* *regarding the Contract, otherwise than in the normal course of the DAB's activities under the Contract and the Dispute Adjudication Agreement, and except to the extent that prior agreement is given by the Employer, the Contractor and the Other Members (if any). The Employer and the Contractor shall be responsible for compliance with this provision, by the Employer's Personnel and the Contractor's Personnel respectively.*

The Employer and the Contractor undertake to each other and to the Member that the Member shall not, except as otherwise agreed in writing by the Employer, the Contractor, the Member and the Other Members (if any):

(a) be appointed as an arbitrator in any arbitration under the Contract;
(b) be called as a witness to give evidence concerning any dispute before arbitrator(s) appointed for any arbitration under the Contract; or
(c) be liable for any claims for anything done or omitted in the discharge or purported discharge of the Member's functions, unless the act or omission is shown to have been in bad faith.

The Employer and the Contractor hereby jointly and severally indemnify and hold the Member harmless against and from claims from which he/she is relieved from liability under the preceding paragraph.   **20–079**

Whenever the Employer or the Contractor refers a dispute to the DAB under Sub-Clause 20.4 of the Conditions of Contract, which will require the Member to make a site visit and attend a hearing, the Employer or the Contractor shall provide appropriate security for a sum equivalent to the reasonable expenses to be incurred by the Member. No account shall be taken of any other payments due or paid to the Member.

COMMENTARY

Condition 5 is linked to the previous condition as the Contractor and   **20–080**
Employer are, in effect, agreeing here not to try and encourage a Member to break the obligations set out in condition 4. For example, neither can request private advice.

However, although the DAB decision can be used in any arbitration, the DAB members cannot either act as arbitrator or be called as a witness. The extent to which an arbitral tribunal would find a decision of the DAB to be persuasive is difficult to judge. Certainly the members of the DAB would be able to make use of their own specialist knowledge, the DAB would have the advantage of being contemporary and it should also benefit from the members' knowledge of the project. However, this would ultimately be a matter for the arbitral tribunal.

The Employer and Contractor must also provide a joint and several indemnity in respect of any claims that may be made against a Member provided that the Member has acted in good faith. If the member acts in bad faith, then he may be subject to the sanctions set out below at cl.8.

MDB HARMONISED EDITION

20–081   The words "and except to the extent that prior agreement is given by the Employer, the Contractor and Other Members (if any)" in the first paragraph have been deleted.

6 PAYMENT

20–082   *The Member shall be paid as follows: in the currency named in the Dispute Adjudication Agreement:*

   (a)   a retainer fee per calendar month, which shall be considered as payment in full for:

   (i)   *being available on 28 days' notice for all site visits and hearings;*

   (ii)   *becoming and remaining conversant with all project developments and maintaining relevant files;*

   (iii)   *all office and overhead expenses including secretarial services, photocopying and offices supplies incurred in connection with his duties; and*

   (iv)   *all services performed hereunder except those referred to in sub-paragraphs (b) and (c) of this Clause.*

   The retainer fee shall be paid with effect from the last day of the calendar month in which the Dispute Adjudication Agreement becomes effective; until the last day of the calendar month in which the Taking-Over Certificate is issued for the whole of the Works.

20–83   *With effect from the first day of the calendar month following the month in which Taking-Over Certificate is issued for the whole of the Works, the retainer fee shall be reduced by 50%. This reduced fee shall be paid until the first day of the calendar month in which the Member resigns or the Dispute Adjudication Agreement is otherwise terminated.*

   (b)   a daily fee which shall be considered as payment in full for:

   (i)   *each day or part of a day up to a maximum of two days' travel time in each direction for the journey between the Member's home and the site, or another location of a meeting with Other Members (if any);*

(ii)   *each working day on site visits, hearings or preparing decisions; and*

(iii)   *each day spent reading submissions in preparation for a hearing.*

(c) all reasonable expenses incurred in connection with the Member's duties, including the cost of telephone calls, courier charges, faxes and telexes, travel expenses, hotel and subsistence costs: a receipt shall be required for each item in excess of five percent of the daily fee referred to in sub-paragraph (b) of this Clause;

(d) any taxes properly levied in the Country on payments made to the **20–084** *Member (unless a national or permanent resident of the Country) under this Clause 6.*

The retainer and daily fees shall be as specified in the Dispute Adjudication Agreement. Unless it specifies otherwise, these fees shall remain fixed for the first 24 calendar months, and shall thereafter be adjusted by agreement between the Employer, the Contractor and the Member, at each anniversary of the date on which the Dispute Adjudication became effective.

The Member shall submit invoices for payment on the monthly retainer and air fares quarterly in advance. Invoices for other expenses and for daily fees shall be submitted following the conclusion of a site visits or hearing. All invoices shall be accompanied by a brief description of activities performed during the relevant period and shall be addressed to the Contractor.

The Contractor shall pay each of the Member's invoices in full within 56 **20–085** *calendar days after receiving each invoice and shall apply to the Employer (in the Statements under the Contract) for reimbursement of one-half of the amounts of these invoices. The Employer shall then pay the Contractor in accordance with the Contract.*

If the Contractor fails to pay to the Member the amount to which he/she is entitled under the Dispute Adjudication Agreement, the Employer shall pay the amount due to the Member and any other amount which may be required to maintain the operation of the DAB; and without prejudice to the Employer's rights or remedies. In addition to reimbursement of all sums paid in excess of one-half of these payments, plus all costs of recovering these sums and financing charges calculated at the rate specified in Sub-Clause 14.8 of the Conditions of Contract.

If the Member does not receive payment of the amount due within 70 days after submitting a valid invoice, the Member may (i) suspend his/her services (without notice) until the payment is received, and/or (ii) resign his/her appointment by giving notice under Clause 7.

COMMENTARY

20–086   A Member's fee can fall into two categories, a retainer and a daily fee. By accepting a retainer, the Member confirms his availability for site visits and hearings on 28-days notice. The retainer also includes reading in and office overheads. As noted above, DAB members are to be provided with copies of all documents which are pertinent to the performance of the Contract. The retainer fee is reduced by 50 per cent from the date of the Taking-Over Certificate. The daily fee, which includes travel time (of up to two days each way), deals with site visits, hearings and working on submissions. The retainer comes to an end when the Member resigns or the DAB is terminated.

The fees are fixed for two years, another reference to the potential length of the appointment. The Contractor is responsible at first instance for paying DAB member invoices. This must be due within 56 days. The Employer is then invoiced for 50 per cent of these fees. If a Member is not paid, he is entitled to interest and ultimately after 70 days may suspend his services or resign.

MDB HARMONISED EDITION

20–087   There have been changes to sub-para.(c) which now reads as follows:

> *(c)   all reasonable expenses including necessary travel expenses (air fare in less than first class, hotel and subsistence and other direct travel expenses) incurred in connection with the Member's duties, as well as the cost of telephone calls, courier charges, faxes and telexes: a receipt shall be require for each item in excess of the daily fee referred to in sub-paragraph (b) of this Clause.*

A new third paragraph has been added as follows:

> *If the Parties fail to agree on a retainer fees or the daily fee of the appointing entity or official moved in the Contract Data shall determine the amount of the fees to be used.*

7 TERMINATION

20–088   *At any time: (i) the Employer and the Contractor may jointly terminate the Dispute Adjudication Agreement by giving 42 days' notice to the Member; or (ii) the Member may resign as provided for in Clause 2.*

If the Member fails to comply with the Dispute Adjudication Agreement, the Employer and the Contractor may, without prejudice to their other rights, terminate it by notice to the Member. The notice shall take effect when received by the Member.

If the Employer or the Contractor fails to comply with the Dispute Adjudication Agreement, the Member may, without prejudice to his/her other rights, terminate it by notice to the Employer and the Contractor. The notice shall take effect when received by them both.

Any such notice, resignation and termination shall be final and binding on the   **20–089**
Employer, the Contractor and the Member. However, a notice by the Employer or the Contractor, but not by both, shall be of no effect.

COMMENTARY

The termination clause is straightforward. In keeping with the provisions   **20–090**
relating to the DAB, the decision to terminate must be a joint one between
Contractor and Employer.

8 DEFAULT OF THE MEMBER

If the Member fails to comply with any obligation under Clause 4, he/she shall   **20–091**
not be entitled to any fees or expenses hereunder and shall, without prejudice to their other rights, reimburse each of the Employer and the Contractor for any fees and expenses received by the Member and the Other Members (if any), for proceedings or decisions (if any) of the DAB which are rendered void or ineffective.

COMMENTARY

Again the meaning of this clause is straightforward. A member must comply   **20–092**
with the obligations of cl.4. If he does not, he is not entitled to be paid and
may be liable to pay to the Employer and Contractor any fees paid out to
other members of the DAB. This liability, as is made clear in the clause,
would only arise, if the failings made a decision void or otherwise ineffective.

MDB HARMONISED EDITION

20–093   The following paragraph has been added:

> *If the Member fails to comply with any of his obligations under Clause 4(e)-(k) above, he shall not be entitled to any fees or expenses hereunder from the date and to the extent of the non-compliance and shall, without prejudice to their other rights, reimburse each of the Employer and the Contractor for any fees and expenses already received by the Member, for proceedings or decisions (if any) of the DB which are rendered void or ineffective by the said failure to comply.*

The MDB version is not so lenient and the Members' conduct does not need to have caused a decision to become ineffective before a liability to reimburse the Employer and Contractor arises.

9 DISPUTES

20–094   *Any dispute or claim arising out of or in connections with this Dispute Adjudication Agreement, or the breach, termination or invalidity thereof, shall be finally settled under the Rules of Arbitration of the International Chamber of Commerce by one arbitrator appointed in accordance with these Rules of Arbitration.*

COMMENTARY

20–095   This clause requires no comment, although as the FIDIC Guide notes, it is entirely possible for the costs of any dispute which may arise here to be prohibitive. Thus, the parties may wish to consider ADR, which is always a sensible approach whatever the size and nature of any dispute.

MDB HARMONISED EDITION

20–096   The words "*by institutional arbitration. If no other arbitration institute is agreed, the arbitration shall be conducted*" have been added to the end of the first sentence. Thus the parties are free to choose their own arbitration forum and it is only in default of such a choice that the ICC rules will apply.

ANNEX PROCEDURAL RULES[17]

1   Unless otherwise agreed by the Employer and the Contractor, the DAB **20–097**
    shall visit the site at intervals of not more than 140 days, including times
    of critical construction events, at the request of either the Employer or the
    Contractor. Unless otherwise agreed by the Employer, the Contractor and
    the DAB, the period between consecutive visits shall not be less than 70
    days, except as required to convene a hearing as described below.

2   The timing of an agenda for each visit shall be as agreed jointly by the
    DAB, the Employer and the Contractor, or in the absence of agreement,
    shall be decided by the DAB. The purpose of site visits is to enable the
    DAB to become and remain acquainted with the progress of the Works
    and of any actual or potential problems or claims. [and, as far as reason-
    able, to endeavour to prevent potential problems or claims from becoming
    disputes][18]

3   Site visits shall be attended by the Employer, the Contractor and the
    Engineer and shall be co-ordinated by the Employer in co-operation with
    the Contractor. The Employer shall ensure the provision of appropriate
    conference facilities and secretarial and copying services. At the conclu-
    sion of each site visit and before the site, the DAB shall prepare a report
    on its activities during the visit and shall send copies to the Employer and
    the Contractor.

4   The Employer and the Contractor shall furnish to the DAB one copy of **20–098**
    all documents which the DAB may request, including Contractor docu-
    ments, progress reports, variation instructions, certificates and other
    documents pertinent to the performance of the Contract. All communica-
    tions between the DAB and the Employer or the Contractor shall be
    copied to the other Party. If the DAB comprises three persons, the
    employer and the Contractor shall send copies of these requested
    documents and these communications to each of these persons.

5   If any dispute if referred to the DAB in accordance with Sub-Clause 20.4
    of the Conditions of Contract, the DAB shall proceed in accordance with
    Sub-Clause 20.4 and these Rules. Subject to the time allowed to give
    notice of a decision and other relevant factors, the DAB shall:

    (a)   act fairly and impartially as between the Employer and the
          Contractor, giving each of them a reasonable opportunity of
          putting his case and responding to the other's case, and

---

[17] These Rules are the Rules for use where the DAB is set up at the beginning of the project and
    not are not for use where an ad-hoc arrangement is being used.
[18] MDB Harmonised version addition.

(b)   adopt procedures suitable to the dispute, avoiding unnecessary delay or expense.

6   The DAB may conduct a hearing on the dispute, in which event it will decide on the date and place for the hearing and may request that written documentation and arguments from the Employer and the Contractor be presented to it prior to or at the hearing.

20–099   7   Except as otherwise agreed in writing by the Employer and the Contractor, the DAB shall have the power to adopt an inquisitorial procedure, to refuse admission to hearings or audience at hearings to any persons other than representatives of the Employer, the Contractor and the Engineer, and to proceed in the absence of any party who the DAB is satisfied received notice of the hearing; but shall have discretion to decide whether and to what extent this power may be exercised.

8   The Employer and the Contractor empower the DAB, among other things, to:

(a)   establish the procedure to be applied in deciding a dispute,

(b)   decide upon the DAB's own jurisdiction, and as to the scope of any dispute referred to it,

(c)   conduct any hearing as it thinks fit, not being bound by any rules or procedures other than those contained in the Contract and these Rules,

(d)   take the initiative in ascertaining the facts and matters required for a decision,

(e)   make us of its own specialist knowledge, if any,

(f)   decide upon the payment of financing charges in accordance with the Contract,

(g)   decide upon any provisional relief such as interim or conservatory measures, and

(h)   open up, review and revise any certificate, decision, determination, instructions, opinion or valuation of the Engineer, relevant to the dispute.

20–100   9   The DAB shall not express any opinions during any hearing concerning the merits of any arguments advanced by the Parties. Thereafter, the DAB shall make and give its decision in accordance with Sub-Clause 20.4, or as otherwise agreed by the Employer and the Contractor in writing. If the DAB comprises three persons:

(a)   it shall convene in private after a hearing, in order to have discussions and prepare its decision;

(b)   it shall endeavour to reach a unanimous decision: if this proves impossible the applicable decision shall be made by a majority of the Members, who may require the minority Member to prepare a *written report for submission to the Employer and the Contractor, and*

(c) *if a Member fails to attend a meeting or hearing, or to fulfil any required function, the other two Members may nevertheless proceed to make a decision, unless:*

(i) *either the Employer or the Contractor does not agree that they do so, or*

(ii) *the absent Member is the chairman and e/she instructs the other Members to not make a decision.*

APPENDIX

Annex A LETTER OF BID

[All italicized text and any enclosing square brackets is for use in preparing the form and should be deleted, together with any square brackets, from the final product.]

Date: _____

ICB No.: _____

Invitation for Bid No.: _____

To: _____

We, the undersigned, declare that:

(a)   We have examined and have no reservations to the Bidding Document, including Addenda issued in accordance with Instructions to Bidders (ITB) _____ [*Number*];

(b)   We offer to execute in conformity with the Bidding Document the following Works _____ ;

(c)   The total price of our Bid, excluding any discounts offered in item (d) below is: _____ ;

(d)   The discounts offered and the methodology for their application are: _____ ; _____ ;

(e)   Our bid shall be valid for a period of _____ days from the date fixed for the bid submission deadline in accordance with the Bidding Document, and it shall remain binding upon us and may be accepted at any time before the expiration of that period;

(f)   If our bid is accepted, we commit to obtain a performance security in accordance with the Bidding Document;

(g)   We, including any subcontractors or suppliers for any part of the contract, have or will have nationalities from eligible countries, in accordance with ITB _____ [*Number*];

(h)   We, including any subcontractors or suppliers for any part of the contract, do not have any conflict of interest in accordance with ITB _____ [*Number*];

(i)   We are not participating, as a Bidder or as a subcontractor, in more than one bid in this bidding process in accordance with _____ , other than alternative offers submitted in accordance with ITB _____ [*Number*];

We, including any of our subcontractors or suppliers for any part of the contract, have not been declared ineligible by the Bank, under the Employer's country laws or official regulations or by an act of compliance with a decision of the United Nations Security Council;

413

(j)   We are not a government owned entity [*or*]

      We are a government owned entity but meet the requirements of ITB _____ [*Number*];

(k)   We have paid, or will pay the following commissions, gratuities, or fees with respect to the bidding process or execution of the Contract:

Name of Recipient	Address	Reason	Amount

      [*If none has been paid or is to be paid, indicate "none".*]

(l)   We understand that this bid, together with your written acceptance thereof included in your notification of award, shall constitute a binding contract between us, until a formal contract is prepared and executed; and

(m)   We understand that you are not bound to accept the lowest evaluated bid or any other bid that you may receive.

(n)   We hereby certify that we have taken steps to ensure that no person acting for us or on our behalf will engage in bribery.

Name: _____

Signed: _____

      in the capacity of: _____

Duly authorized to sign the bid for and on behalf of:

Dated on _____ day of _____ , _____ .

Annex B LETTER OF ACCEPTANCE

[All italicized text and any enclosing square brackets is for use in preparing the form and should be deleted, together with any square brackets, from the final product.]

_____ [*Name and address of the Contractor*]

This is to notify you that your Bid dated _____ [*Date*] for execution of the
_____ [*Name of the Contract and
identification number, as given in the Contract Data*] for the Accepted Contract Amount of the
equivalent of _____ [*Amount in numbers and words*] _____ [*Name
of currency*], as corrected and modified in accordance with the Instructions to Bidders, is hereby
accepted by our Agency.

You are requested to furnish the Performance Security within 28 days in accordance with the
Conditions of Contract, using for that purpose one of the Performance Security Forms included in
the annexes to the Particular Conditions.

Authorized Signature: _____

Name and Title of Signatory: _____

Name of Agency: _____

Annex C CONTRACT AGREEMENT

This **Agreement** made the _____ day of _____ , _____ ,
between _____ of _____ (hereinafter "the Employer"),
of the one part, and _____ of _____ (hereinafter
"the Contractor"), of the other part:

Whereas the Employer desires that the Works known as _____
should be executed by the Contractor, and has accepted a Bid by the Contractor for the execution
and completion of these Works and the remedying of any defects therein,

The Employer and the Contractor agree as follows:

1.   In this Agreement words and expressions shall have the same meanings as are respectively
        assigned to them in the Contract documents referred to.

2.   The following documents shall be deemed to form and be read and construed as part of this
        Agreement. This Agreement shall prevail over all other Contract documents.

        (i)   the Letter of Acceptance
        (ii)   the Letter of Bid
        (ii)   the addenda Nos _____ (if any)
        (iv)   the Particular Conditions - Part A
        (v)   the Particular Conditions - Part B
        (vi)   the General Conditions
        (vii)   the Specification
        (viii) the Drawings, and
        (ix)   the completed Schedules,

3.   In consideration of the payments to be made by the Employer to the Contractor as indicated
        in this Agreement, the Contractor hereby covenants with the Employer to execute the Works
        and to remedy defects therein in conformity in all respects with the provisions of the
        Contract.

4.   The Employer hereby covenants to pay the Contractor in consideration of the execution and
        completion of the Works and the remedying of defects therein, the Contract Price or such
        other sum as may become payable under the provisions of the Contract at the times and in
        the manner prescribed by the Contract.

In witness whereof the parties hereto have caused this Agreement to be executed in accordance
with the laws of _____ on the day, month and year indicated above.

Signed by: _____ (for the Employer)

Signed by: _____ (for the Contractor)

Annex D DISPUTE BOARD AGREEMENT

[All italicized text and any enclosing square brackets is for use in preparing the form and should be deleted, together with any square brackets, from the final product.]

[For a one-person DB]

Name and details of Contract _____

Name and address of Employer _____

Name and address of Contractor _____

Name and address of Member _____

Whereas the Employer and the Contractor have entered into the Contract and desire jointly to appoint the Member to act as sole Member who is also called the "DB".

The Employer, Contractor and Member jointly agree as follows:

1.　The conditions of this Dispute Board Agreement comprise the "General Conditions of Dispute Board Agreement", which is appended to the General Conditions of the "Conditions of Contract for Construction" MDB Harmonised Edition published by the Fédération Internationale des Ingénieurs-Conseils (FIDIC), and the following provisions. In these provisions, which include amendments and additions to the General Conditions of Dispute Board Agreement, words and expressions shall have the same meanings as are assigned to them in the General Conditions of Dispute Board Agreement.

2.　*[Details of amendments to the General Conditions of Dispute Board Agreement, if any. For example:*
In the procedural rules annexed to the General Conditions of Dispute Board Agreement, Rule _____ is deleted and replaced by: " ...... "]

3.　In accordance with Clause 6 of the General Conditions of Dispute Board Agreement, the Member shall be paid as follows:

　　　　A retainer fee of _____ per calendar month,
　　　　plus a daily fee of _____ per day.

4.　In consideration of these fees and other payments to be made by the Employer and the Contractor in accordance with Clause 6 of the General Conditions of Dispute Board Agreement, the Member undertakes to act as the DB (as member) in accordance with this Dispute Board Agreement.

5.　The Employer and the Contractor jointly and severally undertake to pay the Member, in consideration of the carrying out of these services, in accordance with Clause 6 of the General Conditions of Dispute Board Agreement.

6.　This Dispute Board Agreement shall be governed by the law of _____

SIGNED by: _____　SIGNED by: _____　SIGNED by: _____

for and on behalf of the Employer　for and on behalf of the Contractor　The Member in the presence of
in the presence of　in the presence of

Witness: _____	Witness: _____	Witness _____
Name: _____	Name: _____	Name: _____
Address: _____	Address: _____	Address: _____
Date: _____	Date: _____	Date: _____

Annex E DISPUTE BOARD AGREEMENT

[All italicized text and any enclosing square brackets is for use in preparing the form and should be deleted, together with any square brackets, from the final product.]

[For each member of a three-person DB]

Name and details of Contract
Name and address of Employer
Name and address of Contractor
Name and address of Member

Whereas the Employer and the Contractor have entered into the Contract and desire jointly to appoint the Member to act as one of the three persons who are jointly called the "DB" *[and desire the Member to act as chairman of the DB]*.

The Employer, Contractor and Member jointly agree as follows:

1. The conditions of this Dispute Board Agreement comprise the "General Conditions of Dispute Board Agreement", which is appended to the General Conditions of the "Conditions of Contract for Construction" MDB Harmonised Edition published by the Fédération Internationale des Ingénieurs-Conseils (FIDIC), and the following provisions. In these provisions, which include amendments and additions to the General Conditions of Dispute Board Agreement, words and expressions shall have the same meanings as are assigned to them in the General Conditions of Dispute Board Agreement.

2. *[Details of amendments to the General Conditions of Dispute Board Agreement, if any. For example:*
    In the procedural rules annexed to the General Conditions of Dispute Board Agreement, Rule _____ is deleted and replaced by: " ...... "]

3. In accordance with Clause 6 of the General Conditions of Dispute Board Agreement, the Member shall be paid as follows:

    A retainer fee of _____ per calendar month,
    plus a daily fee of _____ per day.

4. In consideration of these fees and other payments to be made by the Employer and the Contractor in accordance with Clause 6 of the General Conditions of Dispute Board Agreement, the Member undertakes to serve, as described in this Dispute Board Agreement, as one of the three persons who are jointly to act as the DB.

5. The Employer and the Contractor jointly and severally undertake to pay the Member, in consideration of the carrying out of these services, in accordance with Clause 6 of the General Conditions of Dispute Board Agreement.

6. This Dispute Board Agreement shall be governed by the law of _____

SIGNED by: _____ | SIGNED by: _____ | SIGNED by: _____

for and on behalf of the Employer in the presence of | for and on behalf of the Contractor in the presence of | The Member in the presence of

Witness: _____
Name: _____
Address: _____
Date: _____

Witness: _____
Name: _____
Address: _____
Date: _____

Witness _____
Name: _____
Address: _____
Date: _____

APPENDIX

Annex F PERFORMANCE SECURITY

Demand Guarantee

[All italicized text and any enclosing square brackets is for use in preparing the form and should be deleted, together with any square brackets, from the final product.]

_____ *[Bank's name, and address of issuing branch or office]*
Beneficiary: _____ *[Name and Address of Employer]*
Date: _____

Performance Guarantee No.: _____

We have been informed that _____ *[Name of Contractor]* (hereinafter called "the Contractor") has entered into Contract No. _____ *[Reference number of the contract]* dated _____ with you, for the execution of _____ *[Name of contract and brief description of Works]* (hereinafter called "the Contract").

Furthermore, we understand that, according to the conditions of the Contract, a performance guarantee is required.

At the request of the Contractor, we _____ *[Name of Bank]* hereby irrevocably undertake to pay you any sum or sums not exceeding in total an amount of _____ *[Amount in figures]* (_____) *[Amount in words, see 1]*, such sum being payable in the types and proportions of currencies in which the Contract Price is payable, upon receipt by us of your first demand in writing accompanied by a written statement stating that the Contractor is in breach of its obligation(s) under the Contract, without your needing to prove or to show grounds for your demand or the sum specified therein.

This guarantee shall expire, no later than the _____ day of _____ , _____ *[see 2]*, and any demand for payment under it must be received by us at this office on or before that date.

This guarantee is subject to the Uniform Rules for Demand Guarantees, ICC Publication No. 458, except that subparagraph (ii) of Sub-article 20(a) is hereby excluded.

Signature(s): _____

1 *The Guarantor shall insert an amount representing the percentage of the Contract Price specified in the Contract and denominated either in the currency(cies) of the Contract or a freely convertible currency acceptable to the Employer.*
2 *Insert the date twenty-eight days after the expected completion date. The Employer should note that in the event of an extension of the time for completion of the Contract, the Employer would need to request an extension of this guarantee from the Guarantor. Such request must be in writing and must be made prior to the expiration date established in the guarantee. In preparing this guarantee, the Employer might consider adding the following text to the form, at the end of the penultimate paragraph: "The Guarantor agrees to a one-time extension of this guarantee for a period not to exceed [six months] [one year], in response to the Employer's written request for such extension, such request to be presented to the Guarantor before the expiry of the guarantee."*

Annex G PERFORMANCE BOND

By this Bond _____ as Principal (hereinafter called "the Contractor") and _____ as Surety (hereinafter called "the Surety"), are held and firmly bound unto _____ as Obligee (hereinafter called "the Employer") in the amount of _____ , for the payment of which sum well and truly to be made in the types and proportions of currencies in which the Contract Price is payable, the Contractor and the Surety bind themselves, their heirs, executors, administrators, successors and assigns, jointly and severally, firmly by these presents.

Whereas the Contractor has entered into a written Agreement with the Employer dated the _____ day of _____ , _____ , for _____ in accordance with the documents, plans, specifications, and amendments thereto, which to the extent herein provided for, are by reference made part hereof and are hereinafter referred to as the Contract.

Now, therefore, the Condition of this Obligation is such that, if the Contractor shall promptly and faithfully perform the said Contract (including any amendments thereto), then this obligation shall be null and void; otherwise, it shall remain in full force and effect. Whenever the Contractor shall be, and declared by the Employer to be, in default under the Contract, the Employer having performed the Employer's obligations thereunder, the Surety may promptly remedy the default, or shall promptly:

(a)   complete the Contract in accordance with its terms and conditions; or

(b)   obtain a Bid or bids from qualified Bidders for submission to the Employer for completing the Contract in accordance with its terms and conditions, and upon determination by the Employer and the Surety of the lowest responsive Bidder, arrange for a Contract between such Bidder and Employer and make available as work progresses (even though there should be a default or a succession of defaults under the Contract or Contracts of completion arranged under this paragraph) sufficient funds to pay the cost of completion less the Balance of the Contract Price; but not exceeding, including other costs and damages for which the Surety may be liable hereunder, the amount set forth in the first paragraph hereof. The term "Balance of the Contract Price", as used in this paragraph, shall mean the total amount payable by Employer to Contractor under the Contract, less the amount properly paid by Employer to Contractor; or

(c)   pay the Employer the amount required by Employer to complete the Contract in accordance with its terms and conditions up to a total not exceeding the amount of this Bond.

The Surety shall not be liable for a greater sum than the specified penalty of this Bond.

Any suit under this Bond must be instituted before the expiration of one year from the date of the issuing of the Taking-Over Certificate.

No right of action shall accrue on this Bond to or for the use of any person or corporation other than the Employer named herein or the heirs, executors, administrators, successors, and assigns of the Employer.

In testimony whereof, the Contractor has hereunto set his hand and affixed his seal, and the Surety has caused these presents to be sealed with his corporate seal duly attested by the signature of his legal representative, this _____ day of _____ , _____ .

SIGNED ON: _____   on behalf of: _____
By: _____   in the capacity of: _____
In the presence of: _____

SIGNED ON: _____   on behalf of: _____
By: _____   in the capacity of: _____
In the presence of: _____

APPENDIX

Annex H   ADVANCE PAYMENT SECURITY

Demand Guarantee

[All italicized text and any enclosing square brackets is for use in preparing the form and should be deleted, together with any square brackets, from the final product.]

_____ [Bank's name, and address of issuing branch or office]
Beneficiary: _____ [Name and Address of Employer]
Date: _____

Advance Payment Guarantee No.: _____
We have been informed that _____ [Name of Contractor] (hereinafter called "the Contractor") has entered into Contract No. _____ [Reference number of the contract] dated _____ with you, for the execution of _____ [Name of contract and brief description of Works] (hereinafter called "the Contract").

Furthermore, we understand that, according to the conditions of the Contract, an advance payment in the sum _____ [Amount in figures] (_____) [Amount in words] is to be made against an advance payment guarantee.

At the request of the Contractor, we _____ [Name of Bank] hereby irrevocably undertake to pay you any sum or sums not exceeding in total an amount of _____ [Amount in figures] (_____) [Amount in words, see 1] upon receipt by us of your first demand in writing accompanied by a written statement stating that the Contractor is in breach of its obligation under the Contract because the Contractor used the advance payment for purposes other than the costs of mobilization in respect of the Works.

It is a condition for any claim and payment under this guarantee to be made that the advance payment referred to above must have been received by the Contractor on its account number _____ at _____ [Name and address of Bank].

The maximum amount of this guarantee shall be progressively reduced by the amount of the advance payment repaid by the Contractor as indicated in copies of interim statements or payment certificates which shall be presented to us. This guarantee shall expire, at the latest, upon our receipt of a copy of the interim payment certificate indicating that eighty (80) percent of the Contract Price has been certified for payment, or on the _____ day of _____, _____ [see 2] whichever is earlier. Consequently, any demand for payment under this guarantee must be received by us at this office on or before that date.

This guarantee is subject to the Uniform Rules for Demand Guarantees, ICC Publication No. 458.

Signature(s): _____

1   *The Guarantor shall insert an amount representing the amount of the advance payment and denominated either in the currency(ies) of the advance payment as specified in the Contract, or in a freely convertible currency acceptable to the Employer.*
2   *Insert the expected expiration date of the Time for Completion. The Employer should note that in the event of an extension of the time for completion of the Contract, the Employer would need to request an extension of this guarantee from the Guarantor. Such request must be in writing and must be made prior to the expiration date established in the guarantee. In preparing this guarantee, the Employer might consider adding the following text to the form, at the end of the penultimate paragraph: "The Guarantor agrees to a one-time extension of this guarantee for a period not to exceed [six months][one year], in response to the Employer's written request for such extension, such request to be presented to the Guarantor before the expiry of the guarantee."*

APPENDIX

Annex I RETENTION MONEY SECURITY

Demand Guarantee

[All italicized text and any enclosing square brackets is for use in preparing the form and should be deleted, together with any square brackets, from the final product.]

_____ *[Bank's name, and address of issuing branch or office]*
Beneficiary: _____ *[Name and Address of Employer]*
Date: _____

Retention Money Guarantee No.: _____

We have been informed that _____ *[Name of Contractor]*
hereinafter called "the Contractor") has entered into Contract No. _____ *[Reference number of the contract]* dated _____ with you, for the execution of _____ *[Name of contract and brief description of Works]* (hereinafter called "the Contract").

Furthermore, we understand that, according to the conditions of the Contract, when the Taking-Over Certificate has been issued for the Works and the first half of the Retention Money has been certified for payment, payment of *[Insert "the second half of the Retention Money" or if the amount guaranteed under the Performance Guarantee when the Taking-Over Certificate is issued is less than half of the Retention Money, "the difference between half of the Retention Money and the amount guaranteed under the Performance Security"]* is to be made against a Retention Money guarantee.

At the request of the Contractor, we _____ *[Name of Bank]* hereby irrevocably undertake to pay you any sum or sums not exceeding in total an amount of _____ *[Amount in figures]* (_____) *[Amount in words, see 1]* upon receipt by us of your first demand in writing accompanied by a written statement stating that the Contractor is in breach of its obligation under the Contract because the Contractor used the advance payment for purposes other than the costs of mobilization in respect of the Works.

It is a condition for any claim and payment under this guarantee to be made that the payment of the second half of the Retention Money referred to above must have been received by the Contractor on its account number _____ at _____ *[Name and address of Bank]*.

This guarantee shall expire, at the latest, 21 days after the date when the Employer has received a copy of the Performance Certificate issued by the Engineer. Consequently, any demand for payment under this guarantee must be received by us at this office on or before that date.

This guarantee is subject to the Uniform Rules for Demand Guarantees, ICC Publication No. 458.

Signature(s): _____

1   *The Guarantor shall insert an amount representing the amount of the second half of the Retention Money or if the amount guaranteed under the Performance Guarantee when the Taking-Over Certificate is issued is less than half of the Retention Money, the difference between half of the Retention Money and the amount guaranteed under the Performance Security and denominated either in the currency(ies) of the second half of the Retention Money as specified in the Contract, or in a freely convertible currency acceptable to the Employer.*

Annex J F PARENT COMPANY GUARANTEE

Brief description of Contract

Name and address of Employer

(together with successors and assigns).

We have been informed that (hereinafter called the "Contractor") is submitting an offer for such Contract in response to your invitation, and that the conditions of your invitation require his offer to be supported by a parent company guarantee.

In consideration of you, the Employer, awarding the Contract to the Contractor, we
[*Name of parent company*] irrevocably and unconditionally guarantee to you, as a primary obligation, the due performance of all the Contractor's obligations and liabilities under the Contract, including the Contractor's compliance with all its terms and conditions according to their true intent and meaning.

If the Contractor fails to so perform his obligations and liabilities and comply with the Contract, we will indemnify the Employer against and from all damages, losses and expenses (including legal fees and expenses) which arise from any such failure for which the Contractor is liable to the Employer under the Contract.

This guarantee shall come into full force and effect when the Contract comes into full force and effect. If the Contract does not come into full force and effect within a year of the date of this guarantee, or if you demonstrate that you do not intend to enter into the Contract with the Contractor, this guarantee shall be void and ineffective. This guarantee shall continue in full force and effect until all the Contractor's obligations and liabilities under the Contract have been discharged, when this guarantee shall expire and shall be returned to us, and our liability hereunder shall be discharged absolutely.

This guarantee shall apply and be supplemental to the Contract as amended or varied by the Employer and the Contractor from time to time. We hereby authorise them to agree any such amendment or variation, the due performance of which and compliance with which by the Contractor are likewise guaranteed hereunder. Our obligations and liabilities under this guarantee shall not be discharged by any allowance of time or other indulgence whatsoever by the Employer to the Contractor, or by any variation or suspension of the works to be executed under the Contract, or by any amendments to the Contract or to the constitution of the Contractor or the Employer, or by any other matters, whether with or without our knowledge or consent.

This guarantee shall be governed by the law of the same country (or other jurisdiction) as that which governs the Contract and any dispute under this guarantee shall be finally settled under the Rules of Arbitration of the International Chamber of Commerce by one or more arbitrators appointed in accordance with such Rules. We confirm that the benefit of this guarantee may be assigned subject only to the provisions for assignment of the Contract.

Date: Signature(s):

Annex K BID SECURITY

Bank Guarantee

[All italicized text and any enclosing square brackets is for use in preparing the form and should be deleted, together with any square brackets, from the final product.]

[Bank's name, and address of issuing branch or office]

Beneficiary: _____ *[Name and address of Employer]*

Date

Blid Guarantee No.: _____

We have been informed that _____ *[Name of the Bidder]* (hereinafter called "the Bidder") has submitted to you its bid dated _____ (hereinafter called "the Bid") for the execution of _____ *[Name of contract]* under Invitation for Bids No. _____ ("the IFB").

Furthermore, we understand that, according to your conditions, bids must be supported by a bid guarantee.

At the request of the Bidder, we _____ *[Name of Bank]* hereby irrevocably undertake to pay you any sum or sums not exceeding in total an amount of _____ *[Amount in figures]* (_____) *[Amount in words]* upon receipt by us of your first demand in writing accompanied by a written statement stating that the Bidder is in breach of its obligation(s) under the bid conditions, because the Bidder:

(a) has withdrawn its Bid during the period of bid validity specified by the Bidder in the Form of Bid; or

(b) having been notified of the acceptance of its Bid by the Employer during the period of bid validity, (i) fails or refuses to execute the Contract Agreement or (ii) fails or refuses to furnish the performance security, in accordance with the ITB.

This guarantee will expire: (a) if the Bidder is the successful Bidder, upon our receipt of copies of the contract signed by the Bidder and the performance security issued to you upon the instruction of the Bidder; and (b) if the Bidder is not the successful Bidder, upon the earlier of (i) our receipt of a copy your notification to the Bidder of the name of the successful Bidder; or (ii) twenty-eight days after the expiration of the Bidder's bid.

Consequently, any demand for payment under this guarantee must be received by us at the office on or before that date.

This guarantee is subject to the Uniform Rules for Demand Guarantees, ICC Publication No. 458.

Signature(s): _____

INDEX

This index has been prepared using Sweet and Maxwell's Legal Taxonomy. Main index entries conform to keywords provided by the Legal Taxonomy except where references to specific documents or non-standard terms (denoted by quotation marks) have been included. These keywords provide a means of identifying similar concepts in other Sweet & Maxwell publications and online services to which keywords from the Legal Taxonomy have been applied. Readers may find some minor differences between terms used in the text and those which appear in the index. Suggestions to *sweetandmaxwell. taxonomy@thomson.com.*

(All references are to paragraph number)

Termination (cont.)
    valuation at date of,
        15–019—15–022
    force majeure, 19–026—19–032
Testing
    completion
        contractors' powers and duties,
            8–013, 9–001—9–006
        delayed tests, 9–008—9–011
        documents, 9–004
        engineers' role, 9–005
        failure to pass tests,
            9–016—9–020, 10–005
        interference with tests,
            10–016—10–019
        retesting, 9–012—9–015,
            11–024—11–027
    meaning, 9–003
Time
    meaning, 1–009, 1–011

"Unforeseeable"
    meaning, 1–036—1–037
Utilities
    contractors' powers and duties
        provision of services,
            4–154—4–157

Valuation
    measurements
        contractors' powers and duties,
            12–001—12–007
        engineers' obligations,
            12–001—12–007
        method, 12–008—12–011
        omissions, 12–019—12–022
        pricing, 12–012—12–018
Variation
    currencies, 13–020—13–023

employees
    dayworks, 13–029—13–034
legislation
    pricing adjustments,
        13–035—13–039
pricing
    changes, 13–040—13–047
    provisional sums,
        13–024—13–028
procedure, 13–015—13–019
right to vary
    contractors' powers and duties,
        13–005
    contract terms, 13–004
    exceptions, 13–007, 13–008
    initiation by engineer,
        13–001—13–008
    procedure, 13–006
value engineering, 13–009—13–014

Workers
    contractors' powers and duties
        accommodation, 6–031—6–034
        conditions of employment,
            6–009—6–012
        health and safety precautions,
            6–035—6–042
        income tax, 6–013
        labour costs, 6–004, 6–007
        particular locality clauses,
            6–005—6–006, 6–040
    recruitment, 6–001—6–008
        supervision, 6–043—6–047
        wages, 6–009—6–011
        welfare provision, 6–003,
            6–031—6–034
        working time, 6–025—6–030
Works *see* **Engineering operations**